T-LEVELS
THE NEXT LEVEL QUALIFICATION

ENDORSED BY

ONSITE CONSTRUCTION

CORE

Mike Jones, Stephen Jones, Tom Leahy, Peter Tanner

The publisher would like to acknowledge Mike Jones for his permission to reproduce text from *Foundation in Construction and Building Services Engineering: Core (Wales)*.

Every effort has been made to trace all copyright holders, but if any have been inadvertently overlooked, the Publishers will be pleased to make the necessary arrangements at the first opportunity.

Although every effort has been made to ensure that website addresses are correct at time of going to press, Hodder Education cannot be held responsible for the content of any website mentioned in this book. It is sometimes possible to find a relocated web page by typing in the address of the home page for a website in the URL window of your browser.

Hachette UK's policy is to use papers that are natural, renewable and recyclable products and made from wood grown in well-managed forests and other controlled sources. The logging and manufacturing processes are expected to conform to the environmental regulations of the country of origin.

Orders: please contact Hachette UK Distribution, Hely Hutchinson Centre, Milton Road, Didcot, Oxfordshire, OX11 7HH. Email education@hachette.co.uk Telephone: +44 (0)1235 827827. Lines are open from 9 a.m. to 5 p.m., Monday to Friday. You can also order through our website: www.hoddereducation.co.uk

ISBN: 978 1 3983 3290 4

© Hodder & Stoughton Limited, Stephen Jones, Peter Tanner 2022

First published in 2022 by
Hodder Education,
An Hachette UK Company
Carmelite House
50 Victoria Embankment
London EC4Y 0DZ

www.hoddereducation.co.uk

Impression number 10 9 8 7 6 5 4 3 2

Year   2026 2025 2024

All rights reserved. Apart from any use permitted under UK copyright law, no part of this publication may be reproduced or transmitted in any form or by any means, electronic or mechanical, including photocopying and recording, or held within any information storage and retrieval system, without permission in writing from the publisher or under licence from the Copyright Licensing Agency Limited. Further details of such licences (for reprographic reproduction) may be obtained from the Copyright Licensing Agency Limited, www.cla.co.uk

Cover photo © Monkey Business - stock.adobe.com

City & Guilds and the City & Guilds logo are trademarks of The City and Guilds of London Institute.
City & Guilds Logo © City & Guilds 2022

'T-LEVELS' is a registered trade mark of the Department for Education.
'T Level' is a registered trade mark of the Institute for Apprenticeships and Technical Education.
The T Level Technical Qualification is a qualification approved and managed by the Institute for Apprenticeships and Technical Education.

Typeset in India by Integra Software Services Ltd.

Printed by CPI Group (UK) Ltd, Croydon CR0 4YY

A catalogue record for this title is available from the British Library.

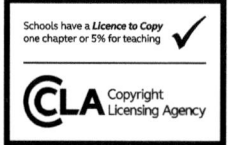

Contents

Acknowledgements ... iv

About the authors ... v

Introduction ... vii

Guide to the book ... viii

1. Health and safety in construction ... 1
2. Construction science principles ... 43
3. Construction design principles ... 83
4. Construction and the built environment industry ... 109
5. Sustainability principles ... 139
6. Measurement principles ... 171
7. Building technology principles ... 186
8. Information and data principles ... 215
9. Relationship management in construction ... 230
10. Digital technology in construction ... 252
11. Construction commercial/business principles ... 265

About the exams and employer-set project ... 284

Answers ... 286

Glossary ... 294

Index ... 299

Acknowledgements

Thank you to Tom Stottor who gave me the opportunity to contribute to this important textbook and my thanks to the copy-editors, proof-readers, illustrators and editors who have put so much time and effort into making this new textbook a reality.

Many thanks to Matthew Sullivan who's been so supportive and provided lots of suggestions and ideas on how to shape the chapters I've produced into something of value for T-Level learners.

Imogen Miles and Sylvia Bukowski have demonstrated their skills in creating a book that is visually appealing as well as of practical value. Thank you.

Finally, thanks to my lovely wife Sue who has patiently supported me through the many hours I've spent at my desk.

Mike Jones

I would like to thank my family and the team at Hodder Education for their continued and relentless support on this project, without which it would not have been possible to successfully complete the book.

I would also like to acknowledge the readers of this book and hope that it provides some support in achieving your personal and professional goals.

Stephen Jones

About the authors

Mike Jones

Working in the construction industry has been the mainstay of Mike's working life over a long period. He has worked as a skilled tradesman, a supervisor and a site manager on projects ranging from small extensions to multi-storey contracts worth millions of pounds. For a number of years, he employed a small team of skilled workers in his own construction company working on contracts for selected customers.

The skills Mike developed over the years allowed him to design and successfully build his own family home in rural Wales, which he has always viewed as a highlight of his construction career.

After over 30 years working out on site, Mike moved into the education sector, first as an NVQ trainer and assessor for three years and subsequently as a college lecturer at Cardiff and Vale College in South Wales for 10 years. After leaving his post at college, he continued to work in education and training, producing teaching and learning resources for City and Guilds, along with his work as a technical author writing textbooks. He is also involved in the writing, reviewing and editing of vocational examinations.

Mike's aim during his time teaching others has been to impart to learners the great job satisfaction that can be gained from becoming a skilled practitioner. Put maximum effort into developing your skills and knowledge, and you will be able to take full advantage of the vast range of career opportunities in the construction industry.

Stephen Jones

Stephen Jones has been writing books about the construction industry, site carpentry and architectural joinery since 2009. Besides being a published author, he is also a lecturer in the construction department at South Devon College. Stephen has Qualified Teaching and Learning Status (QTLS) and is a highly experienced lecturer with a Post Graduate Certificate of Education (PGCE), Higher National Certificate (HNC) in Construction and Nebosh National Certificate in Construction Health and Safety. Prior to teaching in further education, Stephen had his own business and worked in the construction industry as a carpenter, joiner and shopfitter since leaving school at 16 years of age, where he gained a wealth of subject knowledge and skills.

Stephen lives in Torquay on the south coast of Devon with his wife, children and dog, Maisy. His main interest is his family; he is also a football fanatic and enjoys working out in the gym.

Tom Leahy

Tom's journey started over 20 years ago when he served a four-year apprenticeship with a local authority, working on both commercial and domestic plumbing/heating systems whilst attending college for one day a week. This experience would give him the underpinning knowledge and practical skills to make a successful plumber.

In 2005 Tom started work for another local authority, working on a variety of authority-owned buildings, from libraries to town halls and community centres, as part of a reactive maintenance and installation team.

After successfully contributing to the training and development of several apprentices, Tom moved into further education and achieved teaching, assessing and quality assurance qualifications. He continues to work in further education as a Programme Leader training the next generation of Construction and Building Services professionals.

Peter Tanner

Having started in the electrotechnical industry while still at school, chasing walls for his brother-in-law for pocket money, Peter was keen to progress in the industry. After a spell in the armed forces, Peter gained a place as a sponsored trainee on the Construction Industry Training Board training scheme.

On completion of his apprenticeship Peter worked for a short time as an intruder alarm installer, then for a company involved in shop-fitting and restaurant and pub refurbishments. Peter was later seconded to the Property Services Agency, designing major installations within some of the best-known buildings in the UK.

A career-changing accident took Peter into teaching. Peter has worked with City & Guilds for over 25 years and has represented them on industry committees, such as JPEL/64 which is responsible for the production of BS 7671. He is passionate about using his experience to maintain the high standards the industry expects.

Introduction

A T Level in Onsite Construction will provide you with broad and deep foundations for a career in the Onsite Construction industries, including in areas such as bricklaying, carpentry and joinery, plastering and painting and decorating. The Level 3 technical qualification, which was developed by City & Guilds in collaboration with employers and industry bodies, will enable you to progress to a range of careers and further education opportunities. You can find out more about the qualification on the City & Guilds website.

This book will help you to develop the knowledge and understanding you need to complete the core component of your T Level. Each of the chapters in the book follows the topics required for the core component's knowledge outcomes, which you can find in the qualification handbook on the City & Guilds website. Headings in the book follow those in the qualification handbook, so that you can check your learning against the City & Guilds material. Although the book focuses on the core component only, the content that you cover and the skills that you learn will also be useful in your chosen occupational specialism.

The book will introduce you to the Onsite Construction workplace, and the fundamental principles that underpin it, across a range of engaging and real-world examples. The book covers the key content in detail and includes hundreds of photos and technical drawings, as well as industry tips to support you in your T Level work placement. There are also lots of activities and learning features; you can find out more about these and how to use them on the next page.

Your teacher or lecturer might decide to use the book in your classroom, either as a central learning tool or to supplement their teaching. The book will also prepare you for your final exams and the employer set project. (You can find out more about how you will be assessed on page 284 of this book.) If you have any questions about how you should be using the book, you should consult your teacher or lecturer.

Onsite Construction T Level: Core

Guide to the book

The following features can be found in this book.

Learning outcomes

Core knowledge outcomes that you must understand and learn

Key term

Important terms that you should understand

Industry tip

Useful tips and advice to help you in the workplace

Research

Research-based activities: either stretch and challenge activities enabling you to go beyond the course, or industry placement based activities encouraging you to discover more about your placement

Case study

Placing knowledge into a fictionalised, real-life context, to introduce dilemmas and problem solving

Test yourself

A knowledge-consolidation feature containing questions and tasks to aid understanding and guide you to think about a topic in detail

Health and safety

Important points to ensure safety in the workplace

Improve your maths

Short activities that encourage you to apply and develop your functional maths skills in context

Improve your English

Short activities that encourage you to apply and develop your functional English skills in context

Assessment practice

Knowledge-based practice questions to help prepare you for the exam (answers found at the back of the book)

Project practice

Short scenarios and focused activities, reflecting one or more of the tasks that you will need to undertake during completion of the employer-set project

Chapter 1 Health and safety in construction

Introduction

This chapter looks at health and safety legislation and approved standards, and how they protect workers and others who may be affected by work activities.

We will also consider the legal responsibilities of people working in construction and the implications of not following health and safety law.

Learning outcomes

By the end of this chapter, you will understand:
1. construction legislation and regulations
2. public liability and employer's liability
3. approved construction codes of practice
4. implications of poor health and safety performance
5. development of safe systems of work
6. safety conscious procedures
7. safety inspection of a work environment
8. recording and reporting of safety incidents and near misses
9. emergency procedures for unsafe situations
10. types of PPE
11. first-aid facilities
12. warning signs for the seven main groups of hazardous substance
13. safe practices and procedures for the use of access equipment and manual handling
14. safe practices and procedures for working in excavations and confined spaces

1 Construction legislation and regulations

1.1 The role of legislation and regulations in the construction industry

Working in the construction industry can be extremely hazardous. In 1974, the Health and Safety at Work etc. Act (HASAWA) was introduced. It replaced many older laws and enabled new regulations to be passed in Parliament to protect workers and the general public from work activities.

The main objectives of HASAWA are to:
- secure the health, safety and welfare of people at work
- protect people other than those at work (for example, the general public or visitors) from risks to health or safety arising out of or in connection with work activities
- control the possession and use of highly flammable, explosive and dangerous substances.

Everyone has a moral responsibility to protect the health and safety of themselves and others at work. However, under HASAWA some parties – known as **duty holders** – have legal responsibilities. Employers, employees, the self-employed, manufacturers and people in control of premises are all duty holders.

> **Key term**
>
> **Duty holders:** people with a legal responsibility under health and safety law

Shortly after HASAWA was introduced, the UK government established the Health and Safety Executive (HSE). The HSE is an independent regulator that aims to prevent workplace ill health, injury and death by targeting industries with the greatest risks and worst risk-management records, which include construction and manufacturing.

The HSE provides advice and guidance to employers so that they can identify hazards and manage risks correctly. Its emphasis is on the prevention of accidents and ill health by raising awareness in workplaces, so that workers and other people can stay safe and well.

> **Research**
>
> Visit the HSE website (www.hse.gov.uk) to find out the responsibilities of duty holders under HASAWA.

The HSE has legal powers to hold people or companies to account when risks are not managed and legislation is breached. We will look at how the HSE enforces HASAWA later in this chapter.

Legislation made under HASAWA is divided into a number of different regulations. Table 1.1 outlines the main regulations that control health, safety and welfare in the construction industry.

The Health and Safety Executive (HSE) is a reliable source of information regarding current legislation and regulations, and may therefore be used to validate other sources of health and safety information, such as:
- the Construction Industry Training Board (CITB)
- the Royal Society for the Prevention of Accidents (ROSPA)
- the Institution of Occupational Safety for Health (IOSH)
- the British Safety Council.

It is important that people with legal responsibilities to protect the health, safety and welfare of others know how and where to obtain relevant legislation and regulations. They must also understand the importance of ensuring that the information is accurate and current to fulfil their duties under the law. When obsolete legislation or inaccurate information is followed it can compromise the safety of people at work and could result in enforcement action against those concerned by the HSE. Further information on the implications of not adhering to legislation is covered later in this chapter.

▼ Table 1.1 Main regulations that control health, safety and welfare in the construction industry

Regulation	Overview
Reporting of Injuries, Diseases and Dangerous Occurrences Regulations (RIDDOR) 2013	These regulations place legal duties on employers, the self-employed and people in control of premises to report to the HSE serious accidents, dangerous occurrences and occupational diseases resulting from workplace activities.

Chapter 1 Health and safety in construction

Regulation	Overview
Control of Substances Hazardous to Health (COSHH) Regulations 2002	These regulations state that employers must control substances hazardous to health by preventing or reducing workers' exposure. This can be achieved by: • finding out what the health hazards are • deciding how to prevent harm to health by assessing the risks (**risk assessment**) • providing control measures, such as secure storage for chemicals • making sure that control measures are followed • maintaining control measures • providing training, information and instruction for employees and others • monitoring and providing health surveillance when necessary • planning for emergencies. Note: asbestos is not covered by COSHH Regulations.
Control of Asbestos Regulations 2012	There are several different types of **asbestos** that have been used in the construction industry, some of which are more hazardous than others. In most cases, asbestos can only be removed by a licensed contractor. However, small quantities of lower-risk asbestos can be removed by non-licensed contractors, providing they are competent and have effective controls in place. The law states that before doing any building work or maintenance on premises or on plant and equipment that contain asbestos, a risk assessment has to be carried out to manage and control the risks. It is mandatory for anyone who is likely to be exposed to asbestos fibres at work to receive training.
Provision and Use of Work Equipment Regulations (PUWER) 1998	These regulations place duties on people and organisations that operate or have control over work equipment. They state that: • equipment must be suitable for its intended use • equipment must be safe, maintained and regularly inspected • people who use the equipment must have received training, information and instruction • equipment must be used accompanied by suitable health and safety measures, for example guarding or emergency stop devices.
Manual Handling Operations Regulations (MHOR) 1992	These regulations define manual handling as 'any transporting or supporting of a load … by hand or bodily force'. They set out a hierarchy of control measures for dealing with the risks posed by manual handling: 1 Avoid hazardous manual handling operations so far as is reasonably practicable. 2 Assess any hazardous manual handling operations that cannot be avoided. 3 Reduce the risk of injury so far as is reasonably practicable.
Personal Protective Equipment (PPE) at Work Regulations 1992 (Note: the current legislation which refers to the supply of PPE is the Personal Protective Equipment Regulations 2016/425, which is enforced by the Personal Protective Equipment (Enforcement) Regulations 2018)	These regulations relate to the use of personal protective equipment (PPE). Employers have a responsibility to provide PPE free of charge to employees to control the hazards identified in risk assessments. They should also provide information, instruction and training on how to use it and take care of it.
Work at Height Regulations 2005	Under these regulations, people who are in control of others working at height have a duty to make sure work is properly planned, supervised and carried out by competent people. All work at height must be risk assessed.

▼ Table 1.1 Main regulations that control health, safety and welfare in the construction industry (continued)

Regulation	Overview
Control of Noise at Work Regulations 2005	Exposure to high levels of noise can cause temporary or permanent hearing loss. Employers have a duty to reduce the risk by: • assessing the risks • eliminating or controlling the noise • providing hearing protection • providing hearing checks for those at risk • providing employees with information and training • maintaining any noise-control equipment and ensuring it is being used.
Control of Vibration at Work Regulations 2005	These regulations place a duty on employers to reduce the risk to employees' health from exposure to vibration caused by work equipment, machinery and tools, for example when using a core drill. Employers must calculate the amount of vibration that employees may be exposed to at work. At a specific level (referred to as the 'exposure action value'), employers must introduce technical and organisational measures to reduce the risk of personal injury to an acceptable level. The law also prevents workers from being exposed to a higher exposure limit value of 5.0 m/s^2 A(8).
Confined Spaces Regulations 1997	The aim of these regulations is to prevent people working in confined spaces whenever possible, because of the serious risk to health and safety from collapse, drowning, low oxygen levels and exposure to natural gases. If the activity is unavoidable, a **safe system of work** must be followed to control the risk.
Management of Health and Safety at Work Regulations 1999	These regulations apply to every work activity. They explain what employers need to do to manage health and safety at their place of work under HASAWA. The main requirement is for employers to complete a risk assessment and record significant findings when they have five or more employees.
Environmental legislation	There are numerous environmental laws in the UK that protect wildlife, the countryside, listed buildings, national parks and monuments, for example the Control of Pollution Act 1974, the Wildlife and Countryside Act 1981 and the Planning (Listed Buildings and Conservation Areas) Act 1990. The environment is protected further by the Planning and Energy Act 2008, Energy Act 2020 and Climate Change Act 2008. The Climate Change Act 2008 is an environmental law designed to meet the UK government's target to reduce **carbon emissions** recorded in 1990 by at least 80 per cent by 2050.
Waste management legislation	There are several requirements for handling waste, including the Waste Electrical and Electronic Equipment (WEEE) Regulations 2013, as well as regulations on waste carriers. For more information, see Chapter 5, section 8.
Construction (Design and Management) Regulations 2015	The CDM regulations aim to manage health, safety and welfare on all construction projects by defining roles for duty holders. Clients (commercial and domestic), designers, contractors and workers all have legal responsibilities under the Construction (Design and Management) regulations; for example commercial clients that have a project carried out as part of a business must make sure welfare facilities are provided on their project.

Chapter 1 Health and safety in construction

> **Key terms**
>
> **Risk assessment:** a formal process of identifying significant workplace hazards, whom they affect and control measures that could be used to eliminate or reduce risk to an acceptable level
>
> **Asbestos:** a naturally occurring mineral that used to be mixed with other construction materials to create insulation, pipe lagging and flooring. It has been banned in the UK since 1999 because it is extremely hazardous to health; however workers may still be exposed to it in older buildings
>
> **Safe system of work:** a formal set of procedures that must be followed when hazards cannot be eliminated completely
>
> **Carbon emissions:** carbon dioxide released into the atmosphere; scientists believe this is a cause of climate change

Health and safety legislation is regularly reviewed and updated to reflect changes in the construction industry. These changes are often made with guidance and support from the following bodies:
- employers
- unions
- trade associations
- professional bodies
- academics.

> **Test yourself**
>
> How can you reduce the risk of personal injury caused by manual handling?

> **Health and safety**
>
> The use of PPE should be a last resort after all other methods of controlling a hazard have been considered. For example, the use of guardrails, safety nets and airbags should be considered for use while working at height before a lanyard. This is because PPE only protects the user, whereas the other methods protect others as well.

The main roles and responsibilities of people working in the building services engineering industry are no different from those of people working in the construction industry. Building services engineering and construction are considered high-risk industries; therefore the HSE is the authority responsible for enforcing the relevant legislation.

The following people have general duties under health and safety law:
- employers
- employees
- people in control of premises
- manufacturers of products, goods or materials used in the construction industry.

The general duties of employers towards their employees are to:
- protect the health, safety and welfare of their employees so far as is **reasonably practicable**
- ensure **plant** and systems of work are maintained and safe
- ensure the safe use, handling, storage and transport of articles and substances
- provide information, instruction, training and supervision to their employees
- ensure any place of work under their control is safe and without risks to health, including access and egress
- ensure the provision and maintenance of adequate welfare facilities.

> **Key terms**
>
> **Reasonably practicable:** a term used in health and safety law to describe realistic steps that should be taken to comply with the law in terms of time, effort and money
>
> **Plant:** heavy construction machinery, equipment and vehicles, for example a crane or dumper

> **Test yourself**
>
> Make a list of the general duties employees have under HASAWA.

> **Research**
>
> Which other enforcement authorities have the same powers as the HSE?

> **Research**
>
> What responsibilities do other duty holders have, besides employers and their employees?

The implications of not following health and safety legislation were made clear in section 1.3 of this chapter: improvement notices, prohibition notices and prosecution.

> **Research**
>
> Find out the short- and long-term effects that vibration can cause on different parts of the human body. Suggest ways to reduce or control the risks to comply with the Control of Vibration at Work Regulations 2005.

1.2 Regulations relating to the provision of welfare facilities during construction work

Under HASAWA, employers have a duty to provide basic welfare facilities for their employees at their place of work. The Construction (Design and Management) (CDM) Regulations 2015 outline the minimum facilities that should be provided on construction sites:

- Drinking water:
  - An adequate supply of fresh drinking water should be appropriately labelled and accessible in suitable places.
  - An adequate supply of cups or other drinking vessels must be provided, unless the water is supplied from a fountain for workers to drink easily.
- Toilets:
  - These should be separate for men and women, with doors that are lockable from the inside.
  - They should be clean and maintained with adequate ventilation and lighting.
- Washing facilities:
  - These should be located in the vicinity of the toilets and changing rooms, with clean cold and hot or warm running water if possible.
  - Soap or another means of cleaning must be provided, along with towels or another suitable method of drying.
  - Showers may also need to be provided if the nature of the work requires it.
- Rest facilities:
  - These should be equipped with tables and seating with backs for the number of people expected to use them.
  - The rest area must be able to maintain an appropriate temperature and have facilities to prepare and eat meals and boil water.
- Changing rooms with lockers:
  - These must be provided if workers have to wear special clothing for their job, unless they can be reasonably expected to change elsewhere.
  - Changing rooms should be provided with seating and facilities to dry and store clothing.
  - Separate rooms must be provided for men and women.

> **Health and safety**
>
> The CDM Regulations state that, if necessary, employers have to provide facilities for pregnant or nursing mothers to rest lying down.

▲ Figure 1.1 Welfare facilities on a construction site

1.3 Implications of not adhering to legislation

Health and safety law is made up of Acts of Parliament and statutory regulations. Failure to comply with them is a criminal offence. HSE inspectors have a number of powers to enforce the law, including carrying out inspections, issuing simple cautions and notices to duty holders, and sometimes initiating prosecutions.

The law states that HSE inspectors can enter a workplace without notice at any reasonable time to conduct an inspection or investigation. During a visit, they may want to speak to workers, look at possible health risks, and inspect equipment and machinery. They can also take samples, such as sound and dust levels, photographs and measurements, and make copies of records or other documentation needed as part of their investigation.

If the inspector thinks that the employer has broken the law, they will issue a 'notification of contravention'. This document outlines what laws have been broken and how, and what needs to be done to put things right.

An HSE inspector may issue an improvement notice when one or more laws have been breached; this means that faults have to be remedied within a specific period of time, which should be no less than 21 days.

If there has been a serious breach of the law and people are at risk of immediate harm, the HSE will issue a prohibition notice, preventing work from continuing.

Failure to comply with improvement or prohibition notices can result in prosecution, fines and imprisonment.

> **Research**
>
> What are the maximum legal penalties for employers that have not adhered to health and safety legislation?

> **Case study**
>
> Amelia is a self-employed painter and decorator who subcontracts work from medium-sized developers in the local area.
>
> Several months ago, she was visited by an HSE inspector. During the visit, the inspector noticed that a large amount of hazardous paint was not being stored correctly and was a potential source of fuel for a fire; therefore, they issued an improvement notice.
>
> The enforcement action taken by the inspector has been published on the HSE website. What impact do you think this will have on Amelia's business?

When people suffer loss or injury as a result of an accident at work, they may seek compensation. If health and safety legislation has been breached, the duty holder is usually fined because they have broken criminal law. Magistrates and the Crown Court have a discretionary power to award compensation to an injured person. However, claims are usually pursued by the injured person because of a breach of **common law**.

Employers should have liability insurance to cover any claims against them for a breach of common law; however, they cannot insure against fines when they have broken criminal law.

> **Key term**
>
> **Common law:** legislation made in the civil courts rather than statute law that is made in Parliament

> **Research**
>
> Research an employer's common-law duty of care and compare this to their duties under HASAWA.

1.4 Statutory and non-statutory documents in construction

The UK construction industry is subject to a number of laws which cover many different areas, including health, safety and welfare. Legislation comprises Acts of Parliament and regulations (statutory legislation) which have legal status and must be complied with. However, there are also many non-statutory guidance documents, which are not compulsory but offer advice on good practice and compliance with the law, for example the Approved Codes of Practice (ACOPs) produced by the HSE, which describe preferred methods and standards. Unless stated, you do not need to follow the ACOP guidance, but in doing so you will meet the requirements of HASAWA. The HSE states that other practical methods can be used, but they must meet or exceed the standards in the ACOP.

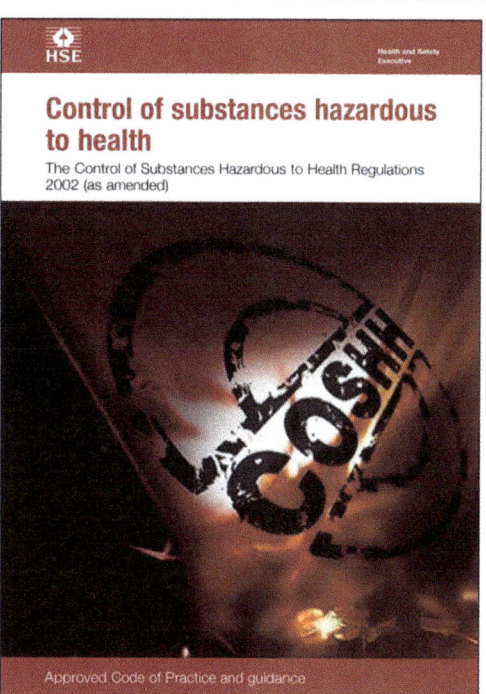

▲ Figure 1.2 An Approved Code of Practice

> **Test yourself**
>
> State one example of statutory legislation that is applicable to a construction activity and an ACOP to support it.

Implications of not adhering to legislation

There are many implications of poor health and safety.

For workers:
- accidents (such as slips, trips and falls) and near misses
- ill health
- injuries
- fatalities.

For employers:
- lower productivity
- higher employee turnover
- an unmotivated workforce
- financial problems, for example due to higher insurance premiums or legal costs
- damage to business reputation.

For the general public:
- accidents and near misses
- ill health
- injuries
- fatalities
- environmental issues.

These implications are covered throughout this chapter. During the design and planning of a building or structure, it is important to consider the health, safety and welfare of workers undertaking the construction phase. However, it is equally important to plan for the protection of people carrying out routine inspections of the building or maintenance work.

2 Public liability and employer's liability

2.1 Current requirements

We have already seen that the HSE has the power to issue fines to employers and other duty holders if they neglect the law. The fines issued by the HSE are intended to punish employers financially; therefore, they cannot insure against them.

However, employers can protect themselves with insurance against claims made by employees through the common-law courts in the pursuit of compensation. In this section, we will look at the different types of insurance that employers may have.

Public liability insurance

Public liability insurance protects a business against claims for compensation for any loss or injury to members of the public, customers or employees in connection with work activities.

There is no legal requirement for employers to have public liability insurance. However, if they engage with the public, or the public and their property are likely to be affected by the employer's work activities, then they should have this insurance to protect themselves from any potential claims made against them.

The level of insurance protection is determined by the nature and scale of the work undertaken. Some clients will stipulate the minimum level of insurance cover needed by a contractor, and have it agreed in writing before awarding any contracts or allowing construction work to start.

Employers' liability insurance

Common law is legislation that places a duty of care on employers to protect their employees from any unnecessary risks, and to ensure a safe system of work. If a person is injured or suffers a loss in the workplace, they may seek financial compensation from the employer if they are **liable**.

> **Key term**
>
> *Liable:* legally responsible

Under the Employers' Liability (Compulsory Insurance) Act 1969, all employers are required by law to insure against liability for injury or disease to their employees arising out of their employment and any financial claims made against them.

The amount of liability insurance an employer needs is usually determined by the number of employees and level of risk involved with their business. However, the minimum amount of cover for most insurance policies starts between 5 and 10 million pounds. If the injured person was successful with their claim against the employer, the compensation would go towards the recovery of their lost earnings, medical treatment, ongoing care and any legal fees.

Chapter 1 Health and safety in construction

> **Research**
>
> Find out the other types of insurance that a contractor or subcontractor may take out to safeguard themselves and their assets.

> **Improve your maths**
>
> The HSE reported that there were 142 work-related fatal injuries in the UK in 2020–21; 39 fatalities occurred in the construction industry. What percentage of fatalities occurred to construction workers?

3 Approved construction codes of practice

3.1 The HSE Legal (L) Series

In 2015, the Construction (Design and Management) (CDM) Regulations came into force. Their aim was to manage health, safety and welfare on all construction projects by defining roles for duty holders. People with responsibilities under the regulations include clients, designers and contractors. If there is more than one designer or contractor working on the same construction project, a **principal designer** or **principal contractor** must be appointed (named). Any other designers or contractors must work under the control of the principal designer or contractor.

The CDM Regulations state that where possible the designer (or principal designer) should plan for health and safety by identifying then eliminating or minimising all foreseeable risks at the planning stage of a project. They also state that all construction work, regardless of size, should have a 'construction phase plan'. This is a key document that identifies all the potential hazards for a project and the measures that will be used to remove or mitigate them.

Contractors and principal contractors have a duty under the CDM Regulations to:
- provide site inductions for all workers and visitors to the site
- provide suitable welfare facilities
- secure the site to prevent unauthorised access and protect non-workers (for example, members of the public).

The HSE has published a range of documents known as the HSE Legal (L) Series (also referred to as the CDM Series). Their primary purpose is to support duty holders in meeting their obligations under the CDM Regulations. They are available online to download for free, or a hard copy can be purchased through the HSE's website. The L Series contains both Approved Codes of Practice (ACOPs) and guidance on various health, safety and welfare regulations.

> **Key terms**
>
> **Principal designer:** a designer appointed by the client to take the lead in planning, managing, monitoring and co-ordinating health and safety during the pre-construction phase of a project involving more than one designer
>
> **Principal contractor:** a contractor appointed by the client to take the lead in planning, managing, monitoring and co-ordinating health and safety during the construction phase of a project involving more than one contractor

> **Improve your English**
>
> Include the word 'mitigate' in a short sentence about health and safety hazards that you may face at your training centre.

> **Test yourself**
>
> How could an employer ensure that ACOPs are implemented in a construction project?

4 Implications of poor health and safety on building performance and individual stakeholders

There are many implications of poor health and safety.

For workers:
- accidents (such as slips, trips and falls) and near misses
- ill health

- injuries
- fatalities.

For employers:
- lower productivity
- higher employee turnover
- an unmotivated workforce
- financial problems, for example due to higher insurance premiums or legal costs
- damage to business reputation.

For the general public:
- accidents and near misses
- ill health
- injuries
- fatalities
- environmental issues.

These implications are covered throughout this chapter.

During the design and planning of a building or structure, it is important to consider the health, safety and welfare of workers undertaking the construction phase. However, it is equally important to plan for the protection of people carrying out routine inspections of the building or maintenance work.

5 Development of safe systems of work

5.1 Development and use of safe systems of work in construction projects

There are different types of management system that can be used to create safe systems of work. These include:
- company management systems
- risk assessments
- method statements
- permits to work
- safety notices (covered later in this chapter)
- CSCS cards.

The HSE favours the following approach, which can be applied to most construction businesses:

- **Plan** – for specific health and safety objectives
- **Do** – implement the plan
- **Check** – that the plan is working and measure performance, for example the number of accidents and near misses
- **Act** – learn from any mistakes and put them right.

The law states that every employer must have a policy for managing health and safety, and that it must be in written form if they have five or more employees. A health and safety policy sets out the general approach an employer has towards managing health and safety in their business and is divided into three sections:
- general statement of intent (the employer's commitment)
- arrangements (what the employer intends to do and how)
- responsibilities (areas of responsibility within the employer's business and defining roles).

The Management of Health and Safety at Work Regulations 1999 contain a schedule known as the 'General principles of prevention'. This provides a hierarchy of control measures for employers to manage risks to health and safety in the workplace (see Figure 1.3).

> **Test yourself**
>
> What is the least effective method of controlling health and safety risks?

> **Research**
>
> Find out about permits to work as part of a safe system of work. Explain:
> - when they are used
> - who issues them
> - how they are controlled
>
> You should also make a list of as many work activities as you can think of that would benefit from a permit-to-work system and justify your reasons for choosing these tasks.

Chapter 1 Health and safety in construction

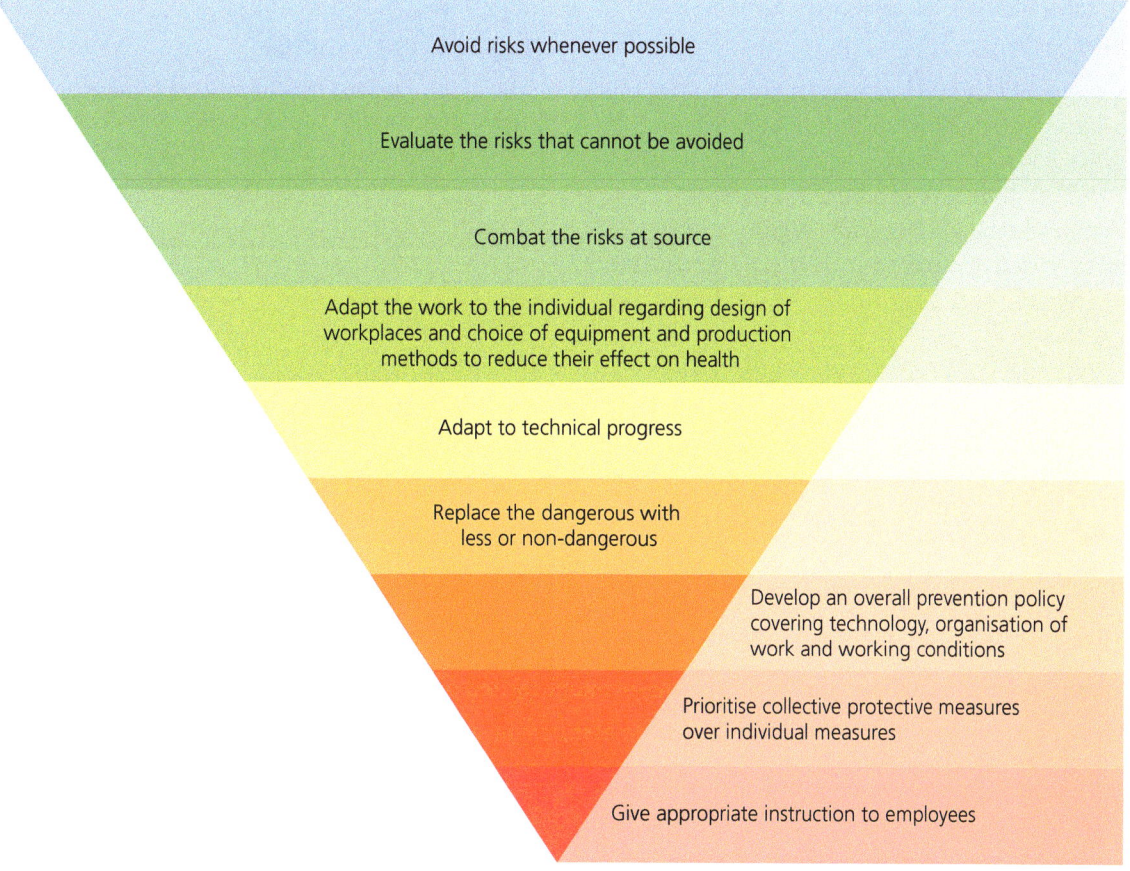

▲ Figure 1.3 Hierarchy of control measures for managing health and safety risks (source: www.legislation.gov.uk)

5.2 How to complete risk assessments

The Management of Health and Safety at Work Regulations 1999 place a legal duty on employers to manage the risk of potential harm to people from work activities. A risk assessment that identifies hazards and determines measures to eliminate or control them is fundamental to reducing work-related accidents and ill health.

Risk assessments should not be designed to produce lots of paperwork; neither do they need to be complicated or difficult to understand.

A risk assessment should be a structured examination of workplace activities, appropriate and proportional to the level of risk and the nature of the hazards. It should be site and task specific; therefore, generic risk assessments would not be considered suitable or sufficient.

Completing a risk assessment is a five-step process:
1. Identify hazards and potential hazards (what could cause injury or illness).
2. Decide who might be harmed and how.
3. Assess the risks by looking at the likelihood of the hazard causing harm and the severity of that harm.
4. Establish control measures to remove or reduce the risk of harm. (Look at what you are already doing to protect people, what future actions need to be taken and who needs to take those actions.)
5. Record any significant hazards found (when there are five or more employees) and review the assessment after the control measures have been put in place to see if they are working, or if anything further could be done.

As part of a good management system, employers should prioritise the highest-level risks before assessing medium-level risks. Once all of these hazards have been identified and eliminated or controlled in some way to a satisfactory level, lower-risk activities should be risk assessed.

Risk assessments should be completed by a competent person with a satisfactory level of knowledge and experience. For more complex activities, the person completing them may seek advice from other sources, such as the HSE or trade organisations. Employers with five or more employees are required by law to record significant findings from their risk assessments.

> **Health and safety**
>
> Some regulations require risk assessments to be carried out for specific tasks, for example the Work at Height Regulations 2005 and the Control of Substances Hazardous to Health (COSHH) Regulations 2002.

Risk assessment template

Company name: M. Bloggs Ltd **Assessment carried out by:** MB
Date of next review: 15/08/2021 **Date assessment was carried out:** 15/05/2021

What are the hazards?	Who might be harmed and how?	What are you already doing to control the risks?	What further action do you need to take to control the risks?	Who needs to carry out the action?	When is the action needed by?	Done
Brick stacks in work area	Bricklayer/General operative; Tripping hazard	Maintain good housekeeping	Remove waste regularly	General operative	Daily	✓

More information on managing risk: www.hse.gov.uk/simple-health-safety/risk/

Published by the Health and Safety Executive 09/20

▲ Figure 1.4 A risk assessment template

Reproduced by kind permission of HSE. HSE would like to make it clear it has not reviewed this product and does not endorse the business activity of Hodder Education.

5.3 How to write method statements

Method statements are documents prepared by employers that describe a logical sequence of steps to complete a work activity in a safe manner. They reflect the hazards identified in risk assessments and describe the way in which the job should be undertaken.

These documents are used as part of a safe system of work to communicate vital health and safety information and guidance to people completing the activity; together with risk assessments they are referred to as RAMS (risk assessments and method statements).

Chapter 1 Health and safety in construction

A typical method statement describes:
- hazards identified
- safe access and **egress**
- supervision needed
- hazardous substances and how to control them
- permit-to-work systems (if applicable)
- personal protective equipment
- emergency procedures
- environmental controls
- health and safety monitoring
- workforce details.

Although method statements are an effective way of keeping people safe at work, there is no legal requirement for employers to produce them. However, they are often requested by principal contractors.

> **Key term**
>
> **Egress:** an exit or way out

METHOD STATEMENT			
Revision Date:	**Revision Description:**		**Approved By:**
Work Method Description	Risk Assessment	Risk Levels	Recommended Actions* (Clause No.)
1.			
2.			
3.			
4.			
RISK LEVELS: Class 1 (high) Class 2 (medium) Class 3 (low) Class 4 (very low risk)			
Engineering Details/Certificates/Work Cover Approvals:		Codes of Practice, Legislation:	
Plant/Equipment:		Maintenance Checks:	
Sign-off			
Print Name:	Print Name:	Print Name:	Print Name:
Signature:	Signature:	Signature:	Signature:
Print Name:	Print Name:	Print Name:	Print Name:
Signature:	Signature:	Signature:	Signature:

▲ Figure 1.5 A method statement

5.4 How to complete a COSHH assessment

Hazardous substances can take many forms, including chemicals, dust and biological agents, and are not just products labelled as hazardous.

People at work are often exposed to hazardous substances and may be at risk of long-term (**chronic**) health problems such as asthma, cancer and skin disease.

Employers and the self-employed have a legal duty under the COSHH Regulations to assess the risk to their employees and others from exposure to products and processes that may contain or create substances harmful to health. If they have five or more employees, these assessments must be written down.

The main legal requirement is to use appropriate control measures to prevent harm from exposure, and this can be achieved by applying the hierarchy of control measures outlined in Figure 1.6.

> **Key term**
>
> **Chronic:** continuing for a long time

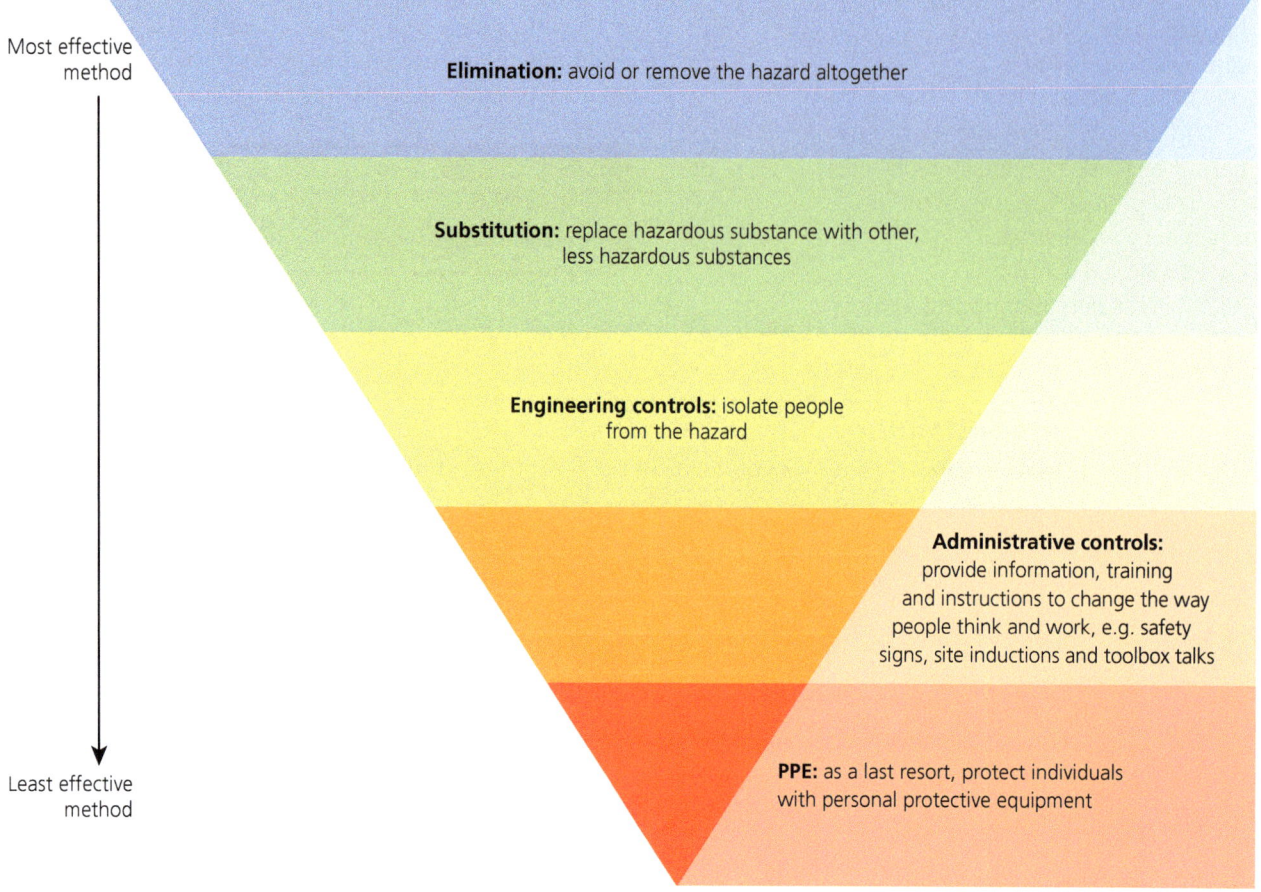

▲ Figure 1.6 Hierarchy of control measures to prevent exposure

By law, suppliers must provide up-to-date **safety data sheets** for any hazardous substances. These contain the supplier's information about a product and should not be confused with COSHH assessments.

> **Key term**
>
> *Safety data sheets:* written documents produced by manufacturers and suppliers of hazardous substances that contain important information about how products should be transported, used, stored and safely disposed of after use, any special conditions you should be aware of and how to deal with the substance in an emergency

Steps to follow to complete a COSHH assessment:
1. Identify the hazardous substance, who is likely to be harmed and how.
2. Evaluate the risk of the hazard causing harm by considering frequency of exposure to the substance and what effects it could have.
3. Decide what reasonably practicable measures are necessary to prevent or control any exposure to the hazard and how these will be maintained; make a plan for emergencies.
4. Record the assessment.
5. Decide if and when the assessment needs to be reviewed, and by whom.

> **Test yourself**
>
> Which substances are covered by the COSHH Regulations?

Chapter 1 Health and safety in construction

Company name: Patio Path and Paving Department:		Date assessment made: Date discussed with employees:				
Step 1 Substance	**Step 2**	**Step 3**		**Step 4 Action**		
What is the hazard?	What harm, and who?	What are you doing already?	What improvements do you need?	Who	When	Check
Breathing in dust from cutting paving	Long-term lung damage, e.g. bronchitis and silicosis. Everyone involved in cutting and anyone working nearby	Water to suppress dust. Protective goggles	Get enough water. Those involved should wear P3 respirators, ear plugs			
Breathing in exhaust fumes from cutter	Carbon monoxide poisoning. Everyone nearby	Always use cutter outdoors	Make sure area is not enclosed and fumes can disperse			
Ready-mix cement – skin contact – splashes	Skin burns when laying concrete or mortar. Anyone – cement on skin	Avoid skin contact. Use protective gloves	Get access to running water. Wash off splashes immediately. Use skin care products			
Moss killer – skin irritation	Mixing concentrate. Anyone – splash on skin	Use protective gloves. Wash out applicator after use	Get access to running water. Use skin care products			
Also:		**Action taken**	**Action needed**			
Thorough examination and test – COSHH		None	Check water suppression			
Supervision		When available				
Instruction and training		Yes	Respirator training for P3 and fit test			
Emergency plans		Mobile phone				
Health surveillance		None	Ask doctor about lung function tests			
Monitoring		None	None			
Step 5 Review date:		1. Review your assessment – make sure you are not sliding back 2. Any significant change in the work? Check the assessment and change it if necessary				
Other hazards needing attention: lifting and handling heavy weights, noise, reversing vehicles						

▲ Figure 1.7 A COSHH assessment

5.5 Construction Skills Certification Scheme

The Construction Skills Certification Scheme (CSCS) is one of many industry-recognised card schemes accredited by the Construction Industry Training Board (CITB). CSCS cards prove that the card holder has a satisfactory level of health and safety awareness; they also show relevant qualifications the holder has achieved to confirm they are working in the correct job role on site.

There are various types of CSCS card, designed to suit people working at different levels in the construction industry, from labourers to academically and professionally qualified managers. CSCS cards last a maximum of five years before the holders have to reapply; however once an apprentice or trainee card has expired it cannot be renewed. There is no legal requirement for workers to hold a CSCS card, although most principal contractors and clients require people to have them as a way of managing health and safety on sites under their control.

To qualify for a basic CSCS card, a candidate must first successfully complete the CITB (Site Safety Plus) Health and Safety Awareness course and test. They must then pass the online CITB Health, Safety and Environment test before they are allowed to apply for an appropriate CSCS card.

5.6 Site Management Safety Training Scheme

Principal contractors and clients usually require people with planning, organising, controlling and monitoring responsibilities (for example, site managers) to hold the Site Management Safety Training Scheme (SMSTS) qualification. This qualification ensures that the holder is up to date with health, safety, welfare and environmental legislation, so that they can fulfil their legal responsibilities. Every five years, the holder has to complete the Site Management Safety Training Scheme Refresher (SMSTS-R) to retain the qualification and to make sure they are aware of any changes in legislation, new guidance and industry best practice.

5.7 Site Supervision Safety Training Scheme

The Site Supervision Safety Training Scheme (SSSTS) is a similar qualification to the SMSTS. It is designed for people with supervisory responsibilities or those preparing to start in this role and provides an understanding of topics such as **site inductions**, risk assessments and method statements. Every five years, the Site Supervision Safety Training Scheme Refresher (SSSTS-R) course has to be completed, to keep up to date with any changes that affect the supervisory role and to retain the qualification status.

6 Safety conscious procedures

6.1 Procedures that promote and support safety consciousness

HASAWA states that employers have a duty to provide information, instruction and training to their employees, so that they know how to work safely without risks to health. Employees and the self-employed should be made aware of:
- the hazards and risks they may face
- control measures that are in place to deal with hazards
- what to do in the event of an emergency, for example assemble at the designated muster point (assembly point).

Employers must pay particular attention to certain groups, such as young people, new recruits and people changing roles, because they are less likely to be aware of potential risks and therefore require additional training and support.

Under the Construction (Design and Management) (CDM) Regulations 2015, principal contractors must engage with workers about their health, safety and welfare, and provide a site-specific induction and any other information and training that they need. Before entering a workplace, people should be made aware of potential hazards and the measures they need to take to protect themselves and others. On construction sites, this is usually achieved by following the site rules displayed on notice boards and safety signs. All workers and visitors to site have to sign in, so that the site manager is aware of who is on site at all times, especially in the event of an emergency. Visitors and new workers will have to attend a site induction, where they will be informed about:
- hazards
- site rules
- personal protective equipment (PPE)
- welfare facilities
- first aid
- fire extinguishers

Chapter 1 Health and safety in construction

- emergency procedures
- key personnel (for example, site manager, assistant manager, supervisors, first aider)
- pedestrian routes
- parking
- working hours
- waste and the environment
- boundaries
- working at height
- traffic-management plans
- **permits to work**
- security
- risk assessments and method statements (RAMS).

> **Key terms**
>
> **Site induction:** an occasion when all new visitors and workers on a construction site are informed about potential hazards and the control measures used to protect people from harm
>
> **Permits to work:** documents issued by site managers to workers undertaking high-risk activities as a method of control; they authorise certain people to carry out specific work tasks within a given timeframe and set out the precautions needed to complete the work safely

▲ Figure 1.8 A hot works permit-to-work activity

> **Research**
>
> Make a list of activities that should be controlled by a permit-to-work system. Choose one activity from your list and create a detailed permit to work using a template from the HSE's website.

The duration of a site induction will depend on the size and nature of the work and the personnel undertaking the training. Experienced workers or designers (architects and engineers) may only need to attend a brief site induction, whereas someone that is newly qualified will need to attend a full induction. Inductions must be completed for every new site that is visited, regardless of whether the person is working for the same employer or principal contractor.

It is likely that the conditions and hazards on a site will evolve as work progresses, after people have attended a site induction. Workers will need to be updated on any matters of health and safety when necessary. Short presentations known as **toolbox talks** are often used to provide this information to workers. These are usually delivered to small groups of workers, in an area of the workplace where they should not be disturbed. These talks usually cover a single aspect of health and safety, such as good housekeeping (keeping areas clean and clear).

> **Key term**
>
> **Toolbox talks:** short training sessions arranged at regular intervals at a place of work to discuss health and safety issues; they give safety reminders and inform personnel about new hazards that may have recently arisen

6.2 Benefits of procedures that promote and support safety consciousness

It is important to remember that employees have duties under HASAWA, as well as employers, and that every effort must be made to comply with the law.

The law states that employees must:
- take care of their own health and safety and that of others
- report any work situations that present a serious and imminent risk
- co-operate with their employer to help them comply with health and safety legislation
- follow any instructions or health and safety training provided by their employer

- inform their employer of any shortcomings they identify in their health and safety arrangements.

When an employee fails in these responsibilities or does not follow their employer's health and safety procedures, they are essentially breaking the law and could be putting themselves and others at risk of injury, illness or death.

Accidents and near misses in the workplace often cost the employer in terms of time, effort and money. There may also be penalties imposed on them by the HSE and civil claims made against them.

A workplace accident could also result in indirect costs for the employer, such as:
- project timescales slipping and missed deadlines
- time spent training new staff
- increased insurance premiums.

Employers with a poor health and safety record usually have difficulty retaining staff, because they do not feel safe in their place of work. It also leads to low morale and a negative health and safety culture, which in turn makes it challenging to recruit new employees.

Poor standards of health and safety can also have a detrimental effect on future contracts and potentially result in losing business.

> ### Test yourself
> 1 What powers does the HSE have to enforce HASAWA?
> 2 What impact do you think an accident at work could have on an injured employee?

> ### Improve your maths
> In 2020, the HSE published a profile of non-fatal workplace injuries by accident kind:
> - falls from a height 8%
> - struck by a moving object 11%
> - slips, trips or falls on the same level 29%
> - acts of violence 9%
> - handling, lifting or carrying 19%
> - other 24%.
>
> Produce a pie chart on a computer to illustrate this data.

7 Safety inspection of a work environment

7.1 Methods used to inspect a workplace to ensure it is safe

Employers have a responsibility to monitor health and safety arrangements in the workplace, to ensure they are effectively controlling risks.

There are two types of monitoring system that are typically used in the construction industry:
- **Active monitoring** is the monitoring of people, procedures, premises and plant in the workplace in order to identify potential hazards before an accident or incident occurs, so that control measures can be taken to prevent harm.
- **Reactive monitoring** is an inspection completed after an incident has taken place, for example a review of accident forms or statistics which might identify an area of concern, such as skin burns from welding equipment or falls from height.

Although reactive monitoring is a review of performance standards, it is a useful method of identifying problems to prevent further accidents and incidents occurring and should be used alongside active monitoring.

> ### Key terms
> *Active monitoring:* monitoring people, procedures, premises and plant in the workplace in order to identify potential hazards before an accident or incident occurs, so that control measures can be taken to prevent harm
>
> *Reactive monitoring:* an inspection completed after an incident has taken place to prevent further incidents occurring

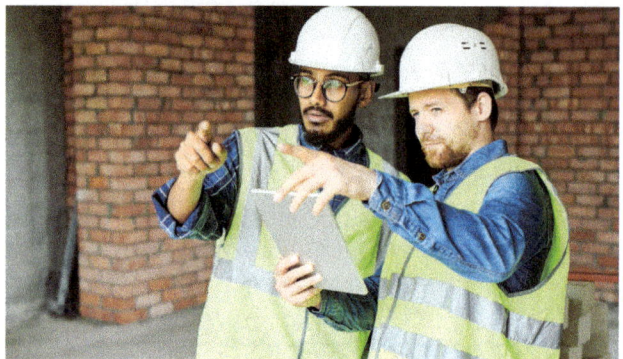

▲ Figure 1.9 A site manager inspecting work on a construction site

7.2 Monitoring health and safety in the workplace

As part of a good health and safety management system, employers should plan a programme of statutory, routine, periodic and pre-use inspections of equipment and their workplace. The frequency of the inspections will be determined by a number of factors, for example pre-use inspections of safety equipment may be completed daily by employees, whereas a periodic inspection of lifting equipment could be every six months, as determined by the Lifting Operations and Lifting Equipment Regulations (LOLER) 1998.

▲ Figure 1.10 A construction crane covered by LOLER

Active monitoring should involve structured, well-planned and organised examinations carried out by competent, experienced inspectors. Employers must decide:
- what type of inspections need to be completed
- who needs to be involved
- what analytical equipment may be needed, even if this means consulting specialists.

There are various types of health and safety inspection that can be implemented in the workplace:
- Health and safety audits (also known as desktop audits) are used to inspect health and safety documentation used in the workplace.
- Safety sampling is used to focus an investigation on a representative sample of a workplace standard; for example a random selection of portable electrical equipment could be selected to check if portable appliance testing (PAT) is up to date. If a big enough sample is selected, then it is likely to reflect any issues that may be identified on a wider scale within the company.
- Safety surveys are often used to focus a detailed health and safety investigation on a particular topic or issue, rather than the workplace as a whole.
- Safety tours are full inspections of entire workplaces to identify any hazards or shortcomings in employers' or employees' health and safety responsibilities.
- Incident inspections are carried out after an accident, a near miss or a case of reported ill health to the HSE.

Routine health and safety inspections are often guided by checklists and templates of set items that must be covered. These provide excellent prompts and are easy to complete with little training. They are also cheap, easy to reproduce and provide a consistent approach for every inspection, regardless of who completes it. The disadvantage is that hazards can often be overlooked if they are not listed, and therefore this form of inspection is not always preferred.

A random visual or sensory inspection of a work area by a team of representatives is sometimes a better way of identifying all potential hazards, because inspectors are not restricted by a checklist or template.

7.3 Types and use of recording documentation

Employers and their safety representatives are responsible for recording all inspections in a clear written report, as soon as possible after they have taken place. In some cases, this is a legal duty.

The report should contain:
- the date the inspection took place
- any potential hazards or defects found
- when the next inspection is due.

Inspection reports are then used to inform an action plan to put right any problems identified. An action plan should detail:
- any corrective actions needed
- people with responsibilities for carrying out the corrective actions
- the timescale in which the corrective actions must be completed.

> **Research**
>
> Research HSE forms F2534 and F2533 – documents that can be used to record the results of inspections and notify an employer of any unsafe or unhealthy working practices and unsatisfactory welfare facilities.

> **Key term**
>
> *Accident book:* a formal document used to record details of accidents that occur in the workplace, whether to an employee or visitor

Under the Provision and Use of Work Equipment Regulations (PUWER) 1998, employers have specific duties to ensure that equipment provided for use at work is:

> '… maintained in a safe condition and inspected to ensure it is correctly installed and does not subsequently deteriorate.'

Other legislation, such as the Lifting Operations and Lifting Equipment Regulations (LOLER) 1998, the Pressure Systems Safety Regulations (PSSR) 2000, the Work at Height Regulations 2005 and the Personal Protective Equipment Regulations 2002, also give specific duties to employers to examine work equipment and keep an up-to-date register of all inspections.

It is important that employers maintain these inspection records because they may be referred to in the event of an investigation into an accident or near miss in the workplace. Inspection records, maintenance schedules and service plans could also be requested by HSE inspectors or required by an employer's insurance company.

8 Recording and reporting of safety incidents and near misses

8.1 Reporting an incident or near miss in the workplace

Workplace accidents and incidents must be reported to employers so that they can be properly dealt with and investigated to help develop potential solutions and reduce the risk of them reoccurring.

Under the Social Security (Claims and Payments) Regulations 1979, employers must record details of any workplace accident in an **accident book**. It is best to complete the entry as soon as possible after the accident has occurred, so that important information is not left out.

As this book contains sensitive information about the accident and personal details about the injured person, it must be stored safely and securely to prevent a breach of confidentiality.

By making reports and keeping records of accidents on site, it is possible to see patterns that may be emerging, possibly due to bad habits or incorrect work practices. These records can assist in planning future work activities to reduce the occurrence of accidents and may be used when legal matters arise related to an accident or emergency. Accident records can be requested and looked at by an employer's insurance company, if a claim is made against them, or the HSE as part of an investigation into an accident or dangerous occurrence. The law states that employers must keep accident records for at least three years after the date they were completed.

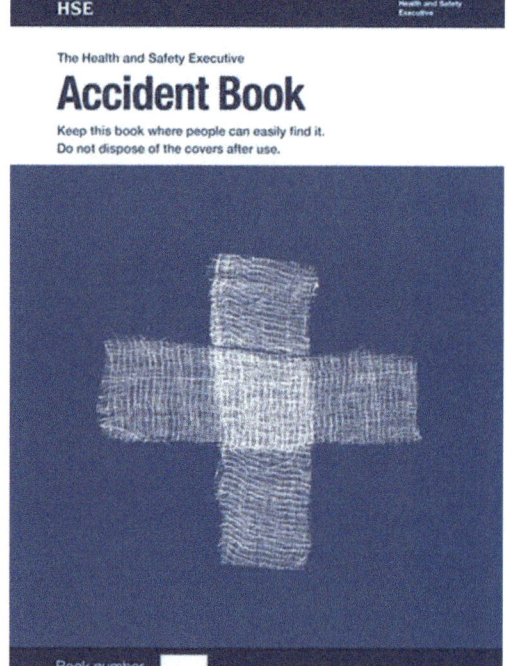

▲ Figure 1.11 An accident book

Chapter 1 Health and safety in construction

Employers have a duty to report the following to the HSE under RIDDOR:
- the death of any person caused by a workplace accident
- dangerous occurrences (also known as 'near-miss accidents')
- gas incidents where someone has died, lost consciousness, or been taken to hospital for treatment
- non-fatal accidents to non-workers, for example a member of the general public; however, this only needs to be reported if the person received treatment at hospital
- occupational diseases:
  - asthma
  - cancer
  - carpal tunnel syndrome
  - dermatitis
  - disease attributed to exposure to a biological agent
  - hand–arm vibration syndrome
  - severe cramp of the forearm or hand
  - tendonitis of the forearm or hand
- specified injuries caused by workplace accidents:
  - amputations
  - injuries arising from working in an enclosed space
  - any scalping requiring hospital treatment
  - crush injuries to the torso or head
  - fractures, other than to fingers, toes and thumbs
  - loss of consciousness caused by asphyxia or a head injury
  - permanent loss of sight or reduction in sight
  - serious burns (including scalding)
- over seven consecutive days' absences when a worker is unable to perform their duties as a result of a workplace accident.

If an employee has been absent from work for three days as a result of an accident, it does not need to be reported to the HSE; however, the employer must keep a record. A copy of a report in an accident book is usually sufficient.

> **Test yourself**
>
> When should an employee be made aware of their employer's accident and incident reporting policy?

▲ Figure 1.12 A dangerous occurrence

> **Research**
>
> Find out how to report a dangerous occurrence to the HSE under RIDDOR.

9 Emergency procedures for unsafe situations

9.1 Procedures to follow in unsafe situations

An emergency situation in the workplace can often result in delays with construction work, missed deadlines and additional costs for the principal contractor, not to mention the risk to workers' lives. To reduce the potential impact of an unforeseen incident, employers have a duty under the CDM Regulations to plan for emergencies on construction sites as part of their pre-construction work. Under the Regulatory Reform (Fire Safety) Order 2005, employers also have a duty to plan for emergencies on other sites such as offices, factories and warehouses.

▲ Figure 1.13 A fire on a construction site

Examples of unsafe or emergency situations in a workplace include:
- fire, for example welding pipes using a blow torch providing a source of ignition
- gas leaks, for example groundworks with a multi-purpose excavator causing damage to a buried gas pipe
- terrorist threats, for example while undertaking construction work on government buildings
- water leak, for example boring (drilling) into the ground and coming into contact with a mains water pipe
- carbon monoxide, for example from a poorly installed or maintained boiler
- electric shock, for example an operative cutting into a masonry wall with a disc cutter coming into contact with buried services such as electrical wires.

As part of their health and safety preparations, an employer (as a **responsible person**) must nominate a competent person to make an emergency plan, complete a fire risk assessment and inform workers about the findings.

> **Key term**
>
> **Responsible person:** (in law) usually the employer (assuming control of the workplace), a person with control of the premises or an owner; that person can then nominate other competent people to perform legal responsibilities, for example to manage fire safety

Fire risk assessments are intended to identify fire hazards in a workplace and those at risk of harm. The level of risk should then be evaluated and a plan should be prepared by the responsible person. Where an employer has five or more employees, the fire risk assessment should be recorded.

Workers on construction sites are usually informed about plans for emergencies and fire risk assessments during their site induction. The actions to follow are often reinforced on safety notices and information boards displayed in work areas.

An emergency plan should be prepared at the start of every project and cover the following areas:
- safe evacuation of the site in the event of an emergency
- emergency escape routes
- safety signage (for example, directional signs and assembly points)
- safe storage of hazardous and flammable materials, including waste
- site security to minimise the risk of **arson**
- assembly points and registers of workers on site
- fire wardens
- fire detection and alarms
- designated spill kits to deal with hazardous chemical, fuel or oil spillages and leaks
- training for workers on emergency equipment and fire safety (for example, use of fire extinguishers, spill kits, and how to raise the alarm)
- procedures for calling the emergency services
- emergency lighting
- fire extinguishers
- maintenance of escape routes and fire doors to keep them clear at all times
- **hot works** permit systems
- arrangements for smoking
- arrangements for vehicles and plant, for example refuelling
- emergency plant shut down and isolation to make processes safe.

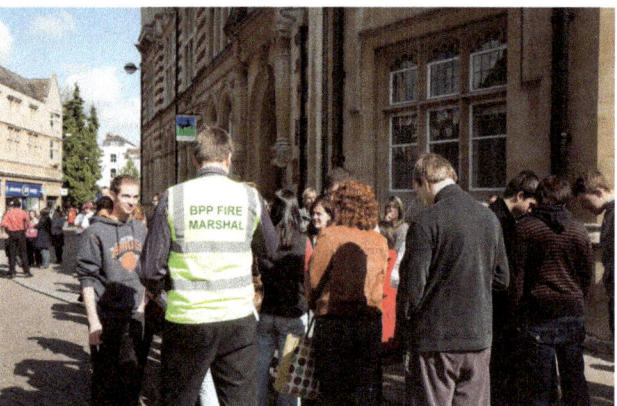

▲ Figure 1.14 A fire marshal

> **Key terms**
>
> **Arson:** the criminal act of deliberately setting fire to property
>
> **Hot works:** any construction work involving open flames or generating heat or friction; hot works are particularly high-risk activities that have the potential to cause fires if they are not controlled properly with a permit-to-work system

> **Research**
>
> Research permits to work and list as many construction activities as possible that may need one.

> **Research**
>
> Research the role of a fire warden and list their responsibilities in the event of a fire in a workplace.

9.2 Actions to be taken when dealing with fire situations

In the event of a fire in the workplace, workers must follow their employer's procedures. The main steps in a fire procedure are outlined in Figure 1.15.

Raise the alarm and inform others.

↓

Walk quickly, following the directional signs, to the closest available emergency exit. Make sure you close all the fire doors behind you. Do not use any lifts between floors.

↓

Only attempt to tackle a small fire if it is blocking your safe exit and if you are trained to use the equipment.

↓

Report to the assembly point and stay there until you are told to leave.

↓

Call the emergency services.

▲ Figure 1.15 Fire procedure

Employers should prepare for an emergency by practising evacuation procedures, without notice, to make sure everyone understands their roles to act promptly and appropriately. The control measures should be monitored, maintained and reviewed by the appointed person to make sure they are effective and do not deteriorate over time; for example a fire drill that is over-rehearsed could be disregarded by workers in a real emergency situation.

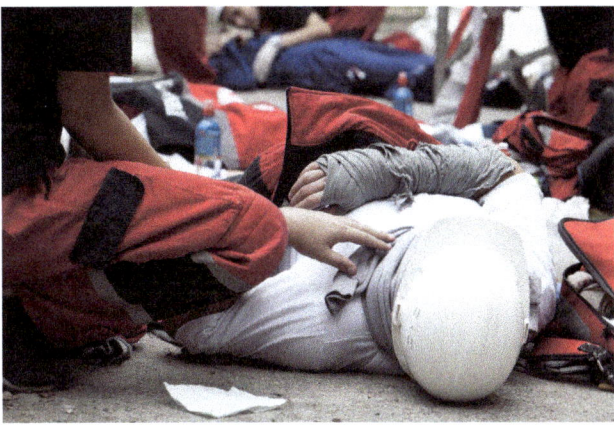

▲ Figure 1.16 A first aider attending to a casualty on a construction site

> **Research**
>
> What are the emergency procedures to deal with electric shock?

9.3 Fire extinguishers and their uses

Fires need three elements to burn:
- heat – such as a spark or naked flame
- fuel – any material or substance that is combustible
- oxygen – a gas that occurs naturally in the air.

If all three are present, a fire is unavoidable. If one element is missing, a fire cannot occur. This is often referred to as the 'fire triangle'.

▲ Figure 1.17 The fire triangle

Fire extinguishers are designed to put out fire by removing one or more of these elements; however, if the wrong extinguisher is used, it could make the fire worse or increase the risk of harm.

The fire risk assessment should identify the appropriate fire extinguishers for a workplace and where they should be situated (known as a fire point). They are usually placed close to exits and where there is an increased risk of a fire, for example hot works.

Fires are classified according to the type of fuel that is burning:
- Class A (flammable solids) – paper, cardboard, fabric
- Class B (flammable liquids) – paints, adhesives, petrol
- Class C (flammable gases) – liquid petroleum gas (LPG), natural gas
- Class D (flammable metals) – magnesium, lithium, aluminium
- Electrical (not classed as E because electrical fires are a source of ignition, not fuel)
- Class F (cooking fats and oils) – chip pan or grease fires.

The CDM Regulations state that everyone at work should be instructed on the safe use of fire extinguishers and understand the colour-coding system used to distinguish different types. All fire extinguishers should have a red body with a colour-coded panel to identify their content and the type of fire that they can safely be used to extinguish.

Table 1.2 describes the range of fire extinguishers.

▼ Table 1.2 Fire extinguishers

Type of extinguisher	Colour of label	Fire classification	Special considerations	How it works
Water	Red	Class A	Do not use on Class B or electrical fires	Cools the fire to remove the heat
Dry powder	Blue	Class A, Class B, Class C, Class D, Electrical	Extinguishers leave a residue that may be harmful to sensitive electronics	Smothers the fire to remove the oxygen
Foam	Cream	Class A, Class B	Do not aim the extinguisher directly at liquid fires. Do not use on electrical or Class F fires	Floats on the surface of the burning liquid to create a seal, starving the fire of fuel

Chapter 1 Health and safety in construction

Type of extinguisher	Colour of label	Fire classification	Special considerations	How it works
Carbon dioxide (CO_2)	Black	Class B Electrical	Horn can become very cold when it is discharged and may freeze, although frost-free versions are available Not to be used in a confined space	Displaces (shifts) the oxygen in the air
Wet chemical	Yellow	Class A Class F	Prevents splashing of hot fats and oils	Cools the fire to remove the heat

Research

Research other types of equipment that can be used to extinguish a fire and explain why they would be preferred to fire extinguishers.

10 Types of PPE

10.1 The purpose and correct use of PPE

When the principles of prevention are applied to mitigate the risk of harm, personal protective equipment (PPE) is always considered a last resort because it only protects the user. Only after employers have considered all other methods of protecting their employees, for example eliminating or controlling risks at source, should they provide suitable PPE or **respiratory protective equipment (RPE)**.

▲ Figure 1.18 A respirator

Key term

Respiratory protective equipment (RPE): personal protective equipment that protects the user's respiratory system

Under the Personal Protective Equipment (PPE) at Work Regulations 1992, employers have legal duties to provide adequate information, instruction and training to their employees on the safe use of PPE. The law also states that employers should provide PPE free of charge, and that their employees should understand how to use and take care of it.

PPE is a last line of defence and will only work properly if the correct type has been selected, it fits correctly and it is in good condition. Therefore, employees must:
- wear PPE when instructed to do so by their employer
- wear and adjust PPE correctly to work safely as trained
- store PPE after use and maintain it
- report any defects in, damage to or loss of PPE.

If more than one item of PPE has to be worn at the same time, you must make sure that they are compatible with each other and that this does not affect the level of protection they provide. PPE that is uncomfortable to wear and does not suit the user can sometimes create further hazards, and discourage workers from using it.

Table 1.3 illustrates types of PPE and the hazards they protect against.

▼ Table 1.3 Types of PPE

Body part protected	Hazards	Types of PPE	Correct use
Ears	Noise – average weekly exposure above 85 dB and other stated levels.	• Ear defenders • Ear muffs • Ear plugs • Canal caps/semi-insert ear plugs	Ear protection should reduce noise levels to an acceptable level, so that you are still able to communicate while wearing it. Ear protectors are manufactured with a single number rating (SNR) system, which allows the acoustic pressure on your ears to be calculated. For example, ear plugs with an SNR of 38 dB (decibels) will reduce a noise of 87 dB by 38 dB to a safe level of 49 dB. To put this into context, a hammer drill produces in excess of 100 dB. Disposable foam ear plugs should be fully inserted in the ear to work properly and disposed of after each use.
Eyes	• Sparks • Dust • Chemicals • Debris	• Goggles • Safety spectacles • Face screens • Face shields • Full-face visors • Sunglasses	Eye protection should be compatible with other PPE worn, adjustable and stored correctly to prevent damage to the lenses.
Feet and legs	• Slips • Falling objects • Objects penetrating the sole (for example, nails sticking out of timber)	• Safety trainers, shoes, boots and wellingtons with toecaps and protective mid-soles • Chainsaw and foundry boots • Knee pads • Kneeling pads	Footwear should have a good grip for different surfaces, for example oil and mud, and be replaced when it becomes damaged. The risk assessment will identify which footwear should be worn. The type of knee protection selected should reflect the type of work undertaken. Kneeling pads protect the user's knees whilst working in the same position for extended periods. They are not worn by the user therefore they are more comfortable to use compared with knee pads. If the work is likely to involve kneeling in different locations, then knee pads may be the better option because they do not rely on the user moving the PPE each time it's needed. There are two different types of knee pads: • Trouser insert knee pads, generally universal fitting, are fitted into work trousers designed with knee pad pockets • Adjustable strap knee pads can be worn with shorts but must be adjusted to suit the user. Poorly fitted adjustable knee pads may reduce the level of comfort and protection offered if they move from their intended position.

Chapter 1 Health and safety in construction

Body part protected	Hazards	Types of PPE	Correct use
Hands and arms	• Cuts and abrasions • Impacts • Chemicals • Temperature extremes • Biological agents • Hand-arm vibration syndrome (HAVS)	• Anti-vibration gloves • Nitrile foam coated gloves • Gloves with cuffs • Gauntlets • Protective arm sleeves • Elbow pads	Care should be taken to select the correct type of gloves to protect against hazards. They must not create further risks, such as entanglement in machinery, when they are being used. There is a variety of protective arm sleeves designed to protect against different hazards (welding, cutting, liquids...) therefore it is important that the correct type is selected and worn. Elbow pads protect the user from the impact of fly debris or hard surfaces. They are often secured and adjusted to suit the user, with elasticated hook and loop straps.
Head and neck	• Falling objects • Hair entanglement • Chemicals • Adverse weather	• Hard hats • **Bump caps** • Snoods • Hair nets	PPE should be worn as directed by the manufacturer. In general, hard hats should be square on your head with the peak facing forwards. Avoid wearing caps or beanies underneath hard hats, unless they are designed for this. Avoid marking hats with paint or pens, because the chemicals damage them. Bump caps are no substitute for hard hats and should only be worn when there is a very low risk of bumping your head.
Lungs (respiratory system)	• Dust • Vapours • Mists • Gases • Atmospheres with low or no oxygen	Respiratory protective equipment (RPE): • disposable half-mask respirators • reusable half-mask respirators with a filter • full-face mask respirators • powered respirators with a mask/hood or helmet • breathing apparatus (BA)	Masks should form a good seal around the user's face to protect them properly. Employers often assess the performance and suitability of masks provided for their employees by undertaking face-fit testing. The type of masks and filters used should reflect the hazards. Employees should understand when and how to replace respirator filters. When masks are not in use, they should be stored correctly to prevent them being contaminated with hazardous substances.
Whole body	• Chemicals • Temperature extremes • Adverse weather • Dust • Metal splashes	• Aprons • Overalls • Boiler suits • Chemical suits • High-visibility clothing	Whole-body protective equipment must be worn according to the manufacturer's instructions and should not cause a risk of entanglement with equipment or machinery. Contaminated PPE should be cleaned or disposed of properly and never mixed with personal clothing.

> **Key term**
>
> **Bump caps:** a type of PPE designed to protect the user's head when there is a low risk of bumping it; it is not designed to take an impact from an object falling from height

Emergency equipment may be required for those working near water, in confined spaces, at height or in oxygen-deficient atmospheres. This includes life jackets, safety harnesses, safety ropes and breathing apparatus. You must be trained in the safe use of emergency equipment and report any damage to it.

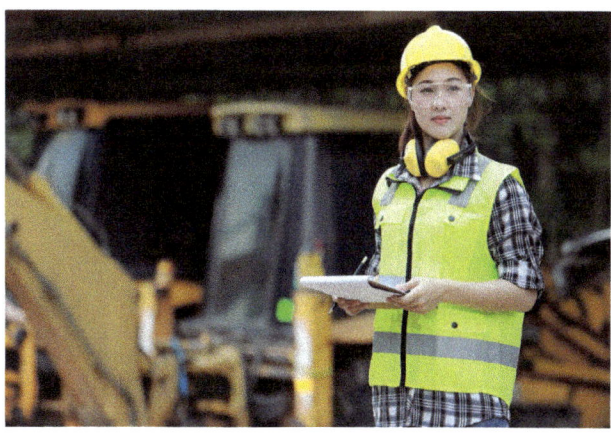

▲ Figure 1.19 A construction worker wearing PPE

Onsite Construction T Level: Core

> **Health and safety**
>
> Always check the expiry dates of safety helmets to ensure they offer the correct level of protection in the event of an accident. Your employer should tell you how to find this information as part of your PPE training.

> **Health and safety**
>
> The Personal Protective Equipment (Enforcement) Regulations 2018 state that PPE must be carefully selected to make sure it is manufactured to European standards. Always check your PPE before using it to make sure it bears **CE marking**. If it does not, inform your employer immediately and replace it. Note that rules on CE marking on PPE are due to be replaced with UK Conformity Assessed (UKCA) marking on 1 January 2023.

> **Key term**
>
> **CE marking:** a mark on a product that shows it has been designed and manufactured to meet EU safety, health or environmental requirements; CE is an abbreviation of a French term '*conformité européenne*', meaning 'European conformity'

11 First-aid facilities

11.1 First-aid facilities that must be available in the work area

The Health and Safety (First Aid) Regulations 1981 place legal duties on all employers to provide adequate and appropriate first-aid equipment, facilities and people to assist their employees if they are injured or taken ill at work. To establish what is considered adequate and appropriate, employers must first carry out a workplace-specific first-aid assessment to determine their needs.

A first-aid assessment should take into consideration:
- the nature of work being carried out and any specific hazards
- the number of staff on site at any one time
- the spread of the workforce
- **lone workers**
- the amount of first aiders and cover needed for holidays/absences
- the number of first-aid boxes and their contents
- first-aid resources needed, for example a first-aid room, stretcher and defibrillator
- staff training, information and instruction.

> **Key term**
>
> **Lone workers:** those who work by themselves without close or direct supervision by their employer or work colleagues; they may work in a remote location or in a workplace with members of the public

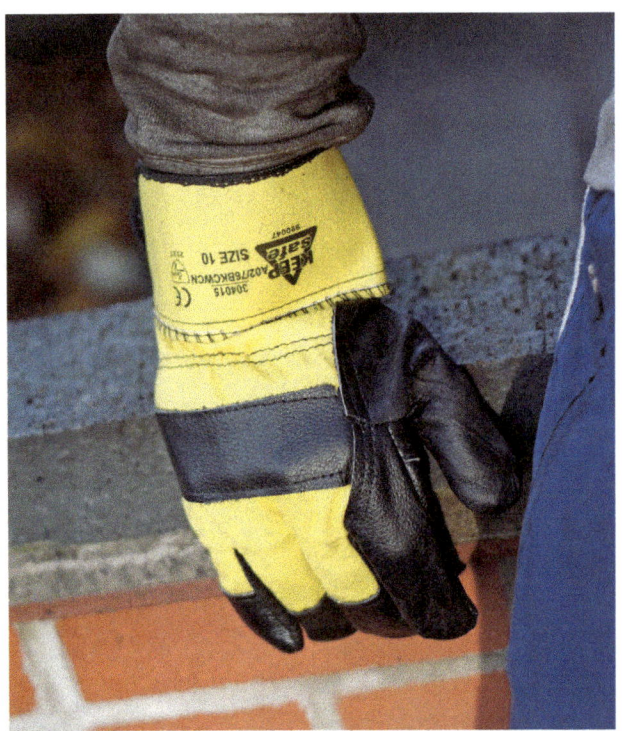

▲ Figure 1.20 PPE with CE marking

▲ Figure 1.21 First-aid facilities on a construction site

Employers responsible for work activities in low-risk environments have to provide a first-aid kit for their workers. They must also appoint a person to take charge of their first-aid arrangements, for example checking first-aid boxes are adequately stocked and calling the emergency services when necessary.

In workplaces where there is a significant risk to employees' health and safety, employers must also have a trained first aider. In the event of the first aider's absence, due to any unforeseen circumstances, the appointed person should be able to provide emergency cover for them.

The HSE publication 'First aid at work' recommends the contents of a basic first-aid box. However, an employer's first-aid assessment will usually determine their exact needs.

The first-aid equipment and facilities provided by employers should be kept clean and dry and made easily accessible. They should also be clearly signposted, in accordance with the Health and Safety (Safety Signs and Signals) Regulations 1996.

Employers have a legal duty to inform their employees about their first-aid arrangements, including who their first aiders are, how to contact them and the location of the first-aid provision. This information is usually shared with workers when they first arrive at work, as part of their site or workplace induction.

> **Research**
>
> What provision has your industry placement employer made for lone workers?

> **Health and safety**
>
> The HSE recommends that tablets and medicines are not kept in a first-aid box.

12 Warning signs for the main groups of hazardous substance

12.1 Categories of safety sign

Safety signs and notices are often displayed around work areas and construction sites to convey important health and safety information. The Health and Safety (Safety Signs and Signals) Regulations 1996 categorise safety signs by their geometrical shape, colour and a simple pictogram. They may also be supported with supplementary text in the same colour as the sign.

The main categories of safety sign are listed in Table 1.4.

▼ Table 1.4 Categories of safety sign

Type of safety sign	Description
Mandatory	A circle with a blue background and white symbol or text Tells you that something *must* be done, for example eye protection must be worn
Safe condition	A square or rectangle with a green background Shows directions to areas of safety and medical assistance in case of emergency

▲ Figure 1.22 First-aid eyewash kit

▼ Table 1.4 Categories of safety sign (continued)

Type of safety sign	Description
Prohibition	A circle with a red outline and a red line from the top left to the bottom right. Tells you that something *must not* be done, for example no smoking or no pedestrian access
Warning	A yellow triangle with a black outline. Makes you aware of nearby danger
Fire fighting	A red rectangle or square with a white pictogram. Marks the location of fire-fighting equipment and fire-alarm activation points

The law states that safety signs should only be used when there is a significant risk to health and safety that cannot be removed or controlled in other ways, or when they can reduce a risk further. If too many signs are displayed in close proximity to each other, they can cause confusion and the main points may be overlooked; therefore, employers must select and position them carefully.

> **Health and safety**
>
> Safety signs should not be used as a substitute for other control methods, for example engineering controls such as safety guards.

> **Test yourself**
>
> What are the five main classifications of safety sign?

> **Case study**
>
> The principal contractor for a construction project is in the pre-construction stage of planning for work on a new building site. The project is expected to last longer than 30 working days and have more than 20 workers on site at any one time during the construction phase.
>
> Explain what safety signs and notices would legally have to be displayed around the site under the Construction (Design and Management) Regulations 2015.
>
> What do you think are the advantages of using safety signs and notices, rather than other safety control measures?

12.2 Symbols for hazardous waste

Manufacturers, importers, distributors and other users of chemicals in the UK have legal duties under the Classification, Labelling and Packaging (CLP) Regulation 2010 to use appropriate safety signs for the classification, labelling and packaging of hazardous substances and waste. These signs are explained in Table 1.5.

▼ Table 1.5 CLP Regulations safety signs

Safety sign	Meaning	Encountered when using ...
Explosive	Explosive, self-reactive	Gas

Safety sign	Meaning	Encountered when using ...
Flammable	• Flammable gases, solids, liquids and aerosols • Self-heating, self-reactive • Contact with water creates flammable gas	• Expanding foam • Nail-gun canisters • Solvent cement • Paint stripper
Oxidising	• Oxidising gases, liquids and solids • May cause fire or explosion • May intensify fire	Chemicals
Gas under pressure	• Contains gas under pressure • May explode if heated • Contains refrigerated gas which may cause cryogenic burns	Carbon-dioxide cylinders used in welding
Corrosive	• Corrosive to metals • Causes severe skin burns and eye damage	• Portland cement • Hydrated lime • Brick cleaner • Batteries

▼ Table 1.5 CLP Regulations safety signs (continued)

Safety sign	Meaning	Encountered when using ...
Acute toxicity	• Toxic from single or multiple exposure • Toxic/fatal if swallowed, in contact with skin or inhaled	• Materials containing formaldehyde • Hazardous air pollutants
Health hazard/hazardous to the ozone layer	• May cause respiratory, eye or skin irritation • May cause drowsiness or dizziness • Harmful if swallowed, inhaled or in contact with skin • Harms the environment by destroying the ozone layer	• Expanding foam • Grab adhesive • Wood adhesive • Solvent cement • Portland cement • Paint stripper
Hazardous to the environment	Toxic to the surrounding natural environment, especially aquatic life	• Wood preservative • White spirit • Diesel, petrol and paraffin oils • Epoxy resin • Bitumen paint
Serious health hazard	• May be fatal if swallowed or enters airways • May cause damage to organs • May damage fertility or cause genetic defects • May cause cancer • May cause allergy, asthma or breathing difficulties if inhaled	• Expanding foam • Grab adhesive • Paint stripper • Wood dust • White spirit • Asphalt • Silica dust

Research

Find one example for each of the CLP Regulation classifications listed in Table 1.5 and describe how they should be safely handled and disposed of as hazardous waste.

The Carriage of Dangerous Goods and Use of Transportable Pressure Equipment Regulations 2009 deal with potentially hazardous substances that are transported by road or rail. These regulations have a slightly different set of diamond-shaped safety signs, designed to suit their classifications of predominant hazards.

> **Research**
>
> Research the classifications of safety sign under the Carriage of Dangerous Goods and Use of Transportable Pressure Equipment Regulations 2009.
>
> Make a list of some of the materials you may have used that would require these signs to be displayed on the vehicles transporting them, and explain how waste materials and packaging were disposed of after use.

13 Safe practices and procedures for the use of access equipment and manual handling

13.1 Types of access equipment

The HSE reported that 111 workers died as a result of falling from height in 2019–20 in the UK, and 40 of those people worked in the construction industry.

If there is a risk of people falling any distance above or below ground that could result in an injury, the employer must take necessary precautions to eliminate the hazard completely or reduce the risk of harm to an acceptable level by assessing the risks.

Where a risk remains, employers should use equipment or other measures to minimise the distance and consequences of a fall, for example safety nets or air bags, or PPE such as a safety harness and lanyard. In the event of an accident, employers should have a well-rehearsed plan and equipment designed to rescue and evacuate people from height.

Equipment specifically designed for working safely at height is referred to as access equipment. It should only be used by trained, competent and authorised people in accordance with the manufacturer's instructions.

Ladders

▲ Figure 1.23 Ladder

A ladder should be:
- set at an angle of 75° or a ratio of 1:4 (the distance between the wall and the base of the ladder should be one quarter of the ladder's height)
- placed on firm, level ground
- placed against a stable surface
- extended 1 m above a working platform
- secured to prevent slipping/moving
- used for light work and short durations.

The user should have three points of contact with the ladder at all times and never overreach.

Mobile scaffold towers

▲ Figure 1.24 Mobile scaffold tower

A mobile scaffold tower should:
- be set up on firm, level ground
- be erected in accordance with the manufacturer's instructions
- not be overloaded
- use guardrails and toe boards
- have brakes applied on castors before use
- have its outriggers correctly positioned and secured when necessary to gain extra height
- not be repositioned with people, materials or equipment still on it.

Scaffolding

▲ Figure 1.25 Scaffolding

Scaffolding should:
- only be erected, inspected, adjusted and dismantled by trained and competent scaffolders
- be designed and erected in accordance with British Standards and the Work at Height Regulations 2005
- have handrails 950 mm high, with no more than a 470 mm gap between guardrails
- have toe boards 150 mm high
- have platforms kept clean and clear.

Chapter 1 Health and safety in construction

Steps

▲ Figure 1.26 Steps

Steps should be:
- fully opened
- placed on firm, level ground
- positioned facing the work, not sideways
- used for light work and short durations.

Podiums

▲ Figure 1.27 Podiums

Podiums should:
- be set up on firm, level ground
- be erected in accordance with the manufacturer's instructions
- have the gate locked while working
- have brakes applied on castors before use.

Staging boards

▲ Figure 1.28 Staging boards

Staging boards are designed to create a safe working platform between two supports, with guard rails on one or both sides.

You should never exceed the safe working load of staging boards with materials and people.

Trestles

▲ Figure 1.29 Trestles

Trestles should:
▶ be set up on firm, level ground
▶ be erected in accordance with the manufacturer's instructions
▶ not be overloaded
▶ have staging boards kept clean and clear
▶ allow safe access and egress.

Mobile elevating work platforms (MEWPs)

▲ Figure 1.30 Boom lift (top) and scissor lift (bottom)

There are a number of different types of mobile elevating work platform, each one designed for access, safety and work at height. Only suitably trained, experienced and competent workers should operate a MEWP.

13.2 Safety checks on access equipment

Under the Provision and Use of Work Equipment Regulations (PUWER) 1998, employers must ensure that access equipment is suitable for its intended use, safe and well maintained.

They must also regularly inspect the components of access equipment for signs of wear and damage, to ensure they have not deteriorated. Records should be kept of weekly, monthly and annual inspections by the people that supply the access equipment, for example employers and hire companies. A tagging system may be adopted, with tags attached to different types of access equipment as a way of recording inspections and identifying when it is safe to use.

Chapter 1 Health and safety in construction

▲ Figure 1.31 A tag on access equipment

Some work equipment may be subject to other regulations which have specific requirements regarding inspections, such as LOLER. Scaffolding, for example, should be inspected before use and every seven days after that. However, further inspections may be necessary if there has been adverse weather or an accident/incident.

Daily visual pre-use inspections of access equipment should be completed by competent employees, to ensure there are no defects and the equipment is safe to use. These inspections should include checking that the non-slip feet on ladders and step ladders are not missing or damaged, and the stiles, steps and rungs are not bent or missing. Even the smallest amount of damage to access equipment can compromise its strength; therefore it should always be reported as soon as possible and damaged equipment should never be used.

13.3 Selecting access equipment

Access equipment must be appropriate for the work intended. This will be determined by the risk assessment and method statement, not its availability and cost.

Before working at any height, employers must consider the safest way to complete the task, its duration and the frequency with which access equipment will be used.

Other points to consider when selecting access equipment include:
▶ whether it is for internal or external use
▶ the ground conditions, for example uneven, sloped or soft
▶ the weather conditions, for example high winds or snow/ice
▶ safe access and egress
▶ the height at which it will be used
▶ the space needed for people and materials
▶ how materials and equipment will be loaded onto it
▶ the weight it will carry
▶ nearby hazards, for example overhead power lines and fragile roofs
▶ the type of work being undertaken
▶ training, information, instruction and supervision.

The risk of a fall from ladders and steps is relatively high; therefore they should only be used for a short duration and no longer than 30 minutes at a time. If this is not possible, the risk must be reduced by using a safer type of equipment, such as a podium.

Certain types of access equipment have action points for heights; for example a tower scaffold needs outriggers to support the equipment when it exceeds heights specified in the manufacturer's instructions. Further advice and guidance on the safe use of access equipment can be found on the HSE website.

13.4 Manual handling operations

Manual handling is defined as any lifting, carrying, supporting or moving of a load using bodily force. Employees working in the construction industry will inevitably be involved in carrying out manual handling operations and therefore have an increased risk of an injury or chronic muscular damage. These types of injury are referred to as musculoskeletal disorders (MSDs) and affect different parts of the body, including the back, shoulders, arms, legs, feet and toes. The HSE reported in the year 2019–20 that half a million people were suffering with a new or existing work-related musculoskeletal disorder, and this figure does not include workers who have not reported injuries under RIDDOR.

Where employees are expected to undertake manual handling operations in the workplace, their employers must take reasonably practicable measures to protect them from harm. The Manual Handling Operations Regulations 1992 state that this should be done by:

- avoiding carrying out manual handling if possible – consider completing the task another way, for example with a forklift, telehandler or crane
- assessing the hazards when it is not possible to avoid manual handling activities by completing a suitable and sufficient manual handling risk assessment
- reducing the risk as much as is reasonably practicable.

If this is not possible, workers should use good **kinetic lifting** techniques or be provided with mechanical lifting aids, such as a pallet truck or sack truck. Where this is not possible, consider changes to the task, load and working environment.

> **Key term**
>
> *Kinetic lifting:* the act of manual handling

Manual handling injuries often occur because activities are badly performed or planned. Employers have a duty to make sure their workers have the necessary information, instruction and training. They must consider the following points when assessing manual handling activities:

- **Task:**
  - What does the task involve?
  - Does the task involve more than one person?
  - How far will the load have to be carried?
  - How often will the task take place?
- **Individual:**
  - Is the worker physically and mentally able to carry out the task?
  - Does the worker have the necessary knowledge to complete the task safely?
  - Has the worker received the necessary training?
  - Does the worker have the right PPE, for example gloves or footwear with a good grip?
- **Load:**
  - Consider the size, shape, weight, centre of gravity and temperature of the load.
  - Is the load sharp?
  - Is the load likely to move?
- **Environment:**
  - Consider the weather and ground conditions.
  - Will the route be clear or are there likely to be obstructions or obstacles, such as stairs or scaffolding?

▲ Figure 1.32 Safe manual handling sequence for a single person

▲ Figure 1.33 Manual handling as a team

Employees have a legal duty to follow their employer's safe system of work, to protect themselves and others from the risk of injury from manual handling operations. They must also use lifting equipment and machinery provided by their employer, if they are trained to use it, and report any unsafe conditions they identify to their supervisor as soon as possible.

▲ Figure 1.34 Examples of mechanical lifting aids: a pallet truck (left) and a sack truck (right)

> **Improve your English**
>
> Think of a word that can be used to describe a condition that develops suddenly or in a short space of time. Use this word in a sentence to explain a particular risk to health in your industry.

> **Health and safety**
>
> There are *no* safe working loads when you are manual handling. Even light loads carry a risk of injury.

14 Safe practices and procedures for working in excavations and confined spaces

14.1 Dangers associated with excavations

Excavations are often created on construction sites to form trenches and holes for building foundations, or to gain access to underground services and drainage. Pedestrians, materials and vehicles are all at risk of falling into exposed excavations unless satisfactory preventative measures are taken to keep them at a safe distance.

Working in an excavation can be extremely dangerous because of the risk of flooding or collapse causing people to be crushed. There could be reduced levels of oxygen and poisonous or explosive atmospheres.

Excavations are most likely to fail if they are too deep and unsupported, or if the soil conditions are weak. The walls of an excavation may also become unstable if there is vibration in the ground caused by vehicle movement around the site or nearby public roads. Additionally, they can weaken and collapse if the soil of an excavation dries out in hot weather, expands after a heavy frost or becomes saturated with rain water.

▲ Figure 1.35 An excavation on a construction site

Plant operators must be competent and take precautions while excavating with machinery to ensure they do not come into contact with buried services, such as gas pipes or live electricity cables. Special care should also be taken when digging next to adjacent structures and buildings, to make sure they are not disturbed or undermined, causing them to become unstable or collapse.

If the construction site has previously been used for manufacturing, industrial or agricultural purposes, the soil may have been contaminated with hazardous waste or chemicals that have been buried or have naturally occurred in the ground.

Under the CDM Regulations, the client has a duty to provide information to the contractor about the location of buried services, any underground structures or any other relevant information before work starts on site. This will allow the contractor to prepare a pre-construction health and safety plan, reducing the risk of an accident or dangerous occurrence.

> **Key term**
>
> *Plant operators:* people in control of heavy construction machinery and equipment

▲ Figure 1.36 A plant operator

14.2 Safety when working in excavations

All excavations should be individually planned and carried out in a safe way by competent workers and supervisors. Wherever possible, working in an excavation should be avoided and only considered as a last resort.

Employers have a duty to reduce the likelihood of an accident or injury in an excavation by assessing the risks and introducing control measures such as the following:
- Isolate the hazard by erecting suitable edge protection barriers, for example Heras-style fencing (temporary metal fence panels), to prevent unauthorised access and to protect the general public.
- Install 'stop blocks' a safe distance away from the excavation to prevent vehicles getting too close and reduce the risk of hazardous exhaust fumes entering the excavation.
- Display safety signs and notices to raise awareness of the hazards and the precautions to be taken.
- Provide adequate lighting.
- Provide safe access (entry) to and egress (exit) from the excavation, for example using ladders or temporary stairs.
- Use a temporary support system to prevent the excavation from collapsing, for example a trench box (metal sidewalls held apart with struts).
- Provide safe designated crossing points.
- Test the soil for contamination.
- Adopt a permit-to-work system to dig and work in the excavation.
- Provide adequate ventilation to workers in the excavation.

- Identify and protect exposed services.
- Carry out daily inspections of the excavation.
- Ensure a competent person carries out atmospheric testing at various points within the excavation before anyone enters, at the start of each shift and after breaks, and continuously thereafter if indicated by the risk assessment and permit to work. Personal monitors may also be worn by individuals working in the excavation.
- Use appropriate portable gas-detection equipment.
- Ensure good communication methods between workers in the excavation and the supervisor.
- Put in place a rescue plan to recover an injured person and provide appropriate rescue equipment.

▲ Figure 1.37 A trench box supporting an excavation

14.3 Dangers associated with confined spaces

The definition of a confined space is a workplace which may be substantially but not always entirely enclosed, where there is a foreseeable serious risk of injury because of the conditions or from hazardous substances. Excavations, loft spaces, sewers or wells could be described as confined spaces, because they are enclosed with restricted access and egress.

Confined spaces may already contain hazards, such as a lack of oxygen or a risk of an explosion. Hazards can also be created by workers with electrical equipment, machinery, materials and substances such as petrol and solvents. Some confined spaces may also contain toxic atmospheres created by gases such as naturally occurring methane, nitrogen or carbon dioxide. Carbon monoxide and carbon dioxide are particularly dangerous because they are colourless, odourless and tasteless, making them more difficult to detect without specialist testing equipment.

Chapter 1 Health and safety in construction

The air we breathe contains a mixture of different gases, with oxygen making up about 21 per cent. If the oxygen in the atmosphere is reduced just below this level, it affects our ability to function normally; if depleted further to 6 per cent or below, it will result in almost certain death.

Additional hazards that may be found in confined spaces include:
- fire
- extremes of heat and cold
- dust, fumes and vapours
- flooding resulting in drowning
- free-flowing solids causing suffocation, for example sand
- entrapment.

▲ Figure 1.38 A confined space

Research

Which solvents that you use in your trade would be particularly hazardous in a confined space? Consider what alternative substances you could use.

Test yourself

Which regulations control the use of hazardous substances?

14.4 Safety when working in confined spaces

The Confined Spaces Regulations 1997 state that working in a confined space should be avoided wherever possible, and work should be completed in another way without entering the space. If this cannot be done and there is still a significant risk of injury, then the work must be properly planned and organised with appropriate control measures in place before it starts.

The Management of Health and Safety at Work Regulations 1999 state that employers and the self-employed must complete a suitable and sufficient risk assessment for work in a confined space. This is used to identify hazards and determine what precautions need to be taken to reduce the risk of injury. It will also help employers to prepare a safe system of work.

Everyone involved in working in a confined space must be competent and specifically trained to undertake their tasks. This includes managers, supervisors and emergency personnel.

If a permit-to-work system is adopted, it would limit the number of people and time spent in a confined space and safely control work activities. Before a permit is issued, workers are told about:
- the risks to health and safety and how to control them
- methods used to communicate from the inside to their supervisor or **sentry** on the outside
- how to raise the alarm in an emergency
- testing and monitoring of the atmosphere
- PPE or RPE to be worn
- arrangements to recover them from the confined space if necessary (a rescue plan).

Key term

Sentry: a person who supervises workers from the access/egress points of a confined space

Assessment practice

Short answer

1. Which organisation enforces the Health and Safety at Work etc. Act 1974?
2. What are the five main categories of safety sign?
3. Which regulations require specific accidents, injuries and near misses to be reported to the HSE?
4. What should not be kept in a first-aid box?
5. When should you be informed about the site rules at work?

Long answer

6. Explain the duty every employer has under the Control of Vibration at Work Regulations 2005 to protect people from the harmful effects of exposure to vibration.
7. Explain the term 'safe system of work'.
8. Explain what a toolbox talk is and when it might be used.
9. Give a definition of 'confined space'.
10. Explain the best way to avoid the hazards of working in a confined space.

Project practice

Your industry placement employer has reported a significant rise in the number of employees absent from work in the last 12 months due to musculoskeletal disorders (MSDs). At the last health and safety meeting, the management team decided to conduct a health and safety audit of the reported accidents and injuries over the past year and a safety survey, with a view to reducing the risk of MSDs.

Your task is as follows:
▶ Work collaboratively with your workplace mentor or employer to gather as much information as you can about the number of people at work that may have had an MSD. This could be achieved by studying health and safety records, interviewing workers or conducting a survey.
▶ Review the current manual handling risk assessment and method statements and identify any areas that could be improved.
▶ Produce a digital toolbox talk for a manual handling activity and present this to a small group of workers.

Chapter 2 Construction science principles

Introduction

In this chapter, we will explore scientific principles that underpin all the career pathways you may choose. Later in your studies, you will use these scientific foundations and build on them with specialised principles.

Learning outcomes

By the end of this chapter, you will understand:
1. materials science principles
2. mechanical science principles
3. electricity principles
4. structural science principles
5. heat principles
6. light principles
7. acoustics principles
8. earth science principles.

The International System of Units (SI)

Imagine a job building a house where the carpenter measures everything in inches, the plumber measures in centimetres, the bricklayer uses feet, and the electrician uses metres. What do you think the outcome would be?

Equally, in science, if someone performed a calculation using one unit of measurement, and another person used a different unit, they would produce different results which could lead to confusion.

As a result, the International System of Units, known as SI units, was introduced. The system sets out what units of measurement, symbols and indices are used to calculate quantities.

Before we look at the SI units, it would be a good idea to understand what indices are. Indices are used to replace repetitive multiplications. For example, $10 \times 10 \times 10 = 1000$. This calculation can be written easily as 10^3, which means ten multiplied by itself three times, or three lots of ten multiplied together.

Where indices are negative, the value becomes a fraction, because the value moves to the right of the decimal point, becoming smaller. For example:

$$5^{-1} = \frac{1}{5} \text{ or } 0.2$$
$$5^{-2} = \frac{1}{25} \text{ or } 0.04$$
$$5^{-3} = \frac{1}{125} \text{ or } 0.008$$

Most calculators have an x^2 button to square a number, and scientific calculators also have an x^y button, which allows a number to be raised to any power or index. For example, to calculate 5^5, use buttons 5 x^y 5 $=$ 3125. This is much easier than keying $5 \times 5 \times 5 \times 5 \times 5$ into a calculator.

Generally, in construction science and principles, large values are used, such as thousands of **watts** or millions of joules. Other aspects of building services engineering or construction work deal with tiny amounts, such as millionths of a metre or thousandths of an ohm. This can become a problem in calculations, as errors may occur if the correct number of zeros is not entered into the calculator.

Instead of inserting the actual number with lots of zeros, we use 'to the power of ten'.

The 'power of' numbers are given names that are explained in Table 2.1. There is less chance of making an error using this method.

▼ Table 2.1 Numbers expressed as indices (to the power of 10)

Actual number	Number shown to the power of 10	Prefix used
1 000 000 000 000	10^{12}	tera (T)
1 000 000 000	10^9	giga (G)
1 000 000	10^6	mega (M)
1000	10^3	kilo (k)
100	10^2	hecto (h)
10	10^1	deca (da)
0.1	10^{-1}	deci (d)
0.01	10^{-2}	centi (c)
0.001	10^{-3}	milli (m)
0.000 001	10^{-6}	micro (μ)
0.000 000 001	10^{-9}	nano (n)
0.000 000 000 001	10^{-12}	pico (p)

Test yourself

What is 30 kilometres multiplied by 500 millimetres?

Improve your maths

Calculators are different depending on the manufacturer and when they were made. Using the internet, your calculator instructions, or by simply experimenting with your calculator, find out what the following functions do. Note that what appears on the button may differ depending on the calculator.

- x^{-1} or $\frac{1}{x}$
- ENG and shift ENG
- $\times 10^x$ or EXP
- x^2 and x^3

SI units are broken down into two different categories:
- base SI units
- derived SI units.

Base SI units are the main category used, whereas derived units require the base functions to determine them. Base SI units are shown in Table 2.2.

Key term

Watt: SI unit of power

Chapter 2 Construction science principles

▼ Table 2.2 Base SI units

Quantity	Unit of measurement	Identification symbol
Mass	kilogram (kg)	m
Length	metre (m)	l
Time	second (s)	t
Temperature	kelvin (K)	T
Electric current	ampere (A)	I
Luminous intensity	candela (cd)	I

Note: the amount of a substance in moles is also a base unit but not relevant to the subjects in this book.

As well as knowing the base SI units of measurement, you may be required to carry out simple calculations using them.

Calculations using base SI units obtain other values called derived SI units. For example, length multiplied by width of a rectangle gives the rectangle's area (or m × m = m²).

Derived SI units

There are many derived SI units of measurement. Table 2.3 shows those relevant to construction-based activities, including the formulae used to calculate them.

▼ Table 2.3 Derived SI units of measurement and associated calculations

Quantity	Unit of measurement	Identification symbol	Base formula (where relevant)
Area (or cross-sectional area)	Square metre (m²)	A	For squares or rectangles: $$length \times width$$ For circles: $$\pi \times radius^2$$ or $$\pi r^2$$
Volume	Cubic metre (m³)	V	For cuboids: $$length \times width \times depth$$ or $$area \times depth$$ For cylinders: $$\pi r^2 \times depth$$ or $$area \times depth$$
Flow	Kilograms per second (kg/s) or commonly litres per second (l/s)	mdot ($\dot{m}$)	$$\frac{volume}{time} \text{ or } \frac{litres}{second}$$
Velocity (speed)	Metres per second (m/s)	v	$$\frac{distance}{time}$$ or $$\frac{metres}{second}$$
Acceleration	Metres per second squared (m/s²)	a	$$\frac{velocity}{time}$$

▼ Table 2.3 Derived SI units of measurement and associated calculations (continued)

Quantity	Unit of measurement	Identification symbol	Base formula (where relevant)
Electromotive force	Volt (V)	ε	$\varepsilon = BLv$ Based on the length of conductor (L) in a field of magnetic flux density (B) and the velocity of movement of the conductor through the field (v)
Electrical resistance	Ohm (Ω)	Ω	$\dfrac{\rho L}{A}$ Based on material resistivity (ρ), length of the conductor (L) and cross-sectional area of the conductor (A)
Illuminance (light on a surface)	Lux (lx)	E	Lumens per m², where lumens are the measure of visible light emitted by a source
Internal pressure	Pascal (N/m²)	Pa	$\dfrac{N}{m^2}$ or $\dfrac{J}{m^3}$
Atmospheric pressure	Bar	bar	Equal to 100 000 Pa
Energy (work)	Joule (J)	E	force × distance or f × d or power × time or watts × seconds
Force	Newtons (N)	F	mass × acceleration
Density	kg/m³	ρ	$\dfrac{mass}{volume}$
Power	Watts (W)	P	$\dfrac{energy}{time}$ or $\dfrac{joules}{seconds}$ In electrical circuits: volts × amperes or V × I
Specific heat capacity	Joules per kilogram per degree Celsius (J/kg/°C)	C_p	$J \times kg \times (t_2 - t_1)$ Total energy (J) required to raise the temperature of a mass (kg) from temperature 1 (T_1) to temperature 2 (T_2)

Chapter 2 Construction science principles

1 Materials science principles

1.1 Materials and their properties

Construction work and BSE installations use many different materials based on their properties. Some of these materials are described in Tables 2.4 and 2.5.

> **Industry tip**
>
> You will need to apply indices when calculating room sizes from plan drawings, as plans use mm ($\times 10^{-3}$) and area and volume SI units are m^2 and m^3 respectively.

> **Improve your maths**
>
> A room measures 5 m wide by 8 m long and has a ceiling height of 2.5 m. The air mass in the room is required to calculate heating needs. If air has a density of 0.946 kg/m^3, what is the mass of air in the room?

▼ Table 2.4 Properties and common uses of metals

Material	Properties	Common uses	Reasons for use	Other information (where relevant)
Iron	• Produced by melting mined iron ore • Very heavy **ferrous metal** with a high melting point	• Versatile material used for decorative fences and gates, as well as stoves • Widely used for structural supports before steel	• Common element which is mined worldwide • Resistant to high temperatures but can be formed and shaped when heated, providing decorative structures	• Prone to oxidising (rusting) when exposed to the weather if not suitably treated, such as by painting
Steel	• Technically an alloy, made from iron and carbon • Ferrous metal that is very rigid	• Structural supports • **Rebar** • **Catenary wires** • Support cables • Sheet materials, such as stainless-steel sinks or splashbacks • Cable reinforcement	• Very strong dense material which provides long-lasting structural support and expands at the same rate as concrete, so used widely for reinforcement	• Prone to rusting if not treated during or after production
Copper	• Fairly soft but rigid **non-ferrous** metal which is mined as ore • Often not pure and mixed to make an alloy • Corrosion resistant • Good electrical conductor	• Pipes • Cable conductors • Architectural features	• Durable, corrosion-resistant metal with low **thermal** expansion • Anti-microbial, meaning it remains clean from bacteria	• Common material which is easily recycled (which is why it may be an alloy, due to impurities during the recycling process) • Can form a green coating known as patina
Aluminium	• Very lightweight and durable non-ferrous metal • Corrosion resistant to weather	• Sometimes used as a cable conductor where lightweight heavy-duty cables are needed • Used widely in construction for facias/cladding and window frames • Used extensively for flexible ducting systems	• Low thermal expansion • Easily formed or fabricated • Lightweight material	• Can react with other materials • Not suitable for applications such as piping, due to reaction with salt and other chemicals

▼ Table 2.4 Properties and common uses of metals (continued)

Material	Properties	Common uses	Reasons for use	Other information (where relevant)
Alloys	• Customised metals made by adding different metals together • Common examples include nickel, tungsten and manganese • Can be ferrous if iron or steel is used in the process	Made specifically for an application by mixing metals with the desired properties, for example adding chromium to steel produces stainless steel	Ability to produce a material suitable for an intended application, such as low thermal expansion materials or corrosion-resistant metals	

> **Key terms**
>
> **Ferrous metal:** a metal that contains iron and is magnetic
>
> **Rebar:** reinforced steel bar commonly used in concrete to act as a frame to stop it moving and cracking
>
> **Catenary wires:** strong wires which are tied at each end and used to support other objects, such as cables which may stretch or break under their own weight when hung between two buildings
>
> **Non-ferrous:** a metal that does not contain iron and is non-magnetic
>
> **Thermal:** related to heat or temperature

Table 2.5 identifies other common materials used in construction, together with their properties and common uses. None of these contain metal, so they are not ferrous.

▼ Table 2.5 Properties and common uses of non-metals

Material	Properties	Reasons for use	Other information (where relevant)
Aggregates	Mineral materials in the form of loose particles or grains. Examples include crushed stone or sand. Aggregates are defined as coarse or fine dependent on them passing through a 4-mm sieve. Sand (fine aggregate) for use in construction must be clean and free from contaminants. A silt test is carried out to confirm suitability	Commonly mixed with cement for mortar or concrete depending on the mixture used. (See mortar and concrete detailed separately below)	Aggregates can also be used as stand-alone materials such as **hardcore** or gravel
Lime	Usually bagged as a fine white powder. It is an alkaline which is produced by heating crushed limestone	Used in mortars as it retains water well, acting as a plasticiser by forming a paste which allows grains of sand to move over each other more freely. This increases workability. Displaces air from mortar, which increases bond strength	Liquid chemical plasticisers can be used as an alternative to lime in mortar

Chapter 2 Construction science principles

Material	Properties	Reasons for use	Other information (where relevant)
Cement	A mixture of crushed limestone and other materials such as clay and chalk. For Portland cement, the most common type, the materials are heated to 1500°C in a kiln (a type of high-temperature oven) then ground to a fine powder. Other types of cement such as Rapid Hardening, Sulphate Resisting and High Alumina are produced for specialist purposes	When added to other aggregates, cement acts as a binder in mortar or concrete	Portland cement was so named because when used as a mortar, it resembles Portland stone
Mortar	A mixture of cement and an aggregate such as sand, with water. The amount of water used depends on the type of masonry material being used and the working conditions on site, such as excessively high or low air temperature and wet weather. The mortar mix will usually include an additive to improve workability (plasticiser) or occasionally to vary the hardening and curing time (accelerator or retarder)	Mortar is used to bond together masonry materials such as brick, block and stone.	Ratios of cement to sand in a mortar mix will change according to the type of masonry material specified and the type of wall finish required. • 1:3 Exposed external use • 1:4 General external use • 1:5 Sheltered internal use • 1:7 General internal use
Concrete	Concrete is a versatile material consisting of a mix of specified proportions of cement, sand (fine aggregate) and chippings (coarse aggregate). It is widely used in foundations and solid floor slabs for buildings. When used with suitable formwork (essentially temporary moulds often constructed from sheet timber), it can be reinforced with steel bars or mesh to form the load-bearing structure of a building. It can be used to form roadways and paths. For efficiency, it is often mixed off site and transported to the work location to be poured into position	Concrete can withstand high compressive loadings, making it suitable for applications like foundations. It can be designed to withstand tensional forces when used in beams or girders, by being reinforced with steel bars or mesh	Mixed ratios of cement, fine aggregate and coarse aggregate vary depending on use. Concrete is specified by 'M' (for 'Mix') categorisations. For example: • M15 will be mixed to a ratio of 1:2:4 which would be suitable for use when securing fence posts. • M20 will be mixed to a ratio of 1:1.5:3 which will be suitable for foundations for a garden wall. There are many other specified mixes. Removing air bubbles in the mix creates a denser, stronger material when the concrete cures. This can be achieved by vibrating the laid concrete with specialist equipment before hardening, or by tamping the surface.

▼ Table 2.5 Properties and common uses of non-metals (continued)

Material	Properties	Reasons for use	Other information (where relevant)
Bricks	Bricks fall into three main categories: • Facing bricks • Engineering bricks • Common bricks They can be manufactured from a range of materials such as clay, clay mixed with shale, concrete, and calcium silicate (also known as 'sand and lime' bricks). Clay and clay/shale bricks are shaped in moulds or 'wire-cut' using steel wire that cuts through a slab of soft material of the correct thickness. The soft brick is then converted to a durable component by a process known as 'firing' in a large kiln. Concrete bricks are moulded and allowed to cure at an even temperature. Calcium silicate bricks are moulded to shape under high pressure in a hydraulic press and then subjected to high pressure steam in large piece of equipment called an autoclave	• Facing bricks can be formed in a range of colours and textures making them suitable for use as the 'face' or finish of the external walls of a building. They can also be used internally as a decorative feature. • Engineering bricks are often used where they must withstand high loadings and continuous saturation such as in an underground inspection chamber. • Common bricks do not have a decorative face and are used where they may be hidden out of sight such as in a partition wall that is rendered or plastered over	Bricks are specified for a project according to their characteristics. Consideration will be given to the location and level of exposure to weather, the type of finish required, the loadings that will be created in the building and other factors. The brick description will include characteristics such as: • Dry weight • Compressive strength • Soluble salt content • Water absorption • Thermal conductivity • Absorption rate Bricks are tested to ensure that their properties are fully suited to the requirements of a project and that consistent manufacturing standards are maintained
Timber	Softwood timber used in construction is normally from species such as spruce, pine, Douglas fir or larch, as these trees grow tall and straight. Softwood timber used for structural purposes is treated to protect it from fungal attack and insect infestation. Hardwood timber is often used where bespoke components, such as a feature staircase, are specified or an external architectural feature, such as period window frames, is installed	Timber is used throughout the construction of buildings. It can be used for structural frames and supports, floors, internal partition walls, roof structures, door and window frames and much more. It is a renewable sustainable construction material which, when used for structural components in a building, can speed up the construction process and reduce drying times allowing earlier occupation of the building	• Timber has two principal gradings: GS for general structure and SS for special structure • Timber strength is given using C gradings such as C14 or C30; the higher the C number, the stronger the timber is structurally • Timber structural strength can be very different depending on the direction of grain, presence of knots or **fissures** in the timber
Sheet materials – wood based	These are factory-manufactured sheets: • Plywood – made from gluing together multiple layers of wood sheets with the grain of each layer at right angles. The stronger the sheet becomes with more layers as shown in Figure 2.1 • Blockboard – softwood timber strips which are glued and sandwiched between two thin outer sheets • Chipboard or particle board – formed by pressing and gluing together small wood chippings or waste pieces of wood as shown in Figure 2.2 • MDF (Medium Density Fibreboard) – an engineered wood product made by breaking down wood fibres and compressing and heating them in combination with wax and a resin binder	Sheet materials are used in construction for flat roof decking, floor coverings, sheathing partition walls and load-bearing external walls in timber frame construction, and much more. Wood-based sheet materials add strength and allow fast completion of work tasks	Plywood boards for external use are referred to as 'marine ply' and are manufactured with waterproof glue to bond the layers of wood sheets. While the glue is waterproof, the timber layers are not, so the sheets of material need protection from continuous saturation

Chapter 2 Construction science principles

Material	Properties	Reasons for use	Other information (where relevant)
Plasterboards	Plasterboard is a sandwich of dried and hardened gypsum plaster between two sheets of paper. By changing the mixture of gypsum, these boards can be water resistant, and provide acoustic insulation or fire resistance.	Used extensively to provide a smooth internal wall or ceiling surface ready for decorating	
Paints	Paints can be categorised as water or oil based and these act as the binder, or main body of the paint. Pigments are added to the binder to give different colours or durability to the paint. Solvents are also added to alter the viscosity or thickness/density of the paint, which alters its workability and ability to penetrate surfaces or blemishes in the surface	Paint is generally used for two reasons: • appearance • surface/materials protection	Solvents that change the viscosity of paint include turpentine for oil-based paints or water for emulsions, which are water-based
Thermoplastics and thermosetting plastics	Soft plastics which are more flexible and suitable for higher temperatures than general PVC	• Flexible • Can operate at high temperatures while remaining good insulators • Used for cable insulation and sheathing	
Unplasticised polyvinyl chloride (uPVC)	More rigid material than PVC but has a lower thermal expansion rate	• Does not expand at the same rate as PVC • Durable, non-corrosive product which can be moulded into any shape during production • Used for window and door frames; guttering, facias and cladding	Shatters easily when hit, as far less flexible than plastic
Rubber	Very flexible material which can maintain electrical insulation at higher temperatures, such as 125°C	Although a natural resource, can be replicated using a manufacturing process to provide a highly durable yet very flexible material that is corrosion and water resistant. Used for: • Cable insulation • Flexible joints • Water seals	
Ceramics	• Includes a broad range of materials manufactured from **minerals**, sometimes with added metals to provide high degrees of strength • Hard, durable and resistant to heat and fire	• Heat-resistant bricks (firebricks), tiles, sinks and worktops • Retain and slowly release heat in storage heaters • Electrical insulators in high-temperature applications	• High durability when exposed to heat • Corrosion resistant, tough materials

Onsite Construction T Level: Core

▼ Table 2.5 Properties and common uses of non-metals (continued)

Material	Properties	Reasons for use	Other information (where relevant)
Plaster	Made from gypsum, lime or cement mixed with water and sand	• Provides a smooth finish to walls and ceilings • Can be moulded to make decorative features • Can be premanufactured into sheets or boarding and partitioning • Cheap to manufacture • When in a wet state, easy to apply to uneven and out-of-line surfaces to provide a smooth, level finish • Can provide a reasonable fire barrier, slowing the spread of fire in a building	• Only suitable for interior uses • Requires cladding or coating if used in generally wet areas, such as shower rooms

▲ Figure 2.1 Chipboard

▲ Figure 2.2 Plywood

Key terms

Hardcore: made from waste broken bricks, stone and/or blocks, used to provide a base

Fissures: in timber these are areas where the grain opens up leaving a void, normally caused by resin or sap. Fissures can degrade the timber strength

Mineral: a solid, naturally occurring, inorganic substance

Tamped: pressed down by a succession of blows

Site concrete testing

The quality of concrete must be monitored to ensure a uniform mix and consistency through different batches that are made, either onsite or delivered as ready-mixed.

Slump testing

Slump testing involves using a slump test cone (frustum), a tamping rod and a rule.

The cone is filled by one third using a sample from a batch of concrete. This is tamped into the cone 25 times using the tamp rod, ensuring the concrete has filled evenly without air gaps. The cone is then filled to two thirds and tamped again to ensure no air gaps.

Chapter 2 Construction science principles

▲ Figure 2.3 Concrete slump test

Now fill the cone completely and tamp again, ensuring the concrete remains perfectly level with the top of the cone.

The cone is lifted allowing the concrete to 'clump down'. The amount of slump is measured as shown in Figure 2.4 from the top of the cone.

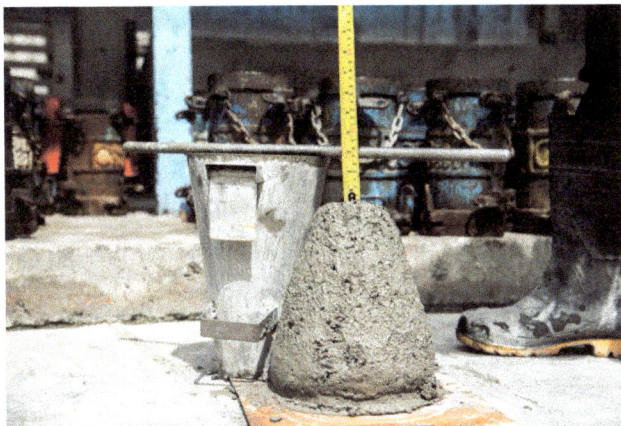

▲ Figure 2.4 Measuring the amount of slump from the top of the cone

The slump measurement should be consistent through batches.

Compaction factor test

This testing method is normally used in a laboratory but can be employed on site. It is more sensitive than a slump test when used for concrete mixes that have low workability.

This is a concrete mix that does not flow easily and will be compacted by vibration when placed in position on site.

The test requires specialist equipment consisting of three cylindrical hoppers mounted vertically in line, through which the concrete is fed and weighed in carefully controlled stages.

Methods of mixing

There are two ways to get the materials ratio correct before mixing. These are:
- ▶ volume batching – where a container such as a gauge box or bucket is used to measure a consistent volume of material quantities. When mixing small amounts, quantities of materials are often measured by the 'shovel full', but this can be inaccurate since the amount loaded onto a shovel can vary depending on the strength of the operative and the techniques used. If materials are supplied in bags rather than bulk loads, the ratio of a mix can be established if the volume of the bagged materials is known – for example, one 25 kg bag of cement is 0.033 m^3.
- ▶ weight batching – where the ratios are mixed by weight. So, a mortar mix of 1:3 cement to sand would be, as an example, 25 kg of cement to 75 kg of sand.

Small mixes can be done either by hand on a board or in a powered mixer such as that in Figure 2.5

▲ Figure 2.5 Using a powered mixer

Mixers can vary in size and where very large batches are required, mixes are made offsite and transported in mixer lorries, such as that in Figure 2.6

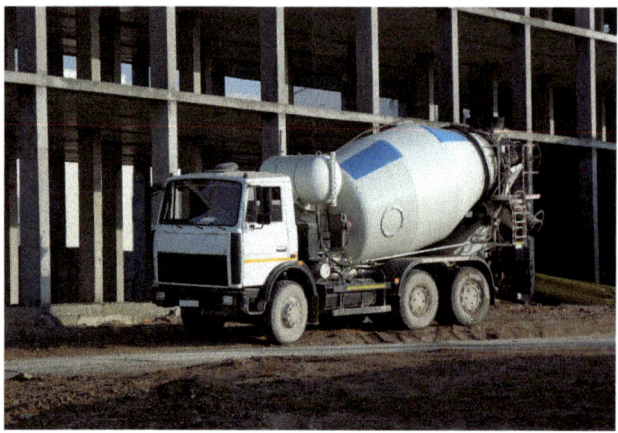

▲ Figure 2.6 Transporting mixes in a mixer lorry

Other environmental effects

Other environmental effects on materials to consider in more detail include:
- electrolytic and galvanic corrosion (metals)
- dissimilar metals
- thermal expansion
- UV radiation.

Electrolytic and galvanic corrosion (metals)

Corrosion is a chemical reaction where two metals are separated by an electrolyte which causes one metal to corrode at an accelerated rate. Without suitable protection, these types of corrosion can cause structural parts of a building to fail.

Electrolytic corrosion occurs where electrical current is induced into nearby metals from other influences, such as buried high-voltage cables. This causes a **voltage** difference between the metals, which in turn creates a circuit. As current is a flow of electrons, the electrons from one metal are drawn to the other, which means the metal is losing electrons and corroding. The electrolyte is whatever is between the two metals, such as soil or clay, and allows electric current to flow through it.

Galvanic corrosion is where two dissimilar metals are in close proximity and separated by an electrolyte or the two metals are touching. If one metal is more chemically active than the other, such as stainless steel to zinc, the proximity is similar to creating a battery and this causes current to flow between them, which again results in loss of electrons and corrosion.

Key term

Voltage: the amount of potential energy between two points in an electrical circuit, expressed as volts or V

Research

Aluminium can be used as a cable conductor as it has good conductive properties but is a lightweight material, meaning that the cables are significantly lighter than equivalent copper cables. However, aluminium also has some disadvantages; for example, it is prone to electrolytic corrosion if preventative measures are not taken.

Research where this problem of electrolytic corrosion occurs and why, and suggest measures to reduce the risk of it occurring.

Thermal expansion

Different materials expand at different rates. This can cause failure of building structures and services if correct measures are not taken.

One example is PVC conduit used to protect electric cables. It is usually fixed to a wall using saddles. The amount of expansion, in metres, for PVC conduit is, on average, 52×10^{-6} for every metre length of conduit, per °C rise in temperature (52×10^{-6} m/m/°C). This is known as its linear temperature expansion coefficient.

If a 6 m run of conduit was installed at 20°C but the temperature of the space was increased to 45°C (as might occur in a loft space), by how much would it expand?

$$52 \times 10^{-6} \times 6 \text{ m} \times (45° - 20°) = 0.0078 \text{ m or } 7.8 \text{ mm}$$

If the conduit was installed without taking the expansion into consideration, the additional 7.8 mm would cause it to bend between supports, making it look wavy instead of straight.

As a result, PVC conduit systems should have expansion couplers evenly spaced in the installation to provide a space for the conduit to expand into and stop it from buckling or waving.

▲ Figure 2.7 PVC conduit with expansion coupler to stop it buckling

Concrete and steel expand at similar rates. If they did not, many structures made of concrete and reinforced with steel would crumble, with one material expanding more quickly than the other.

Other linear temperature expansion coefficients for common materials are shown in Table 2.6.

▼ Table 2.6 Linear temperature expansion coefficients for common materials

Material	Linear temperature expansion coefficient (m/m/°C)
Aluminium	$21 - 24 \times 10^{-6}$
Brick (masonry)	5×10^{-6}
Cast iron	10×10^{-6}
Concrete	$12 - 14 \times 10^{-6}$
Copper	$16 - 17 \times 10^{-6}$
Plate glass	9×10^{-6}
Plaster	17×10^{-6}
PVC	$54 - 110 \times 10^{-6}$
Steel	$10 - 12 \times 10^{-6}$
Wood across its grain	30×10^{-6}
Wood along its grain	3×10^{-6}

Ultra-violet radiation

Ultra-violet (UV) radiation comes from direct sunlight and can cause some plastic-based materials such as white PVC to degrade.

The radiation breaks down polymers in the plastic which normally allow it to be flexible. When the polymers break down, the plastic becomes brittle and can break easily. This is why white PVC pipes or cables should never be used outdoors in direct sunlight.

Some plastics are specially treated to withstand UV radiation, such as black plastics. This is why most outdoor plastic pipes are black or grey. Unplasticised PVC (uPVC) is also UV resistant and is commonly used as a plastic material where white colouring and strength are required, such as for guttering or windows and window sills.

2 Mechanical science principles

It is important to understand the ways in which mechanical loads are measured. You need to know the difference between mass and weight, as well as energy and power. Energy and power can be mechanical as well as electrical.

The fundamental relationship between mass and weight is defined by Newton's second law. According to this law:

$$F = ma$$

where:
- ▶ F = force (N)
- ▶ m = mass (kg)
- ▶ a = acceleration (m/s²).

Mass is a measure of the quantity of matter in an object. It is not dependent on gravity and is therefore different (but proportional) to weight.

Weight is the downwards gravitational force acting on a mass. Newton's second law can be transformed to express weight as a force by replacing acceleration (a) with acceleration due to gravity (g):

$$W = mg$$

where:
- ▶ W = weight (N)
- ▶ m = mass (kg)
- ▶ g = acceleration due to gravity (on Earth, this is 9.81 m/s²).

Acceleration

Acceleration is the rate of change of velocity with time:

$$a = \frac{v}{t}$$

where:
- ▶ a = acceleration (m/s²)
- ▶ Δv = change in velocity (m/s)
- ▶ t = time (s).

As acceleration is a measurement of the rate of change of speed in metres per second (m/s) each second (s), the unit is metres per second per second, or metres per second squared (m/s²).

How levers work

Levers are tools that exert a large force when applying a lesser force. They come in all shapes and sizes, such as crowbars, claws on claw hammers, wheelbarrows or even seesaws in playgrounds.

While levers give an advantage in terms of the force applied, the distance travelled is increased, meaning the energy used is essentially the same. This is because the turning moments (torque) applied to the ends of a lever must be equal but opposite. Therefore:

$$\text{force} \times \text{distance} = \text{force} \times \text{distance}$$

Torque is measured in Newton metres (Nm). This is the product of the force (N) applied to a lever and the distance (m) of the force from the **fulcrum**.

> **Key term**
>
> **Fulcrum:** the pivot point of a lever

Levers are categorised into three classes.

Class 1 levers

In a class 1 lever, the force and the load are on different sides of the fulcrum. The effectiveness of the force depends on its distance from the fulcrum; the greater the distance, the greater the effect of the force.

In Figure 2.8, the force applied downwards can be four times less than the load, as the lever on the force (effort) side of the fulcrum is four times longer than the lever on the load side.

▲ Figure 2.8 Class 1 lever

Examples of class 1 levers are seesaws, crowbars and claw hammers when used to lift nails. The advantage gained by a lever is referred to as mechanical advantage (MA), which is calculated as:

$$MA = \frac{load}{effort}$$

So, for the lever in Figure 2.9, the mechanical advantage is:

$$\frac{400}{100} = 4$$

Class 2 levers

In a class 2 lever, the load is between the force and the fulcrum.

The calculations remain the same, but the load is more limited because it is between the two points, and the force and the desired movement are both in the same direction. An example of a class 2 lever is a wheelbarrow.

▲ Figure 2.9 Class 2 lever

In the example in Figure 2.9, the force of effort of 100 N is applied 2.5 m from the fulcrum, and the load is applied 0.5 m from the fulcrum.

The load that can be lifted is calculated as follows:

force of load × distance from fulcrum =
force of effort × distance from fulcrum

Therefore, the force of the load in Figure 2.10 can be calculated by:

force of load × 0.5 m = 100 N × 2.5 m

This can then be represented as:

$$force\ of\ load = \frac{100\ N \times 2.5\ m}{0.5\ m} = 500\ N$$

Class 3 levers

In a class 3 lever, the force is between the load and the fulcrum. Examples of class 3 levers include fishing rods, tweezers, tongs and the human arm.

The advantage with class 3 levers is not so much the effort needed but the distance travelled.

▲ Figure 2.10 Class 3 lever

Pulleys

A pulley is an effective way of gaining a mechanical advantage when lifting an object. By running a rope through a four-pulley system, a mechanical advantage of four is gained.

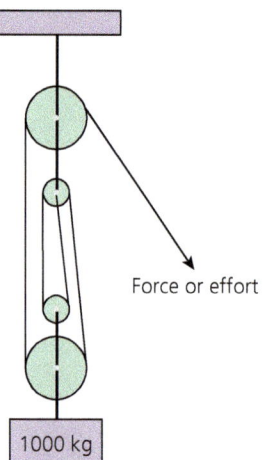

▲ Figure 2.11 A pulley system lifting a mass of 1000 kg against gravity

Looking at Figure 2.11, we can determine the force required to raise the mass of 1000 kg. As the load has been described as a mass, first determine the downward force of the load:

$$\text{force} = \text{mass} \times \text{gravity}$$

$$f = 1000 \times 9.81 = 9810 \text{ N}$$

Remember: gravity is acceleration, which averages 9.81 m/s².

As the system has four pulleys compared to one pulling rope, the mechanical advantage is 4:1.

Remember:

$$MA = \frac{\text{load}}{\text{effort}}$$

Rearrange the formula for mechanical advantage to find the effort required:

$$\text{effort} = \frac{\text{load}}{MA}$$

$$\frac{9810}{4} = 2452 \text{ N}$$

Therefore, a downward force (or effort) of 2452.5 N is required to raise the load.

Although the pulley gives a mechanical advantage, the pulling rope will need to be pulled four times further than the load is raised. This means to raise the load by 1 m, the pulling rope needs to be pulled 4 m.

> **Test yourself**
>
> A mass of 2575 kg needs to be raised 2 m by a pulley with a mechanical advantage of six. How much effort is required?

Work

If the force applied to a body results in movement, then work has been done. This applies to forces that lift, push or twist objects.

When an object moves in the same direction as the force exerted, the work done is equal to the force exerted multiplied by the distance moved:

$$\text{work} = \text{force} \times \text{distance}$$

Or, to include the values used to determine force:

$$\text{work} = \text{mass} \times \text{gravity} \times \text{distance}$$

Mechanical work is measured in **joules** (J). (Newton-metres (Nm) can be used for mechanical work, but are also used as a measurement for torque.) Other units of work or energy, which are not SI units but commonly used for specific applications, include:
- kilowatt hour (kWh), used by electricity supply companies to measure electrical energy
- calorie, often used as a measure of food energy
- BTU (British Thermal Unit), often used for heat source applications such as burning gas.

> **Key term**
>
> **Joules:** the unit of measurement for energy. Where energy is expressed as mechanical energy, it is known as work

Let us work through an example. If a mass of 100 kg is lifted 10 m, calculate the work done:

$$\text{Force} = \text{weight} = m \times a = 100 \times 9.81 = 981 \text{ N}$$

$$\text{Work done} = \text{force} \times \text{distance} = 981 \times 10 = 9810 \text{ J}$$

If the mass is doubled, the work done is:

$$\text{Force} = \text{weight} = m \times a = 200 \times 9.81 = 1962 \text{ N}$$

$$\text{Work done} = \text{force} \times \text{distance} = 1962 \times 10 = 19{,}620 \text{ J}$$

Energy can exist in many forms but is categorised into two main groups:
- kinetic energy (the energy an object has due to its motion, such as a rotating machine)
- potential energy (stored energy held by an object because of its position or state, for example a coiled spring).

The potential energy of gravity keeps a mass on the ground. If the mass is raised, then a machine uses kinetic energy. If the input of kinetic energy ceases, the potential energy tries to bring the mass back down to the ground.

Power

Power is defined as the rate of doing work. It is calculated by dividing work done by the time taken to carry out that work:

$$\text{average power} = \frac{\text{work done}}{\text{time taken}}$$

The unit of power is joules per second (J/s), and 1 J/s is equivalent to 1 watt (W).

The output (mechanical) power required for a motor to raise a mass of 1000 kg to a height of 5 m above the ground in one minute is calculated as:

$$P = \frac{m \times a \times d}{t}$$

So:

$$P = \frac{1000 \times 9.81 \times 5}{60} = 817.5 \text{ W}$$

If the same motor raised the same load in 10 seconds, the output power required by the motor would be:

$$P = \frac{1000 \times 9.81 \times 5}{10} = 4905 \text{ W}$$

The amount of energy used is the same, no matter how quickly the task is carried out, but more power is required to do the work in a shorter time.

> **Test yourself**
>
> How much power is needed to raise a load of 90 kg a distance of 150 m in 20 seconds?

Efficiency

The law of conservation of energy states that energy cannot be created or destroyed, only transformed from one form to another. However, during the transformation process, some of the energy may be turned into unwanted forms, such as noise or heat, known as energy losses. Such losses are common in mechanical processes.

The **efficiency** of a mechanical system is the ratio of output power compared to input power expressed as a percentage:

$$\% \text{ efficiency} = \frac{\text{output power}}{\text{input power}} \times 100$$

> **Key term**
>
> **Efficiency:** the ratio of output power compared to input power expressed as a percentage

It is more common for efficiency to be expressed in terms of power rather than energy, although energy could replace power in this formula.

If a machine with a 200 kW output has an input power of 220 kW, the machine efficiency is:

$$\% \text{ efficiency} = \frac{200}{220} \times 100 = 90.9\%$$

> **Test yourself**
>
> An electric motor drives a hoist which is to raise a weight of 400 N to a height of 25 m in 120 seconds.
>
> If the motor and mechanical losses are 20 per cent, calculate the electrical power input to the motor.

As well as occurring in mechanical systems such as machines, losses in efficiency can also occur in specific BSE systems:

- Electrical systems: the passage of an electric current represents a flow of power or energy. When current flows in a circuit, power loss occurs in the conductors due to resistance. This causes heat dissipation and voltage drop, which are also losses.
- Heating systems: pipework can lose heat. Long systems can dissipate a lot of heat if measures such as lagging are not used. Pipe systems with small diameters or many bends can also lose pressure, meaning more power is required to pump fluids through the system.
- Forced air systems: these can lose pressure if ducting contains many bends or reduced diameters, which in turn will require more power to force the air through.

Centre of gravity

The centre of gravity of an object is an imaginary point where the weight of the object is concentrated.

Figure 2.12 shows a uniform object, where the centre of gravity is in the centre of the object. If a pivot is placed under the centre of gravity, the object remains balanced, but if the pivot is placed under any other part of the object, the object is unstable and falls.

When lifting or manually handling, it is easier to carry an object if it is supported under the centre of gravity. This state, where the object is perfectly balanced, is known as equilibrium.

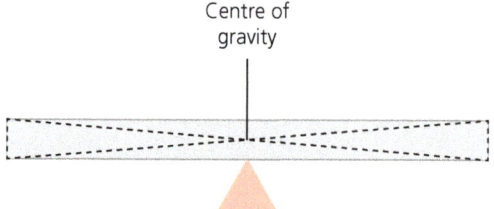

Centre of gravity over the pivot means the object remains balanced

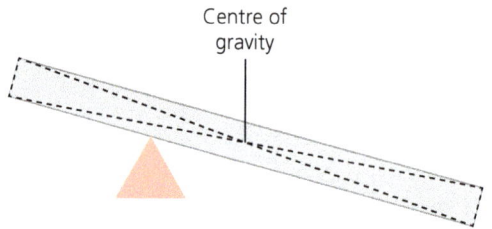

If the centre of gravity is moved away from the pivot, the object falls or becomes unstable

▲ Figure 2.12 Effects of centre of gravity

Chapter 2 Construction science principles

A building is constructed from materials and components that add up to a potentially significant weight – a multi-storey building may weigh thousands of tonnes, creating what are referred to as 'loadings'.

The weight of the building itself and the permanent fixtures and systems within it are called 'dead loads'. The weight of occupants, furniture and other non-fixed items are referred to as 'imposed loadings' or 'live loads'.

The design of the building ensures that the forces generated by these loads or loadings are transferred safely through the structure to the designed foundation and dispersed into the natural foundation beneath the building – the ground that it sits on.

Certain components of a building are designed and tested to withstand loadings at specific locations in a building. For example, openings in walls are spanned by load-bearing components called lintels, and floors are supported by joists or beams that span from wall to wall.

Components like these are designed to withstand a range of forces under load. For example, the points at which the ends bear on walls are subject to a force referred to as 'shear'. The bottom of the component is subject to tension, while the top of the component is subject to compression. (More on these forces in section 4 of this chapter.)

The whole length of the component must be designed to minimise bending and deflection caused by the weight the component is supporting above it.

Earlier in this chapter, the relationship between mass and weight was explained using Newton's second law. Structural engineers use formulae based on scientific laws and principles to design components and structures that are load-bearing. An example is the method of quantifying pressure in Newtons per square millimetre (N/mm²).

Moments or torque

Torque is a term often used in building services and mechanics. It refers to how tight something needs to be, such as a nut, bolt or screw. Too tight and it may break the thread or snap, not tight enough and it may come loose.

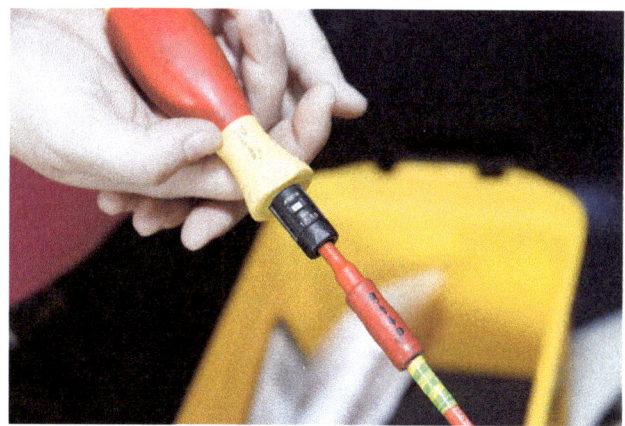

▲ Figure 2.13 A torque screwdriver with adjustable settings

Torque, or moment, is measured in Newton metres (Nm) and determined by:

$$M = \text{force } (F) \times \text{distance } (d)$$

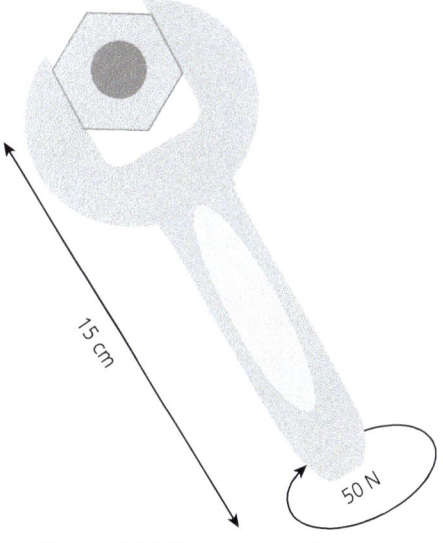

▲ Figure 2.14 Force applied to a spanner in the form of torque

The torque applied by the spanner in Figure 2.14 can be calculated, first ensuring the 15 cm is converted to 0.15 m:

$$\text{Moment or torque} = 50 \text{ N} \times 0.15 \text{ m} = 7.5 \text{ Nm}$$

> **Test yourself**
>
> A nut requires 12 Nm torque by a wrench that has a length of 320 mm. How much force needs to be applied?

Onsite Construction T Level: Core

3 Electricity principles

Whichever career path you take in construction, you will work on or near electricity, particularly in building services engineering. Knowing basic electrical principles will help you to work safely.

There are three key areas for delivering an electrical supply to an installation:
- generation
- transmission
- distribution.

3.1 Generation

Electricity can be generated in many ways due to a large-scale move towards **renewable resources**. The use of **fossil fuels** in the UK is now in decline, due to the need to protect the environment, although they are better suited to times of peak demand.

Peak demand refers to times when consumer demand for electricity is at its highest, for example on cold, dark days when more heat and light are needed. Another example would be at half-time during a big televised football match, when large numbers of people boil a kettle to make tea.

Fossil-fuel sources

Fossil fuels all work in the same way. Fuel is burned, which heats water to form high-pressure steam. This steam turns a **turbine**, which rotates a generator at high speeds to produce electricity.

Fossil fuels are not renewable. They are very polluting and release carbon dioxide and other greenhouse gases when burned, leading to **climate change**.

> **Key terms**
>
> **Renewable resources:** resources that can be replaced over time by natural processes, for example wind energy or solar energy
>
> **Fossil fuels:** fuels such as coal, oil and gas that are mined from the earth and burned to produce energy
>
> **Turbine:** a machine that uses a moving stream of air, water, steam or hot gas to turn a wheel and generate power
>
> **Climate change:** a large-scale, long-term change in the Earth's weather patterns and average temperatures

Gas

Gas is the most widely used fossil fuel. It can produce instant heat, so gas generators can heat water to steam much more quickly than coal generators.

While its use is in decline, due to the need to reduce greenhouse gases produced when it is burned, it remained the largest energy source in the UK in 2021. Location of the power plant is not critical, as gas can be piped to most mainland areas of the UK.

Oil

Oil is used for many regional generators, which power local areas at times of peak demand. The generators can deliver electricity immediately.

Oil is also used for private standby generators, for buildings such as hospitals that require a continuous power supply during mains power cuts.

Traditionally, oil-fired power stations are located near coastal oil refineries, to minimise transportation of the oil.

Coal

Coal used to be the most widely used fuel to produce high-pressure steam for turbines. However, it is highly polluting and its use has therefore been scaled down in the UK, with only a few coal-fired power stations remaining. Most of these are only used during periods of peak demand. Coal-fired power stations were traditionally located near to UK sources of coal, such as South Wales, Yorkshire and the North East of England.

In 2017, the UK went 24 hours without using a single coal-fired power station, for the first time since 1882. In 2020, the entire summer months went without coal being burned. This is a huge step forward in reducing pollution and climate change.

Renewable fuel sources

Wind

Wind power turns a propeller, which directly drives a generator. Wind is a natural resource, so no pollution is created or resources are required after production of the turbine.

The use of wind energy is ideal for the UK, as an island with a lot of wind from the sea. Electricity generation from wind power is increasing each year, with many offshore wind farms being constructed several miles out to sea.

Chapter 2 Construction science principles

Wave
The sea produces a huge amount of energy, both in the form of waves and as the tide moves in and out each day. This movement of water can be used to turn electrical generators through a rotating waterwheel effect or using a back-and-forth motion.

Hydro
There are three types of hydro generation:
- Run of river: this uses the natural downward flow of rivers and harnesses the water's energy by using existing **weirs**.
- Storage: watercourses are held back by a dam and released through pipes that divert the water through turbines.
- Pumped storage: a large volume of water is held in a reservoir high up on hills or mountains until needed. When there is high demand for electricity, the water is released from the reservoir through turbines. When the station is offline, the water is pumped back up to the reservoir, ready for the next time it is needed.

> **Key term**
>
> *Weirs:* low dams across a river, which increase the force of the water as it flows over the top; sections of a weir can be raised or lowered to regulate the force of the water

Photovoltaic
Photovoltaic (PV) cells convert solar energy into electricity without any moving parts. You can see these on the roofs of some houses and in fields called solar farms.

> **Research**
>
> Biofuel is increasingly being used to generate electricity and provide a fuel source for boilers in offices and houses. What is biofuel and is it carbon neutral?

Electricity is generated at power stations using a three-phase system. A three-phase system is one where the rotation of the generator is used to maximum effect by having three windings, so one turn of the generator gives three times the current distributed over three wires. A single-phase generator would only distribute current in one wire with a neutral return path. This means much of the magnetic field is unused during one revolution of a large generator. To maximise output, most generators are three-phase, with only small-scale portable generators being single-phase.

3.2 Transmission systems

In the UK, the transmission system is called the **National Grid**. It is a network of mainly overground cables used to send electricity all around the UK from the generator stations. As transmission systems are hundreds of miles long and send vast amounts of electrical power, they use a range of very high voltages:
- 400 kV (known as the super grid)
- 275 kV
- 132 kV.

> **Key terms**
>
> *National Grid:* the network of power lines supplying electrical energy around the country

There are two main reasons why high voltages are used:
- High voltages mean reduced current, so smaller conductor sizes can be used.
- As the cables travel vast distances, the voltage lost due to cable resistance has less of an impact on high voltages than it would on lower voltages.

As an example, a locality consumes 80 MW of electricity. Calculate the current demand at 400 kV and 400 V.

To calculate the current demand based on power and voltage values, use:

$$I = \frac{P}{V}$$

So, at 400 kV and remembering that 80 MW is 80×10^6 watts:

$$\frac{80 \times 10^6}{400000} = 200 \text{ A}$$

So, the cables need to be big enough to carry 200 A to the area of demand.

To calculate the current demand at 400 V, the same process is used, so:

$$\frac{80 \times 10^6}{400} = 200000 \text{ A or 200 kA}$$

At this level of current, the cables would need to be huge. The benefit of increasing the voltage is to reduce the current, making cable sizes much more realistic.

Where cables have resistance (see section 3.5), this resistance leads to power loss in the form of voltage drop. The amount of voltage drop is directly proportional to the amount of current and the value of resistance as:

$$\text{voltage drop} = \text{current} \times \text{resistance}$$

So, if a section of transmission cable had a resistance of 20 Ω and carried a current of 200 A, the voltage drop would be:

$$200 \times 20 = 4000 \text{ V}$$

> **Test yourself**
>
> If a section of a transmission system has a resistance of 30 Ω and supplies a locality with a demand of 25 MW, what is the voltage loss at 400 kV and 132 kV?

There are issues with transmitting at higher voltages:

- High voltages can break down insulation. When cables are overground, air is used as an insulator between conductors. In underground cables, materials such as PVC are used as insulators between conductors.
- Conductors need to be suspended high above the ground and several metres away from each other, as high voltages can jump across or break down air, especially when the air has a high water content (i.e. relative humidity).

The towers used to carry electricity cables are known as pylons. The higher the voltage used in the transmission system, the bigger the pylons must be. Pylons used for the 400 kV super grid are as high as a tower block and have six cables suspended from them (three on each side). There is a single cable running between the tops of pylons, which acts as a common earth.

Transformers are used to change voltages within the transmission system, by either stepping up the voltage (for example from 132 kV to 400 kV) or stepping it down. If voltage is increased, current will decrease; if voltage is decreased, current will increase. This relates to the number of primary (input) windings in the transformer compared with the number of secondary (output) windings, shown by:

$$\frac{N_p}{N_s} = \frac{V_p}{V_s} = \frac{I_s}{I_p}$$

> **Key term**
>
> **Transformers:** devices that convert voltages and current, proportionately, to different values

All transmission voltages are three-phase, so the transformers used to step up or step down the voltage in the transmission system need to be three-phase as well. This means they have three incoming wires into three windings, which step up or step down the voltage into three outgoing wires.

Where transmission systems enter urban areas or pass through the countryside but need to be hidden, they may need to run underground. In this case, the voltage will be stepped down (for example to 132 kV) so it does not damage the insulation between the conductors in the underground cable.

> **Test yourself**
>
> Why are high voltages used over long distances?
>
> Why are high voltages, such as 400 kV, not used all the way up to the local substation?

3.3 Distribution systems

At points around the National Grid, electricity is tapped off to be distributed to the user. These systems are known as distribution systems. They are looked after by distribution network operators (DNOs).

As distribution is much more localised, the voltages can be stepped down to lower values. This keeps the pylon sizes smaller. Lower voltages mean cables in urban areas can be run underground, keeping the supplies invisible. As underground systems are more expensive, rural areas normally use cheaper overhead supplies.

Depending on how far the electricity needs to be distributed, distribution voltages may be:

- 33 kV (three-phase)
- 11 kV (three-phase)
- 400 V (three-phase)
- 230 V (single-phase).

In most cases, the underground cables in towns and cities are 11 kV. These supply the many **substation** transformers.

> **Key term**
>
> **Substation:** equipment that transforms voltage to a suitable level for consumers

On the outgoing side of the substation (to consumers' installations), the supplies are 400 V three-phase or 230 V single-phase, depending on overall current demand. As a general rule:

- buildings with a demand over 100 A will have a three-phase supply
- buildings with a demand below 100 A will have a single-phase supply.

▲ Figure 2.15 Generators (G) supplying the consumer via transmission/distribution systems

Large buildings and facilities may have their own substation transformer. These places may be supplied at 11 kV or 33 kV.

▲ Figure 2.16 Three-phase and single-phase distribution to consumers

3.4 Basic electrical circuit principles

Electricity is the flow of electrons from one atom to another. Materials that are good **conductors** have electrons which move out of orbit from atom to atom. When the material is connected to an electromotive force (emf) such as a battery, the flow can be controlled in one direction. This is because the electrons are attracted to the positive plate of the battery.

This flow of electrons is called **charge** and happens in materials that are good conductors, such as copper, iron and steel.

Materials whose atoms keep their electrons in orbit make good **insulators**, such as rubber or PVC.

What makes materials or elements different is the structure of their atoms. The number of electrons orbiting the nucleus is different for different types of material. For example, copper has 29 electrons but iron has 26. The number of electrons in an atom equals the number of protons.

When 6.24×10^{18} electrons flow in one direction, this is one coulomb of charge.

> **Key terms**
>
> **Conductors:** materials that have atoms less densely packed together and allow electron flow
>
> **Charge:** the measure of electron flow (Q) in a material, measured in coulombs (C)
>
> **Insulators:** materials that have atoms which are densely packed together so that electrons cannot readily move

3.5 Properties of an electrical circuit

Table 2.7 shows key electrical values, their SI units of measurement and formulae used to calculate them.

▼ Table 2.7 Properties of an electrical circuit

What needs calculating	SI unit of measurement	Symbol	Associated formulae
Charge	Coulomb (C)	Q	$1\,C = 6.24 \times 10^{18}$ electrons $Q = It$
Current	Ampere (A)	I	$I = \dfrac{Q}{t}$ or $I = \dfrac{V}{R}$
Electromotive force or circuit voltage	Volt (V)	V	$V = I \times R$
Resistivity	Ohm-metre (Ω-m)	ρ	See resistance below
Resistance	Ohm (Ω)	R	$R = \dfrac{V}{I}$ or $R = \dfrac{\rho L}{A}$
Power	Watt (W)	P	$P = V \times I$ or $P = I^2 \times R$

Conductors and resistivity

Before we study **Ohm's law** in detail, we need to look at conductors and insulators, and how much resistance a circuit has based on the material used as the conductor. This is based on the resistivity (ρ) of the material.

> **Key term**
>
> **Ohm's law:** a law that states the relationship between current, voltage and resistance in an electrical circuit

As different materials have different numbers of electrons, they conduct electricity differently. Each material has a resistivity value based on the measurement of the resistance of a 1 m³ block of the material at 20°C. Table 2.8 shows some common resistivity values.

> **Research**
>
> The elements in the periodic table are all given an atomic number. What do these numbers represent?

> **Key term**
>
> *Resistance:* the measure of how well a material conducts electricity in ohms (Ω); the lower the value of resistance, the better it conducts

▼ Table 2.8 Common materials and their resistivity

	Material	Resistivity value (ρ)
Common materials used as conductors	Copper	0.0172×10^{-6} Ω-m
	Aluminium	0.028×10^{-6} Ω-m
	Gold	0.024×10^{-6} Ω-m
	Steel (used in cables)	0.46×10^{-6} Ω-m
Common materials used as insulators	Hard rubber	1×10^{13} Ω-m
	Glass (average value)	1×10^{12} Ω-m
	Dry wood	1×10^{14} Ω-m
	PVC (average)	1×10^{15} Ω-m

To work out the value of resistance, we use the material's length (L) and cross-sectional area (CSA) (A) and apply the resistivity calculation:

$$R = \frac{\rho L}{A}$$

Remember, when you see a formula with no sign between two values, multiply them. So ρL is actually $\rho \times L$.

When calculating resistivity, if you always use a resistivity value in $\mu\Omega$-m (micro-ohm-metres or $\times 10^{-6}$) and the CSA in mm² ($\times 10^{-6}$), then the two values of $\times 10^{-6}$ cancel out. You can simply input the values without using the $\times 10^{-6}$.

What is the resistance, at 20°C, for 30 m of copper cable having a CSA of 2.5 mm²?

$$R = \frac{\rho L}{A}$$

So:

$$R = \frac{0.0172 \times 10^{-6} \times 30}{2.5 \times 10^{-6}} = 0.21 \, \Omega$$

> **Test yourself**
>
> Calculate the resistance of an aluminium cable at 20°C with a length of 100 m and a CSA of 25 mm².

Ohm's law

Ohm's law explains the relationship between current, voltage and resistance in any electric circuit. It is applied to work out the quantities of a DC circuit and can be expressed as:

$$V = IR$$

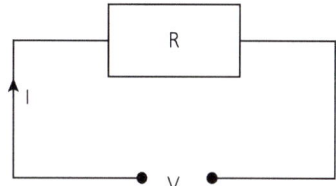

▲ Figure 2.17 Ohm's law

To remember Ohm's law, use the triangle in Figure 2.17 and cover the value you wish to find. It will leave the two other values and tell you whether they need to be divided or multiplied.

▲ Figure 2.18 Simple DC circuit

Figure 2.18 shows a simple circuit with a resistor (R), a current (I) and an electromotive force (emf) (V). We can apply Ohm's law to calculate values based on those we know. If the circuit in Figure 2.14 has a voltage of 12 V and a resistance of 8 Ω, what is the current?

$$V = IR$$

So, transposed to find the current:

$$I = \frac{V}{R}$$

So:
$$\frac{12}{8} = 1.5 \text{ A}$$

> **Test yourself**
>
> If the circuit in Figure 2.14 has a current of 6 A and a voltage of 120 V, what is the circuit resistance?

Series circuits

When resistors in a circuit are connected one after the other, they are connected in series. The total resistance is found by adding all the resistances together:

$$R_{total} = R_1 + R_2 + R_3 \ldots \text{ and so on}$$

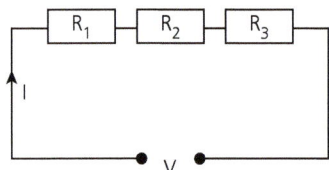

▲ Figure 2.19 Circuit with three resistors in series

The circuit shown in Figure 2.19 has a supply voltage of 200 V and three resistors of 10 Ω, 15 Ω and 25 Ω in series. Let us calculate the circuit current.

▲ Figure 2.20 Measuring voltage in a series circuit

As the current is based on the total circuit resistance, this needs to be determined first:

$$R_t = 10 + 25 + 15 = 50 \text{ Ω}$$

Now Ohm's law can be applied to determine the circuit current:

$$I = \frac{200}{50} = 4 \text{ A}$$

> **Test yourself**
>
> What is the circuit current if a circuit has a supply voltage of 400 V and 12 Ω, 8 Ω, 16 Ω and 22 Ω resistors in series?

In a series circuit, the current remains the same through each resistor, but the voltage across each resistor is different. The voltage across each resistor can be calculated by applying Ohm's law to that resistor. The value of all the voltages must equal the total circuit voltage.

To calculate the voltage across each resistor in Figure 2.20:

$$V1 = 4 \times 10 = 40 \text{ V}$$
$$V2 = 4 \times 15 = 60 \text{ V}$$
$$V3 = 4 \times 25 = 100 \text{ V}$$

And to check the values equal the circuit voltage:

$$40 + 60 + 100 = 200 \text{ V}$$

> **Test yourself**
>
>
>
> ▲ Figure 2.21
>
> For the circuit in Figure 2.21, determine the:
> ▶ value of resistor X
> ▶ voltages V1, V2 and V3.

Parallel circuits

When resistors are arranged in parallel (see Figure 2.22), the supply voltage remains constant across all resistors, but the current in each branch changes.

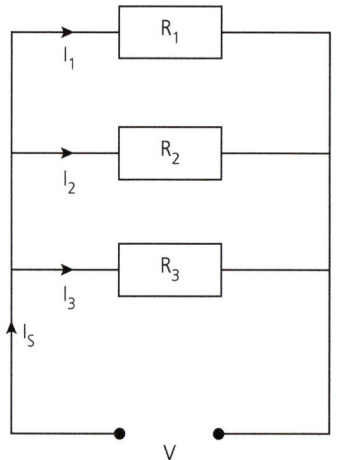

▲ Figure 2.22 Parallel resistors in a circuit

Chapter 2 Construction science principles

The value of current in each branch is based on the value of resistance and voltage in each branch. The total of each branch will equal the supply current (I_s).

The total resistance for the circuit is calculated by:

$$\frac{1}{R_{total}} = \frac{1}{R_1} + \frac{1}{R_2} + \frac{1}{R_3} \ldots \text{and so on}$$

A golden rule is that the total resistance must be less than the lowest resistor in the circuit.

When calculating parallel resistances, do not forget that the total is divided into 1 at the end.

Let us calculate the total resistance and each value of current, if the circuit in Figure 2.22 has the following values:

- $R_1 = 20\ \Omega$
- $R_2 = 30\ \Omega$
- $R_3 = 40\ \Omega$
- $V = 400\ V$.

$$\frac{1}{R_{total}} = \frac{1}{20} + \frac{1}{30} + \frac{1}{40} = \frac{1}{0.108\ldots} = 9.23\ \Omega$$

$$I_1 = \frac{V}{R_1} = \frac{400}{20} = 20\ A$$

$$I_2 = \frac{V}{R_2} = \frac{400}{30} = 13.33\ldots\ A$$

$$I_3 = \frac{V}{R_3} = \frac{400}{40} = 10\ A$$

$$I_s = \frac{V}{R_{total}} = \frac{400}{9.23} = 43.33\ldots\ A$$

And to check:

20 A + 13.33… A + 10 A = 43.33… A

To work out the total resistance on a calculator, use the x^{-1} button. So, for the values in the worked example, push:

> **Test yourself**
>
>
>
> ▲ Figure 2.23

The circuit in Figure 2.23 has the following values:
- $R_1 = 5\ \Omega$
- $R_2 = 100\ \Omega$
- $R_3 = 200\ \Omega$
- $V = 100\ V$

Calculate the total circuit resistance and each current value.

> **Improve your maths**
>
> A 10 m run of steel conduit contains a 1.5 mm² copper circuit protective conductor (CPC), which is connected to the earth of a socket outlet at the end of the conduit run. The conduit has a 20 mm outside diameter and an 18 mm internal diameter (the bit the cables go in).
>
> As the conduit acts as an earth in parallel to the CPC, what is the overall resistance of the conductor?

Power in parallel and series circuits

Let us look at a circuit with series and parallel resistors, then calculate the power.

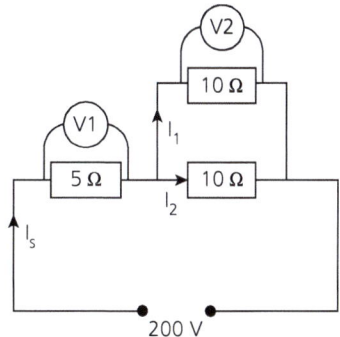

▲ Figure 2.24 Circuit with series and parallel resistors

Looking at the circuit in Figure 2.24:
- The current in the 5 Ω resistor is the same as the supply current (I_s).
- The current through each 10 Ω resistor is calculated in the same way as for a parallel circuit and must total the supply current.
- The voltage V2 is the same across each 10 Ω resistor and V1 + V2 must equal the supply voltage.

So, we can calculate:
- total resistance of the parallel section (R_P)
- total circuit resistance
- circuit current I_s
- voltages V1 and V2
- currents I_1 and I_2.

$$\frac{1}{R_p} = \frac{1}{10} + \frac{1}{10} = \frac{1}{0.2} = 5\ \Omega$$

67

As the parallel branch totalling 5 Ω is in series with the 5 Ω resistance, the total resistance of the circuit is:

$$R_{total} = 5 + 5 = 10 \text{ Ω}$$

So I_s can be found using Ohm's law:

$$I_s = \frac{V}{R_{total}} = \frac{200}{10} = 20 \text{ A}$$

Voltage V1 is then found by:

$$V_1 = I_s \times R = 20 \times 5 = 100 \text{ V}$$

Voltage V2 will also be 100 V, as the total resistance of the parallel section is also 5 Ω.

Current I_1 is found by applying Ohm's law to that section:

$$I_1 = \frac{V_2}{10} = \frac{100}{10} = 10 \text{ A}$$

Current I_2 has the same properties:

$$I_2 = \frac{V_2}{10} = \frac{100}{10} = 10 \text{ A}$$

As the total circuit current is 20 A, the current is split evenly over the two equal resistances in parallel.

Power **dissipated** in a circuit can be calculated in two ways:

$$P = V \times I \text{ or } P = I^2 R$$

> **Key term**
>
> *Dissipated:* energy consumed by converting to heat energy

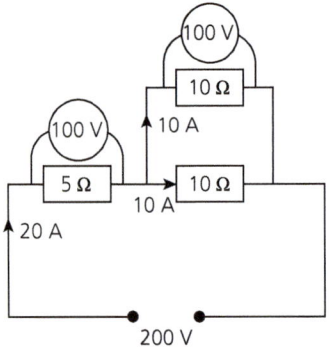

▲ Figure 2.25 Calculating power in a circuit

Figure 2.25 shows the circuit from the previous example with calculated values. Remember that the total circuit resistance was 10 Ω.

To work out the power of the whole circuit, we can use the total circuit or supply values, so:

$$P = V \times I = 200 \times 20 = 4000 \text{ W or 4 kW}$$

Equally:

$$P = I^2 R = 20^2 \times 10 = 4000 \text{ W or 4 kW}$$

So, either formula can be used depending on the values known.

The power dissipated by the 5 Ω resistor is:

$$20 \text{ A} \times 100 \text{ V} = 2000 \text{ W or 2 kW}$$

The power dissipated by one of the 10 Ω resistors is, using the other formula:

$$10 \text{ A}^2 \times 10 \text{ Ω} = 1000 \text{ W}$$

As the other 10 Ω resistor is the same, a total of 2000 W is dissipated in the parallel part of the circuit.

> **Test yourself**
>
>
>
> ▲ Figure 2.26
>
> Using the circuit shown in Figure 2.26, calculate the:
> - total resistance of the circuit
> - current I_s
> - power dissipated by the 35 Ω resistor
> - power dissipated by the parallel section
> - total power dissipated by the circuit.

They also have a thermal trip (normally a bimetallic strip), which causes gradual mechanical movement when heated by lower overload currents. This also causes the switch to trip, disconnecting the circuit.

Residual current devices (RCDs)

RCDs have either electronic devices or a small **toroidal** transformer. They monitor the current entering a circuit through the live wire and the current returning through the neutral wire.

> **Key term**
>
> *Toroidal:* circular or doughnut shaped

If the circuit is healthy, the two currents remain identical. If a small fault happens, current flows to earth; the live wire has more current than the neutral wire.

If the imbalance in current exceeds the device's residual current setting, it trips instantly. The most common residual current setting for an RCD is 30 mA or 0.03 A, so these devices are highly sensitive.

Residual circuit breaker with overload (RCBO)

These are a miniature circuit breaker and RCD in the same body. They have the characteristics of both types of device.

4 Structural science principles

Buildings need to stand up to a lot of stresses and forces. Approved Document A of the Building Regulations 2010 requires a building to be constructed in such a way as to withstand the combined dead, imposed and wind loads without deformation or movement that will affect stability. Therefore, the effects of forces need to be considered.

4.1 Forces acting on a building

Five types of force can act on a building or part of a building. These forces are described in Table 2.9 and shown in Figure 2.29.

▼ Table 2.9 Forces

Force	Description	Example	How to resist it
Tension	Force that tries to stretch a building or its components in opposite directions	The tension on vertical cables used in a suspension bridge, which hang the deck from the main cables	Use steel, which has good tensile strength, especially when wound into cables
Compression	Force that tries to crush a component by pushing on both ends	A column in a building with the weight of the building resting on it and the force acting on the ground below	Use materials such as concrete, stone or masonry, which act well under compression
Bending	Force that acts on the centre of a beam that is supported at each end	Beams that provide horizontal support, such as wooden floor joists, or support steels like rolled-steel joists (RSJs)	Use reinforced concrete for long spans, or steel or wooden joists for shorter lengths
Torsion	Force that tries to twist the component in opposite directions	An entire building when subjected to strong wind	Use a closed hollow section, such as a steel box section, or circular structural elements such as poles
Shear	Force that tries to split or divide a component. Shearing forces work in opposite directions but do not have the same line of action, although they can be close	Wind in different directions at the top and bottom of a building	Use brick or concrete walls. Shearing can occur in cement joints, so steel bracing helps

Health and safety

When you see a beam with 'SWL 500 kg' stamped on it, never suspend any more than 500 kg from it. SWL stands for 'safe working load', and any load beyond this may bend or break the beam.

There are other terms used to describe forces which act on a building:
▶ Vertical forces act upwards or downwards on a building and result in tension or compression.
▶ Horizontal forces act on the sides of a building and result in shear or torsion. They may be:
  – lateral – acting across the width of the building
  – longitudinal – acting along the length of the building.

▲ Figure 2.27 Forces acting on buildings and components

4.2 Structural members

The supporting parts of a building are known as structural members.

▲ Figure 2.28 Structural members used in a building

In Figure 2.28, you can see the following parts:
- **Foundation:** this is the lowest part of the substructure, supporting the building from sinking into the ground. Metal or concrete piles may be required where a foundation is unable to meet a solid base; these are drilled or driven deep into the ground to make sure the building is on a firm substructure.
- **Substructure:** this is the complete section of a building extending below ground-floor level.
- **Footings:** these are the sections of masonry from the foundation to ground-floor level. They are normally linked to a particular member, such as a wall, not to the entire structure.
- **Columns:** these are upright supports and can be steel or concrete. When columns are made from brick, they are often referred to as pillars.
- **Beams or girders:** these are horizontal elements that form the support for floors. In a house, beams are used as main structural supports for first-floor walls that do not have a wall on the ground floor to support them.
- **Joists:** these are similar to beams but are often wooden and many in number. They are used to support a flooring system above them and a ceiling below them. They are supported by walls, beams or girders.
- **Trusses:** these are normally associated with roofing systems and form triangles to provide support along large spans. Where trusses are used in walls, they are called braces.
- **Lintels:** these are used to provide structural support above windows and doors.

For a structure to remain sturdy and withstand all expected forces, it is important that the thickness and strength of all members are carefully calculated by structural engineers.

4.3 Drilling, notching and chasing

Where services such as electrical wires, pipes and ducts are installed in buildings, they need to be run through some structural parts.

Building regulations stipulate where notches or holes can be cut into a joist or beam so as not to weaken it, as shown in Figure 2.29.

▲ Figure 2.29 Notching and cutting holes in a joist or beam

Other things to note when cutting joists include:
- The maximum diameter of holes should be 0.25 × joist depth.
- Holes in the same joist must be at least three hole diameters apart.
- The maximum depth of notch should be 0.125 × joist depth.

Chapter 2 Construction science principles

> **Test yourself**
>
> A 30 mm notch needs to be cut into a wooden beam that is 8 m long. At what maximum length from the end of the beam can the notch start?

To avoid weakening the structure:
- vertical **chases** in walls should be no deeper than one third of the wall thickness
- horizontal chases should be no deeper than one sixth of the wall thickness.

If a wall is 400 mm thick, chases should be no deeper than:

$$\text{Vertical} = \frac{400}{3} = 133 \text{ mm}$$

$$\text{Horizontal} = \frac{400}{6} = 66.6 \text{ mm}$$

> **Improve your maths**
>
> What is the maximum depth of a vertical chase in a wall which is 0.2 m deep?

> **Key term**
>
> *Chases:* cuts in a masonry or plaster wall to contain pipes or cables; once the pipes or cables have been fitted, they are covered with plaster or similar

4.4 Other effects on structures

There are many other factors that can affect a structure and increase the risk of weakening its stability.

Other buildings

Neighbouring buildings can have several effects on each other, such as creating increased wind speeds. Currents of air moving around structures travel at different speeds, and these forces can have damaging effects on buildings.

Foundations from other buildings also have an impact, as the force of the building's weight is transferred into the ground at 45-degree angles. This could cause soil movement beneath or close to adjacent buildings.

Drains and sewers

These underground systems can collapse if buildings are built on top of them, causing buildings to lean as one section of the foundation drops.

Proximity to drains and sewers also influences the depth of a foundation. Building weight is transferred to the ground from the foundation at a 45-degree angle from the footprint. As a result, there is potential for damage to drains and sewers if they are located within that 45-degree area. This means that the foundation is typically excavated to a depth that is at least the same as the deepest part of the drain or sewer.

Trees

The roots of trees, especially fast-growing trees, can damage building foundations. As trees absorb moisture from the ground, this can cause uneven drying of soils under foundations or footings. Shrinkage of soils can cause foundations to collapse or crack.

Different soil structures

Soils shrink and expand depending on rainfall and levels of the water table. Soils which are permeable (allow water to drain through) are more stable than non-permeable soils, such as clay.

Soils can also shift or move. If this happens under part of a building, it can cause **subsidence**, meaning one side of the building drops or moves away from other parts. The initial signs of subsidence are cracks forming in walls.

> **Key term**
>
> *Subsidence:* the sinking of a structure into the ground

5 Heat principles

There is much more to heating a building than just considering the source of the heat. Buildings need to be kept warm in winter and cool in summer. Equally, temperature control and air changes are important to prevent condensation, which can lead to damp and mould growth.

> **Health and safety**
>
> You may need to consider the effects of temperature on particular groups of people, such as elderly people.
> - 21°C is the recommended comfortable living temperature.
> - Respiratory problems can start when subjected to long periods at 16°C.
> - If exposed to a temperature of 12°C for more than two hours, elderly people are at risk from raised blood pressure and heart attack.

Onsite Construction T Level: Core

5.1 Heat transfer

There are three methods of transferring **heat**:
- convection
- conduction
- radiation.

> **Key term**
>
> **Heat:** does not, in BSE terms, mean hot; it simply means heat energy, which can be hot or cold

Convection

Convection occurs when air or water is heated and then moves away from the source of the heat, carrying thermal energy.

For example, a wall-mounted convection heater or radiator heats the air around it, causing the hot air to rise. This is moved into the room by cooler air entering the bottom of the heater or radiator, creating convection currents which rotate around a room.

The term 'radiator' is a misnomer, as most of the heat output is via convection. They are not called convectors, however, because a true convector has a heat source that is not exposed.

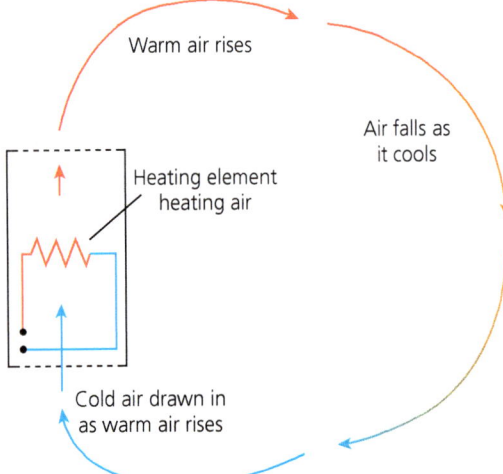

▲ Figure 2.30 The convection cycle

Convection also occurs when an immersion heater conducts heat to the water close to the element. This causes hot water to rise in the vessel, allowing the element to heat cooler water that replaces the hot, rising water.

Convection heaters:
- are a good way to heat most sizes of room
- can be wall mounted
- can use elements or pipes buried in the floor to heat a large space
- are less effective in small spaces with limited air circulation.

Conduction

Conduction is when heat is transferred directly into a material, for example brick in the case of a **storage heater**. When a substance is heated, the molecules gain more energy and vibrate more. They then bump into other nearby molecules and transfer some of their energy to them. Conduction is therefore not appropriate for heating air directly, as the molecules are too far apart.

> **Key term**
>
> **Storage heater:** an electric heater that stores thermal energy; it heats up internal ceramic bricks when electricity is cheaper at night and then releases heat gradually during the day, acting in the same way as a convection heater

The following are examples of heating systems using conduction:
- Floor-heating systems conduct heat to the floor, which causes air convection above it.
- Hot-water immersion systems directly heat water around the element.
- Instantaneous water heaters have an element that is wound around a metal water pipe. Heat is conducted through the pipe into the water. As the amount of water in the pipe is small, it heats up quickly.

Radiation

Radiant heaters:
- use infrared radiation to heat bodies, but not the air around them
- are directional – there must be a line of sight in order to heat a person
- are particularly useful in large areas where heating the air is not required or efficient, for example outdoor patio heaters
- require a large amount of power and are therefore less efficient than convection heaters.

▲ Figure 2.31 A radiant heater

5.2 Characteristics of air

Modern buildings strive to be more energy efficient by keeping heat in, but this can also cause problems if a building does not allow air changes to occur. Air will become stale and contaminated, causing breathing problems.

In addition, if a building is allowed to heat up but moist warm air comes into contact with colder surfaces, such as windows, condensation forms, which in turn can cause mildew or rot. Air contains water vapour, and the amount of water vapour in the air is known as the level of humidity. Warmer air can hold more water vapour, but if that air touches cooler surfaces, or the air is rapidly cooled by mechanical means such as an air-conditioning system, the humid air vapour will condense into liquid water.

▲ Figure 2.32 Mould in the corner of a window

Getting the balance right is difficult, so if a building does not have a forced air change by fans, it needs to occur naturally. This can happen by pressure differences, which can be caused by wind or the convection effect created by differences in temperature or humidity. In either case, the amount of ventilation depends on the size and positioning of openings in the building.

Buildings with a staircase or an open chimney will create air changes by the stack effect, where warm, less-dense air rises, drawing cooler dense air into the base of a building through any opening. As air warms, it expands and becomes less dense, as there is less air occupying a space, but as it cools it contracts, meaning more air occupies a space, so it becomes denser.

Ideally, a normal room in a house should have a minimum of four air changes per hour. A minimum of six changes per hour should occur in bathrooms and toilets, but these are likely to have forced air systems such as extraction fans creating more changes.

In public places, this increases to a minimum of ten changes per hour in normal occupied rooms, and 12 in places such as restaurants.

5.3 Heat loss in buildings

Buildings lose heat through the different construction materials used. Even if walls are well insulated, doors and windows can cause heat loss.

All materials used in construction have ratings of thermal conductivity, known as K-values. These are values given by manufacturers of the materials.

Table 2.10 gives examples of commonly used materials and their average thermal conductivity (K) value in watts per metre Kelvin (W/m K).

▼ Table 2.10 Thermal conductivity of common materials

Material	Average K-value (W/mK)
Light blockwork	0.38
Dense blockwork	1.63
Exposed brick	0.84
Chipboard	0.15
Concrete	1.1
Glass	1.05
Gypsum plaster	0.46
Plasterboard	0.25
Steel	20–40
Granite	2.5
Timber (softwood)	0.14
Timber (hardwood)	0.16
Mineral-wool insulation	0.035
Rigid foam board	0.026
Air in a cavity	0.18

R-values

Once a K-value is known for a material, the amount of heat resistance the material offers can be calculated. This is known as the R-value.

The R-value can be calculated based on the thickness of the material:

$$\text{R value} = \frac{\text{thickness of material (m)}}{\text{K value (W/mK)}}$$

Using Table 2.10, the R-value for 12.5 mm (0.0125 m) thick plasterboard can be determined as:

$$R = \frac{0.0125}{0.25} = 0.05 \text{ W/m}^2.\text{K}$$

Note the units for R-value: watts per metre squared Kelvin or W/m².K.

> **Test yourself**
>
> Using Table 2.10, determine the R-values for:
> - 25 mm thick chipboard
> - 500 mm thick concrete.

U-values

A **U-value** is the rate of heat transfer for a wall, ceiling or floor.

> **Key term**
>
> *U-value:* a unit of measurement used to record thermal transmittance through a building material

To calculate the total heat loss for a wall, determine the U-value based on the materials used.

For example, consider a wall constructed using the following materials:
- 12.5 mm thick plasterboard (R_1) mounted on 100 mm light blockwork (R_2)
- a 50 mm air cavity (R_a)
- 100 mm exposed brick (R_3).

In addition to building materials in the wall, external air resistance and internal air resistance need to be taken into account. These have predetermined R-values of:
- internal air resistance (R_{si}) 0.13 W/m K
- external air resistance (R_{se}) 0.04 W/m K.

So, to calculate the U-value of the wall:

$$R_1 = \frac{0.0125}{0.25} = 0.05 \text{ W/m}^2.\text{K}$$

$$R_2 = \frac{0.1}{0.38} = 0.26 \text{ W/m}^2.\text{K}$$

$$R_a = \frac{0.05}{0.18} = 0.28 \text{ W/m}^2.\text{K}$$

$$R_3 = \frac{0.1}{0.84} = 0.12 \text{ W/m}^2.\text{K}$$

$$\text{Total U-value} = \frac{1}{R_{si} + R_1 + R_2 + R_a + R_3 + R_{se}} \text{ W/m}^2.\text{K}$$

So, for our wall:

$$U = \frac{1}{0.13 + 0.05 + 0.26 + 0.28 + 0.12 + 0.04} = 1.136 \text{ W/m}^2.\text{K}$$

Some manufacturers provide U-values for their products, such as doors and windows. As an example, these could be:

- 3 W/m².K for a double-glazed window that has 12 mm spacing between the panes of glass
- 2.5 W/m².K for an external door that has a standard fire rating.

Using U-values, we can calculate the heat loss in watts through various parts of a building:

$$\text{area of surface} \times \text{temperature difference} \times \text{U-value for surface}$$

The temperature difference relates to the desired room temperature and the average lowest outdoor temperature. So, if a room needs to be heated to 20°C, and the outdoor temperature has an average low of −3°C, the difference is 23°C. Once each surface heat loss has been determined, including the heat loss through air changes, these are added to calculate the overall heat loss.

▲ Figure 2.33 A room with heat loss

Figure 2.35 shows a room with dimensions. Assuming the height of the room is 2.6 m, we can calculate the overall heat loss if the desired temperature is 20°C and the outside average low is 2°C. Let us assume all the R-values have been calculated so that the following U-values are given, and we will assume the room undergoes two air changes per hour:
- Walls: 0.35
- Floor: 0.25
- Ceiling: 0.25
- Windows: 2.9
- Door: 2.9
- Air change: 0.33.

Room volume:

$$6 \times 2.5 \times 2.6 = 39 \text{ m}^2$$

Window heat loss:

$$(1.7 \times 1.5) \times (20 - 2) \times 2.9 = 133.11 \text{ W}$$

Door heat loss

$$(1 \times 1.8) \times (20 - 2) \times 2.9 = 93.96 \text{ W}$$

Wall heat loss:

Wall area:

$$2.5 \times 2.6 = 6.5 \text{ m}^2 \times 2 = 13 \text{ m}^2$$
$$6 \times 2.6 = 15.6 \text{ m}^2 \times 2 = 31.2 \text{ m}^2$$
$$31.2 + 13 = 44.2 \text{ m}^2$$

Less window and door:

$$44.2 - (1 \times 1.8) - (1.7 \times 1.5) = 39.85 \text{ m}^2$$

So, loss is:

$$39.85 \times (20 - 2) \times 0.35 = 251.1 \text{ W}$$

Floor and ceiling heat loss (as both have the same dimension and U-value):

$$6 \times 2.5 \times (20 - 2) \times 0.25 = 67.5 \text{ W} \times 2 = 135 \text{ W}$$

Air-change loss needs to consider the volume of the room, the number of air changes and the U-value for the air loss, which is 0.33:

$$39 \times (20 - 2) \times 2 \times 0.33 = 463.3 \text{ W}$$

So, the total heat loss is:

$$133 + 93.6 + 251.1 + 135 + 463.3 = 1076 \text{ W or } 1.076 \text{ kW}$$

Therefore, the room would require some form of heating system to counter this loss and maintain the required temperature.

5.4 Methods of heating a building

Several types of building services system can be used to provide heat, for example:
- electric wall-mounted convection heaters (including storage heaters)
- electric floor-heating elements
- wet floor-heating systems
- wet heat-emitting radiators
- ground or air source heat pumps.

Electric convection heaters, known as panel heaters, contain electric heating elements. Cool air is drawn into the bottom of the unit, passes over the heating element and rises out of the unit as warmed air.

Electric storage heaters contain bricks that sandwich heating elements. The elements heat the bricks, which then store the heat and gently release it into the room using the convection cycle. The advantage of storage heaters is they can consume electricity at times when costs are lower, such as night time, then release the heat at other times.

Wet heating systems have boilers that heat water and pump it around the building to radiators or floor-heating pipework. The fuel used for the boiler can be:
- gas
- oil
- biomass.

Gas can be delivered by mains pipes into the building, and this is metered at supply. Alternatively, tanks can be used to store gas delivered to the property.

Oil is also delivered to a property and stored in tanks for future use.

Health and safety

Gas components can only be installed and certified by engineers registered with **Gas Safe**.

Key term

Gas Safe: a professional organisation that controls the health and safety of work completed on gas systems in the UK

Biomass heating systems use specially made wood pellets, which are burned to produce the heat source for the pumped water system. They are said to be carbon neutral, or low carbon, as the carbon dioxide (CO_2) released by burning the wood is equal to the CO_2 released by the wood if it were left to decay naturally. We will explore this, ground/air source heat pumps and other technologies used to heat or cool a building, in Chapter 5.

6 Light principles

Lighting design has become a very specialist area because of the vast range of lighting and **luminaires** available. Light-emitting diode (LED) technology is used as the main form of lighting, due to its energy-efficient performance and light quality.

Key term

Luminaires: complete electric lighting units, including the casing, lamp and any internal controlling devices or electronic equipment (known as control gear or drivers)

6.1 Main lighting terms

Table 2.11 considers the main lighting terms.

▼ Table 2.11 Main lighting terms

Term	Definition
Luminous intensity	This is the amount of light emitted per solid angle or in a given direction. It is denoted by the symbol l and measured in candela (cd).
Luminous flux	This is the total amount of light emitted from a source. It is denoted by the symbol F and measured in lumens (lm).
Illuminance	This is the amount of light falling on a surface. It is denoted by the symbol E and measured in lumens per metre^2 (lux).
Efficacy	This is the efficiency of a lamp or luminaire and compares the amount of light emitted to the electrical power consumed. It is denoted by the symbol K and measured in lumens per watt (lm/W).
Maintenance factor (mf) Light loss factor (llf)	These are factors used to derate the light output of a lamp, allowing for dust. The factor used depends on the environment. An average office environment would have a factor of 0.8, whereas a factory where lots of dust accumulates may have a factor of 0.4.
Coefficient of utilisation or utilisation factor (uf)	This is a measure of the efficiency of a lamp in transferring light energy to a surface, such as a wall or ceiling. Emitted light bounces off reflective walls, making more effective use of the light. An average factor for a room is 0.6. The lighter the colour of the room, the higher the factor.
Glare	Glare is a very bright light that is difficult to look at. Poor positioning of luminaires can cause glare, resulting in people squinting, getting headaches or experiencing general discomfort.
Diffuser/louvre	Diffusers are normally translucent plastic or glass covers that refract harsh light, dispersing it to create an even light from a single source. A louvre is polished metal which reflects light in different directions, dispersing a harsh glare into softer dispersed light.
Colour rendering	This is the appearance of light in terms of colour. Lamps produce either orange light or blue light, depending on the gases used in them. This in turn can make objects under the lights look different in colour. For example, sodium lamps were once commonly used for street lighting, but this gives the appearance that everything lit by them is orange. This was never considered much of a problem, as hazards could still be seen and avoided. In more modern times, whiter colours are preferred as they are better for CCTV images and also make people feel safer.

Test yourself

A lamp has a lumen output of 2500 and consumes 50 W. What is its efficacy?

Research

How does the colour rendering of street and utility lighting affect crime?

6.2 Choosing the right type of lamp

Artificial lighting consumes a lot of energy. Even during daylight hours, many offices, shops and public buildings have lights switched on. It is therefore important to choose lamps that:
▶ provide a good-quality light with the correct colour rendering
▶ consume power efficiently.

Table 2.12 shows the different types of lamp available, their basic operating principles and their characteristics including efficacy (an indication of power to light performance).

▼ Table 2.12 Types of lamp

Lamp type	Basic operating principles	Characteristics and applications
General lighting service (GLS)	A filament is suspended in a vacuum. When current flows through the filament, it glows white hot, producing light.	GLS lamps rated above 100 W were banned due to their poor efficacy (approximately 14 lumens/watt). Caps are usually bayonet (BC) or Edison screw (ES).
Tungsten halogen	A filament is suspended in halogen gas, which prevents evaporation. This allows the lamp to run at much higher temperatures than standard GLS lamps, meaning a brighter light.	These were widely available in ratings up to 500 W but have been superseded by LED lighting. Smaller G4 or G9 lamps are still common in table or display lamps. Halogen lamps have an average efficacy of 17 lumens/watt.
High-pressure sodium (SON)	Current strikes across low-temperature sodium gas in a tube, causing it to heat up and ionise, producing light. The lamps are known as discharge lamps.	These were widely used for car park or street lighting, as they are suited to the illumination of large areas. The colour output is slightly orange. Average efficacy is 120 lumens/watt.
Low-pressure sodium (SOX)	These work on a similar principle to high-pressure sodium lamps but they have larger gas-filled tubes, with lower sodium gas pressures, and are usually U shaped.	These were widely used for street lighting due to their very high efficacy but colour rendering is poor, with a dull orange light output. Average efficacy is 180 lumens/watt. Although still around today, they are rarely used in new lighting schemes.
High-pressure mercury (MBFU)	These work on a similar principle to high-pressure sodium lamps, but have tubes filled with mercury gas.	These are commonly used for utility/communal lighting because of their better colour rendering, which is white with a slight blue tinge. Efficacy values are on average 80 lumens/watt.
Metal halide (HID)	These discharge lamps are similar in operation to high-pressure sodium lamps but they use a mixture of mercury and sodium gases.	These produce a bright white light but efficacy is lower as a result and averages 60 lumens/watt.
Low-pressure mercury (fluorescent)	Commonly known as fluorescent tubes, these mercury-filled tubes have an inner coating of phosphor powder, making a blue ultraviolet light inside the tube and various types of white light on the outside.	These are still widely available but are slowly being replaced by LED tubes. Efficacy is on average 100 lumens/watt.
Compact fluorescent (CFL)	These are small versions of fluorescent tubes, where the tube is often bent to occupy smaller areas.	Designed to replace the GLS lamps banned by EU directives, these often have ES or BC caps. Efficacy is on average 65 lumens/watt.
LED	Made up of several light emitting diodes (LEDs), these lamps require control gear, known as drivers, to reduce and rectify voltage and govern current.	LED lamps and luminaires are now widely used for all applications, from general room lighting to street lighting and stadium illumination. Efficacy is on average 120 lumens/watt but this is improving all the time.

LED lighting is becoming the most common form of lighting due to:
▶ good efficacy
▶ the ability to change colour rendering to suit a location, for example warm white, daylight or coloured
▶ longer-lasting lamps.

7 Acoustics principles

7.1 Acoustics in buildings

Acoustics is a term used to describe the study of mechanical waves that travel through gases, solids and liquids causing vibration and sound. This could be from somebody speaking, a loudspeaker playing music or someone banging on something like a drum.

> **Key term**
>
> **Acoustics:** the transmission of sound

Different materials and the shape of those materials affect the way sound waves move. In most construction applications, the desire is to provide acoustic comfort rather than to enhance acoustics, such as in a concert hall. Poor acoustic protection can lead to severe health conditions, such as hearing loss, high blood pressure, headaches and sleep deprivation.

7.2 Factors that affect acoustics

Before we consider methods of providing acoustic comfort, we need to understand some terms used in the study of acoustics.

▼ Table 2.13 Terms used in the study of acoustics

Term	Description
Reverberation	Reverberation is the persistence of a sound following its creation. It is caused when sound reflects off surfaces and decays. Even when the initial source of the sound stops, the reflections can continue, with their amplitude decreasing until zero is reached.
Reverberation time	Reverberation time is a measurement, in seconds, of the time it takes for a sound to decay and stop entirely. For small rooms, it should typically be under a second, and no more than two seconds for larger rooms.
Frequency	Frequency is the number of waves that occur in a period of time. For sound waves, this determines the pitch of a sound. Low-frequency sounds (fewer waves per second) are lower in pitch than high-frequency sounds.
Resonance	Materials have different natural resonant frequencies. This is simply the frequency of sound and pressure which causes the material to move easily and vibrate. For example, if a loudspeaker is placed in front of a glass and the frequency of the sound is increased (low pitch to high pitch), there will be a point where the glass shatters because it vibrates to the point of destruction.
Sound absorption	Sound absorption is where sound waves are suppressed or absorbed by an item or structure, rather than being reflected.
Sound transmission class (STC) rating	An STC rating is a numerical value showing how well a structure reduces sound transmission. It is commonly used to rate doors, walls, windows, ceilings and floors. The higher the value, the better the structure is at reducing sound travel. A well soundproofed wall typically has an STC rating above 50.
Sound barrier	A sound barrier is a material that can be placed on a structure (such as a wall, ceiling or floor) to increase its STC rating.
Attenuation	Attenuation is the gradual loss of intensity. It refers to a structure's ability to limit sound transmission.
Baffle	Acoustic baffles are devices which reduce the strength of airborne sound. They absorb sound, reducing echo and lowering reverberation time. When baffles are designed to be suspended from a ceiling, they are referred to as clouds.
Decibel	The decibel is the unit of measurement for sound.
Flanking sound	Flanking sound is sound that travels between two areas, such as through air vents and ducts to reach another room.
Footfall	Footfall is the sound made by walking on a surface.

7.3 Principles of sound

If you have ever been in a large open space with bare, hard walls, you will have experienced a very echoey environment. Soft materials, such as fabrics and foams, create a sound-absorbent surface where sound does not reverberate.

The following construction methods can help to absorb sound waves and prevent reverberation and control unwanted sound, ensuring privacy of occupiers:
- cavity wall insulation, such as rock wool or fibre wool
- isolation membrane (a thin barrier in walls, ceiling voids and doors)
- flooring materials, such as carpets or wood
- soundproof foam panels (usually shaped into small pyramids).

When considering the installation of building systems, their location will have a bearing on the noise they produce. If a system component, such as a gas boiler or circulation pump, vibrates in operation, this will create sound. Any materials in contact with the component can also vibrate, which in turn can amplify this sound. In some cases, acoustic hangers or fixings may be required to reduce the transfer of vibrations.

Chapter 2 Construction science principles

If equipment is located outdoors, such as air source heat pumps, the sound from the fans can cause sound pollution.

▲ Figure 2.34 Ceiling-mounted, sound-absorbing panels used to reduce echo in large spaces

Health and safety

Noise exposure of 80 decibels (dBA) for two hours can cause hearing damage.

7.4 Laws and regulations on noise restriction

Laws restrict noise levels between 11 p.m. and 7 a.m., and councils must investigate any complaints of noise above permitted levels. As a guide, constant underlying noise levels of 24 decibels (dBA) are permitted, with increases to 34 dBA intermittently.

Approved Document E of the Building Regulations covers resistance to sound and sets out standards for new homes and conversions. While the statutory requirements are barely one page long, the guidance for complying with the requirements is very detailed, including information on:
▶ soundproofing, including the transmission of sound between walls, ceilings, windows and floors
▶ prevention of unwanted sound travel within different areas of a building, including flats and connecting buildings
▶ the structure of materials and formation of elements of the building, including building services, especially where these services penetrate walls, floors and ceilings
▶ the need to lag pipework or soundproof socket outlets recessed into a partition wall.

Source	Level	Description
Firework	140 dB	Threshold of pain
Jet plane	130 dB	
Siren	120 dB	
Trombone	110 dB	Extremely loud
Helicopter	100 dB	
Hairdryer	90 dB	
Truck	80 dB	Very loud
Car	70 dB	Loud
Conversation	60 dB	Moderate to quiet
Rainfall	50 dB	
Fridge humming	40 dB	Faint
Whispering	30 dB	
Leaves rustling	20 dB	
Breathing	10 dB	
	0 dB	Threshold of hearing

▲ Figure 2.35 Comparison of different noise levels

8 Earth science principles

Earth science is a wide-ranging subject, covering ground structures, watercourses, water cycles and earth forces such as earthquakes. For buildings to remain safe and stable for their lifespan, they rely on the ground being stable below them.

In this section, we will look at basic ground structures, watercourses and the water cycle, and how earth forces can be detrimental to buildings and structures in the UK.

8.1 Ground structure

We usually consider changing soil conditions in the context of the ground being muddy, hard or good for growing things. However, this only relates to the first layer of topsoil we encounter in our daily lives. When constructing buildings, we need to consider the soil structure deeper below the topsoil; and in the case of high-rise buildings, we must go deeper still. Figure 2.36 shows the different layers of soil. Although the illustrations show the layers in a relatively small depth, in reality the **bedrock** layer can be very deep.

Key term

Bedrock: solid rock usually found beneath the weaker surface materials of the ground

▲ Figure 2.36 Layers of soil

Low-rise buildings with shallow foundations typically have footings or foundations at least 1 m deep in clay-type subsoils, as these soils are prone to shrinkage or freezing due to their water content up to 0.75 m in depth. Stony subsoils, where water is less likely to be held, may allow for shallower foundations, but this depends on the **water table**.

> **Key term**
>
> **Water table:** the level below which the ground is saturated with water; this level can rise with rainfall and fall with periods of dry weather

Where buildings require more stability, such as high-rise buildings, foundations need to be in contact with the bedrock layer. This is normally achieved using piles. Figure 2.37 shows piles being drilled into the ground, where reinforced concrete is then poured to give a solid connection to the bedrock.

▲ Figure 2.37 Piles being drilled, with reinforcement being dropped into the pile hole ready for concrete to be poured

In addition, soil cleanliness is important. On all proposed development sites, soil samples must be taken and analysed before any building works begin, to ensure the ground has not been contaminated by previous activities. Many years ago, some locations had uses that were never documented or recorded, such as landfill-type waste tips. In these situations, the decomposing waste could have caused pockets of methane gas to build up in the ground, creating a long-term explosion risk. Other sources of contamination include chemical seepage from old factory sites.

8.2 Watercourses

A watercourse is a channel through which water flows. Examples include a wide river, a small stream, a ditch which is dry for most parts of the year, or an underground river.

Watercourses are important as they drain the land. Without them, water would cause soils to become unstable, leading to subsidence or even landslips. It is therefore important that watercourses of any type are kept free of obstruction and not overloaded.

When new developments are built, **surface water** needs to be drained into local watercourses, which can create flood risks if not managed correctly. Surface-water flooding is now the biggest cause of property flooding. Water that falls as rain on fields takes time to filter through the soils into the natural drain. When water falls on hard surfaces, such as roads, pavements and roofs, this water drains into the watercourses much more quickly, causing **flash floods**. With more developments comes more risk of flash floods.

> **Key terms**
>
> **Surface water:** water that collects on the ground or above surface structures and buildings, normally in the form of rain
>
> **Flash floods:** floods that appear suddenly following heavy, above-average rainfall

Measures used to manage surface water, known as attenuation, include:
- underground flood tanks that store water and gradually release it into local watercourses
- the creation of ponds or reedbeds to store or absorb surface water or gradually release it into local watercourses such as streams, brooks or rivers
- soakaways, which disperse surface water into the ground where it slowly drains into watercourses.

Chapter 2 Construction science principles

When considering flood risks, it is worth looking at the water cycle and the way land drains after rainfall. Figure 2.38 shows how rainfall seeps through permeable soils and forms the water table. Water seeps through soils to either aquifers, which are like underground rivers or lakes, or into above-ground rivers or lakes. These eventually drain to the sea, where evaporation starts the water cycle by forming rain clouds.

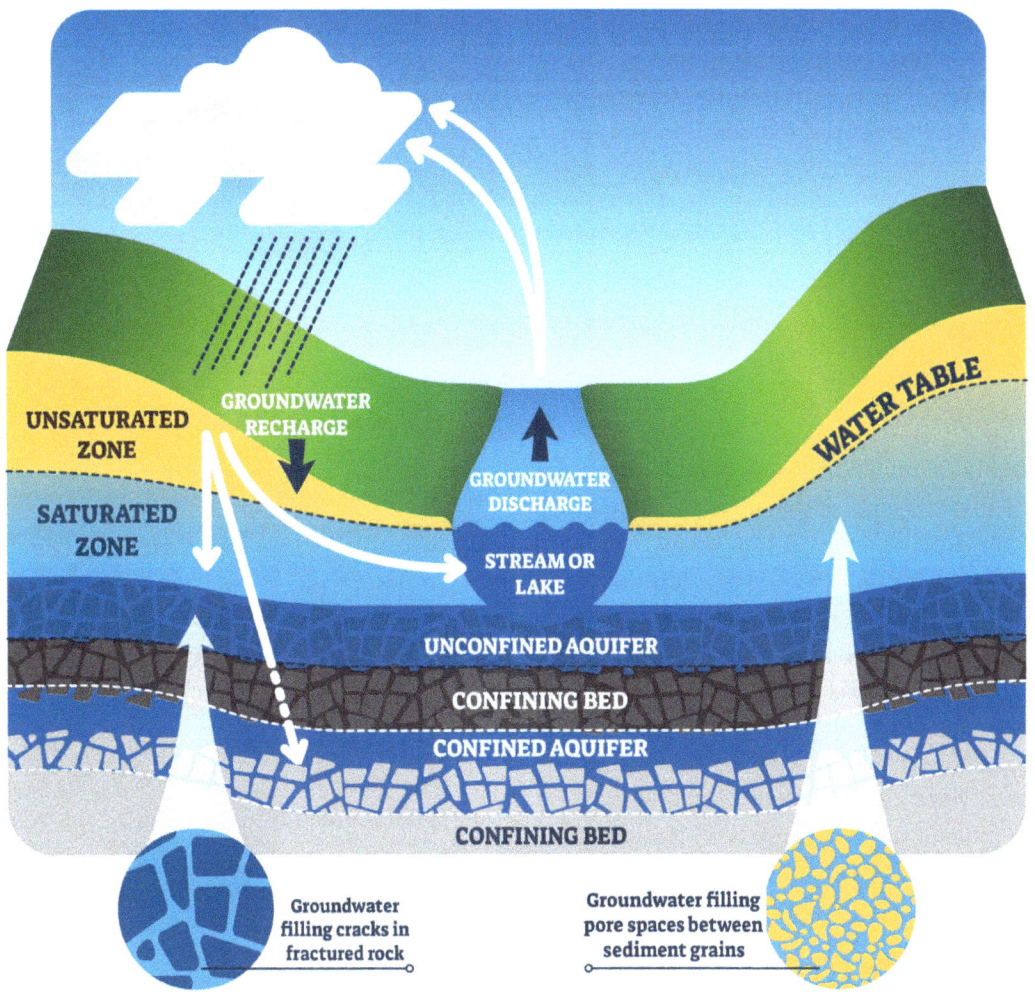

▲ Figure 2.38 How rainfall naturally drains

8.3 Earth forces

The Earth's crust is made up of large, moving pieces known as tectonic plates. At a boundary, plates can move apart, move together or slide past each other. The joins at plate boundaries are known as fault lines, and these are danger zones for earthquakes. On occasion, the moving plates become jammed so tension and forces build. Eventually the forces become too great and the sudden release results in an earthquake, shaking the ground violently. The UK is not close to any fault lines, so the risk of earthquakes is low. That does not mean, however, that they do not happen, as some minor tremors have affected the UK.

Other earth forces that are closer to home are those caused by weather and tides.

While wind can cause serious damage to buildings and structures during storms, persistent wind and rain can also cause slow but damaging erosion to some materials. One example is limestone, which dissolves when it comes into contact with carbonic acid, common in rainwater. Mortar used for pointing between bricks is also prone to weathering, especially in areas such as chimneys. Where building services equipment is located in exposed outdoor areas, it can also become damaged through weathering.

Flooding through storm water is becoming a more frequent occurrence due to global warming.

Storm water is not just a huge amount of rainwater released into drains, as discussed above with watercourses. High tidal conditions, caused by **spring tides** and **storm surges**, also restrict river drainage, because high levels of seawater act as a barrier to the draining river water.

Climate change is increasing the likelihood of heavy rain in the UK, particularly during the winter months, leading to higher risks of flooding and less stable soils.

> **Key terms**
>
> **Spring tides:** exceptionally high tides that occur twice monthly at the time of the new moon and the full moon, when the moon's orbit aligns with the sun to create a greater gravitational pull; they are known as spring tides because they act like a spring
>
> **Storm surges:** large-scale rises in sea level caused by storms; high winds push the seawater towards the coast and low pressure at the centre of the storm pulls up the water level

Assessment practice

Short answer

1. a State the SI unit of measurement for density.
   b Calculate the area of a circle with a diameter of 840 mm.
2. List three pure metals that are non-ferrous.
3. A lever is to raise a load with a mass of 180 kg. The load is placed 0.5 m from the fulcrum and a force acts 3 m from the fulcrum on the other side. Calculate the force required to raise the load.
4. A weight of 5820 N is to be raised 28 m in 1.5 minutes. Determine the power required.
5. A circuit has a total resistance of 58 Ω and is connected to a 20 V DC supply. Calculate the power dissipated by the circuit.

Long answer

6. Explain how pumped storage hydro plants provide additional power during peak times.
7. Explain how an RCD provides earth fault protection.
8. Explain why glare from artificial lighting should be avoided and describe measures that can be taken to minimise it.
9. Describe how sound attenuation is achieved in a large open room.
10. A room measures 3 m × 2 m and has a height of 3 m. There is one door leading into the room, with a dimension of 2 m × 0.9 m.

    The walls have a U-value of 0.4. The room temperature is 28°C, and the outside temperature is −5°C.

    Calculate the heat loss from the room through the walls.

Chapter 3 Construction design principles

Introduction

This chapter compares modern and traditional building methods, and investigates how good design can reduce the impact of construction work on the wider environment. It considers how modern building methods can reduce project durations, lower costs and improve the health and safety of workers by manufacturing off site.

It then explores the role of different disciplines in the design process from conception to completion, together with factors that influence the design of a building project.

Learning outcomes

By the end of this chapter, you will understand:
1. benefits of good design
2. design principles
3. the role of different disciplines involved in design
4. the design process from conception to completion
5. the concept of the 'whole building', including life cycle assessment.

1 Benefits of good design

1.1 Factors of good design

Design function

Before choosing a method of construction, it is important to understand how the building will achieve its design function. All buildings are initially designed to meet the client's **specification** because without them to finance the work, there is no project. However, sometimes the needs of the client have to be compromised to satisfy local restrictions, planning laws, building regulations and environmental requirements. In addition, designers have to consider:

- the aesthetics of the building and how it will fit into its proposed environment
- access and egress
- security
- internal layout and arrangement of rooms/spaces
- energy efficiency.

Once these aspects have been determined, designers and clients must decide what form of construction to use for the **superstructure** of the building. There are several important factors to consider when selecting a method of construction, for example, build speed, familiarity with the building system and cost.

> **Key terms**
>
> **Specification:** a detailed description of the materials and working methods that must be used for a project
>
> **Superstructure:** the part of a building above ground level, built on the basement or foundation

By contrast, the substructure of a building is the part below ground level. For more information on superstructure and substructure, see Chapter 7.

Environmental impact

The way we construct and use buildings for all types of purpose will have an impact on the environment. However, with considerate planning and the use of modern and innovative building materials, this impact can be minimised.

Without tight monitoring systems and control measures, construction work on **greenfield sites** and other types of land such as conservation areas may result in damage or destruction of natural wildlife habitats. (There is further information on greenfield and **brownfield sites** in section 1.3 of this chapter.)

Inconsiderate construction and inappropriate disposal of waste materials can lead to pollution of the land, the air and natural watercourses, such as streams, rivers and lakes. This can damage or destroy whole ecosystems, outside the boundary of the site and beyond the lifetime of the development.

> **Key terms**
>
> **Greenfield sites:** areas of land that have not been previously developed or built on, above or below ground
>
> **Brownfield sites:** areas of land that have been previously developed or built on, even if there is no physical evidence of earlier use

▲ Figure 3.1 Superstructure and substructure of a building

▲ Figure 3.2 A greenfield site

Chapter 3 Construction design principles

▲ Figure 3.3 A brownfield site

The environmental impact of a building's design, construction and use is controlled and monitored by:
- an environmental impact assessment (EIA)
- an energy performance certificate (EPC)
- the Code for Sustainable Homes
- a site waste management plan (SWMP)
- building regulations (Part L).

Environmental impact assessment (EIA)

The Town and Country Planning (Environmental Impact Assessment) Regulations 2017 require developers to identify significant effects on the environment of their proposed construction project at the planning application stage. If the type of development is listed in Schedule 2 of the regulations and exceeds the thresholds set out there, the local authority must screen the proposals to determine the likely environmental impact and decide whether an EIA is required. This also allows the public an early opportunity to be involved with decision-making procedures.

The regulations affect only a small proportion of private and public projects that require planning permission; however, they could also apply to some **permitted development**.

> **Key term**
>
> **Permitted development:** building work that does not require planning permission from the local authority but must still be constructed according to building regulations

> **Research**
>
> Search online for 'Schedule 3 of The Town and Country Planning (Environmental Impact Assessment) Regulations 2017'. You can access it here: www.legislation.gov.uk/uksi/2017/571/schedule/3/made
>
> Use the screening criteria to determine the environmental impact of construction work you have recently been involved in.

Energy performance certificate (EPC)

EPCs are a legal requirement for domestic and commercial properties in the UK that are available to buy or rent. However, some buildings may be exempt, for example, residential buildings that are intended to be used for less than four months a year and listed buildings.

To obtain an EPC, an accredited assessor must evaluate the energy efficiency of a building. They will check that loft and wall insulation has been correctly installed, and they will review records of energy use (also referred to as heat demand) for the past two years. Once satisfied, they will award an energy efficiency rating from A (most efficient) to G (least efficient) and a certificate containing information about the property's energy use and average energy costs. The certificate may also make recommendations about how to reduce energy use and save money on energy bills in the future.

EPC assessments are valid for ten years for domestic properties and are available to view on the Landmark Register in England and Wales, and the Scottish EPC Register in Scotland.

▲ Figure 3.4 Loft insulation

85

▲ Figure 3.5 Energy performance certificates

> **Case study**
>
> You decide to rent out your home. You do not have an EPC, so you appoint an EPC assessor to inspect your property.
> ▶ What energy efficiency rating do you think the assessor would award your property? Explain your answer.
> ▶ What recommendations do you think the assessor would make to improve the energy performance of your property and reduce your energy costs?

Code for Sustainable Homes

Launched in 2006, the Code for Sustainable Homes was a government-led environmental assessment method for rating and certifying the performance of new homes in the UK. It set national standards for the design and construction of new domestic properties, with the aim of improving sustainable building methods.

The government withdrew the code in 2015 and consolidated some of the benchmarks into existing building regulations. Although adherence to the code is no longer a condition for new planning proposals, it still endures where outline planning permission was granted before March 2015 or where there are existing contractual arrangements.

Site waste management plan (SWMP)

A site waste management plan (SWMP) sets out how waste will be managed and disposed of during a construction project.

The plan should be prepared by the client, along with the design team, contractor and subcontractor, and it should identify:
▶ the different types of waste that will be created throughout the project
▶ where the waste is expected to be found
▶ the minimum estimated quantities of each type of waste
▶ the actions that the client and principal contractor intend to take to minimise the amount of waste by reusing and recycling.

For more information on waste management and SWMPs, see Chapter 5, section 8.

▲ Figure 3.6 Construction site waste being recycled in skips

Building regulations (Part L)

Part L of the building regulations is concerned with the conservation of fuel and power in new buildings (Approved Document L1A) and existing buildings (Approved Document L1B). (There is further information on Approved Documents in section 2.1 of this chapter.)

To help the government meet its targets to reduce carbon emissions, the regulations identify measurable rates to determine the energy efficiency of a building.
▶ The calculated rate of CO_2 emissions from a dwelling (the Dwelling CO_2 Emission Rate, DER) must not exceed the Target CO_2 Emission Rate (TER).
▶ The Dwelling Fabric Energy Efficiency (DFEE) rate must not exceed the Target Fabric Energy Efficiency (TFEE) rate.

Chapter 3 Construction design principles

The regulations also provide technical guidance on the quality of materials and standards of work to achieve the TER and TFEE rates in areas such as secondary heating and lighting.

> **Test yourself**
>
> What part of a building is the superstructure?

1.2 The benefits of good design and the potential implications of poor design

A building that has been properly thought out and planned will benefit the end user in many ways. Prospective buyers often look for the following qualities in new homes.

- Aesthetics – the building should have 'kerb appeal' (it should be attractive when viewed from the road) and have modern amenities. It should be in keeping with its surroundings and add value to the community.
- Efficiency – the building should have a good energy efficiency rating, allowing it to be cost-effective to run.
- Sustainability – the use of sustainable materials and alternative energy sources could make the property more desirable.
- Wellbeing and improved quality of life – these can be achieved by maximising natural light and ensuring efficient ventilation within the building, as well as enabling access to outdoor living spaces such as gardens, terraces or balconies.
- Affordable quality – 'affordable' is considered a median household income or lower, determined by the local or national government. Careful selection of building materials, methods and processes without unforeseen delays will bring in the construction project on budget.
- Improved local/community facilities – this could include accessible parks, woodland, play areas, community outdoor gyms, sports clubs, leisure facilities and shops.
- Improved **infrastructure** – this could include footpaths, roads, parking, bus routes and pedestrian bridges.

▲ Figure 3.7 Newly constructed domestic houses for sale

Impact on the local community

A building must meet the needs of both the client financing the project (the 'project sponsor') and the local community, now and for future generations. A development that serves the purpose of the client but is not well designed or managed throughout the construction phase can cause disruption to the local community with noise, light and air pollution, not to mention increased construction traffic and on-street parking.

A poorly planned development could have a negative impact on both the local economy and the way the area functions; for example, consider the effect it might have on local infrastructure. On average, most of the new occupants will drive some form of vehicle that will increase the amount of traffic in the area, particularly at peak times. This may result in delays, wear and tear on the roads, increased pollution and limited parking. The higher population will increase demand on local schools, healthcare facilities, shops and the job market.

> **Key term**
>
> *Infrastructure:* the basic systems and services that a country or organisation needs in order to function properly, including transportation systems (roads, footpaths, railways, bridges, tunnels and airports), water supplies, sewers and telecommunications systems

▲ Figure 3.8 Infrastructure of a community

Building new houses creates short-term employment opportunities for the development team. However, when the building phase is completed, those houses will be sold and occupied by people with social and economic needs.

To ease the shortage of **affordable housing**, it is often a condition of planning approval that a percentage of all newly built homes are allocated to meet this need. The National Planning Policy Framework states that at least ten per cent of most new developments should be affordable housing.

> **Key term**
>
> **Affordable housing:** housing for sale or rent, for those whose needs are not met by the market (including housing that provides a subsidised route to home ownership and/or is for essential local workers); this means that low-income households can buy or rent properties at an affordable price, based on their earnings

> **Improve your maths**
>
> Research the average wages for five different trades in your area, then determine the annual median wage.

Impact on parties in the construction chain

A new housing development needs to enable a local community to function and grow, and to bring employment opportunities. A failure to meet these needs will be reflected in slow sales and reduced profitability.

Large housing developments are usually divided into smaller phases by the building management team. Each phase is completed and sold in turn, to finance further phases of the overall development. If the development and buildings have not been well designed and constructed in the right area at the right time, then slow sales at an early stage in the development can have a major impact on a number of different parties in the construction chain:

▶ The client may be unable to finance further phases or to make repayments to the project sponsor (for example, a bank, a shareholder or an investor). Broken repayment agreements could result in financial penalties for the client. A failed venture could damage the client's reputation and impact on the financing of future developments.
▶ The project sponsor may not finance any further building work or future developments.
▶ Construction work might be paused, leading to a reduction in size of the project team until sales pick up and further phases of construction work are started. This could cause difficulties when building recommences in terms of re-employing contractors and subcontractors at the right time, leading to further delays.
▶ Suppliers may not be paid on time by contractors and subcontractors, because they are no longer employed. This can lead to a loss of confidence and goodwill, making it difficult to find suppliers willing to provide further building materials or credit terms. Material costs could increase the longer the development is delayed.

Impact of poor design and poor quality of work

Poor design and poor quality of work can result in a building failing to perform as expected after transfer of ownership to the end user. This can result in further negative impacts on the customer and client, such as high energy bills and the fabric of the building degrading.

Materials that have been inappropriately specified, selected, stored and used during construction work can often lead to failures at a later stage. Building materials can fail when exposed to certain conditions, as shown in Table 3.1.

Chapter 3 Construction design principles

▼ Table 3.1 Common failures in building materials

Building material	Failure	Cause
Ferrous metal	Corrosion (an electrochemical process)	Exposure to moisture and other chemicals
Timber	Decay/collapse (when the timber contains no chlorophyll)	Fungal attack (dry/wet rot) due to moisture content above 20 per cent Woodworm
	Warps, splits and shakes	Poor storage (unsupported/exposed to the elements/poor seasoning)
	Burning	Unprotected timber positioned too close to a heat source
Brickwork and concrete	Deterioration and cracking of concrete or mortar joints Splitting of bricks and blocks	Frost attack as a result of water expanding as it freezes in porous materials **Sulphate attack** Incorrect ratios of materials used in the mixing of mortar and concrete
	Efflorescence	Water containing salt in the materials migrating to the surface and drying
Paint	Flaking/peeling	Ultraviolet light (sunlight) attacking the paint Inadequate paint preparation and poor choice of primers

89

▼ Table 3.1 Common failures in building materials (continued)

Building material	Failure	Cause
Plastic	Becoming brittle	Ultraviolet light (sunlight) attacking the plastic
	Expansion and contraction	Extreme temperatures (heat/cold)
	Discolouration	Low-grade plastic
Glass	Misting	Failure of seals between double-/triple-glazed units, allowing the thermal gases to escape and moisture to enter between the layers of glass

> **Key terms**
>
> **Ferrous metals:** metals that contain iron and are vulnerable to rust when exposed to moisture
>
> **Sulphate attack:** a chemical process where sulphates cause the cement in concrete to deteriorate, often resulting in cracking

> **Test yourself**
>
> What are the benefits of good design and the implications of poor design?

1.3 Factors that can impact on the profitability of projects

At the planning stage of any development, a **feasibility study** is usually carried out by the client and the management team, to check that the project is both viable and potentially profitable. The study usually considers areas such as:
- finances (profits/losses)
- budgets
- resources (labour, plant and equipment)
- environmental impact
- scheduling
- appraisal of the proposed development site regarding existing services, waterways, boundaries and **topography**
- investigations into the history of the land above and below ground, identifying any unforeseen features that have the potential to make building work more complicated and costly, such as old wells, tunnels and mines
- historical architectural interest or any existing features, structures or buildings that could delay construction work or even prevent it from going ahead.

> **Key terms**
>
> **Feasibility study:** an investigation to establish the likely success of a proposed project
>
> **Topography:** the physical features and shape of land surfaces

Chapter 3 Construction design principles

▲ Figure 3.9 The topography of a piece of land

Greenfield and brownfield sites

Reusing a brownfield site for a new development makes environmental sense. However, from a financial point of view it is more costly, because existing buildings need to be removed and the ground needs to be prepared (for example, by clearing contaminated soil and waste).

For ease of building purposes and to maximise profits, many developers would prefer to use a greenfield site. Building work can usually start quickly on these sites with less preparation, and contractors do not have to contend with any existing hazardous waste or unforeseen development issues. However, building on a greenfield site is controversial, as many people believe we have a moral responsibility to protect the environment. There may also be local planning restrictions in place to preserve greenfield sites and only allow new developments when certain criteria are met.

Over-specification and difficulty of assembly

Developments that are over-specified or over-engineered can result in lower profits, even before construction work has begun. Unique design ideas and complicated assembly of components often result in difficulties during manufacturing and construction, due to the workforce's lack of familiarity, knowledge and skills. These complications can lead to increased timescales, missed deadlines and stretched budgets.

Vernacular construction

The design of the houses in a development may be sympathetic or particular to a region, relying on locally sourced materials and traditional skills that have developed over generations. This method of region-specific house building is referred to as vernacular construction. Traditionally, construction methods did not follow contemporary mainstream architectural designs, until the Victorian period when building materials and styles became increasingly standardised.

Given the distinct regional characteristics of such buildings, it is problematic to design and specify this form of construction anywhere other than where it is usually found. It would be difficult to achieve the same standards of work without employing labour and sourcing materials from further away.

▲ Figure 3.10 This stone house in the Cotswolds, built with local materials, is an example of vernacular construction

Corporate social responsibility (CSR)

Although construction businesses need to make a profit, this should not be at the expense of the world around them. **Corporate social responsibility** (CSR) refers to strategies adopted by businesses to ensure they monitor and manage their social, economic and environmental impact on all aspects of society.

The CSR business model is self-regulated and can take many different forms, depending on the type of industry. It supports companies in being socially accountable to themselves, their stakeholders and the general public, in order to have a positive influence on the world. Businesses using this model not only benefit society but also themselves, by boosting morale between employers and employees and improving the reputation of their corporate brand.

> **Key term**
>
> *Corporate social responsibility:* the commitment of an organisation to carry out its business activities in a socially and environmentally responsible way

Typical examples of CSR in construction include:
- providing services for the community
- supporting charitable activities
- using ethically and sustainably sourced building materials
- reducing the environmental impact of construction work.

> ### Research
>
> Search online for the International Organization for Standardization's guidance document 'Discovering ISO 26000'. You can access it here: www.iso.org/files/live/sites/isoorg/files/store/en/PUB100258.pdf
>
> Research what social and ethical responsibilities could be adopted by your work-placement employer to manage their impact on society and the environment.

> ### Test yourself
>
> What factors could have an impact on the profitability of a building project?

2 Design principles

2.1 Factors to be considered during the design of building services

A client must consider many factors as part of their **concept** for a building, and some of these are more obvious than others, for example:
- building capacity
- number of bedrooms and bathrooms
- internal arrangement of rooms and other spaces
- parking and garages
- aesthetics (design features, choice of materials and use of colour).

> ### Key term
>
> **Concept:** a principle or idea

One of the most important factors to consider when planning a building project is the availability of mains services, such as electricity, water and drainage. If all these services are not easily accessible at the site (for example, in rural locations), then it could be expensive to connect to them. This usually involves paying service providers to dig long trenches in the ground to run pipes and cables to the boundary of the development, which is not always feasible and may cause damage to the natural environment. Special permission may also need to be obtained if this involves working on someone else's land.

Even if the proposed new development is on a brownfield site, there is no guarantee that the services provided will be in good order, for example, older water pipes may have low pressure or the water supply may be contaminated if the pipes are made of lead or corroded steel.

In remote locations where mains water cannot be accessed, a borehole may have to be drilled into the ground to find a natural water source. This can be very expensive, because the depth of the hole and the ground conditions are relatively unknown until the drilling starts, and any water that is found will have to be filtered and purified before use.

A client may choose to have a self-contained site using alternative energy sources, such as photovoltaic panels for electricity and lighting, ground and air source heat pumps to supply heating and hot water, and storage tanks for sewage. The main benefit of this approach is that it does not rely on the use of finite fossil fuels to generate power, and although the initial installation costs can be quite high, these will be offset with reduced energy bills over a period of time. For more information on energy sources, see Chapter 5.

▲ Figure 3.11 Services being connected

Chapter 3 Construction design principles

Approved Documents

In many ways, the design of a building and the choice of construction method are influenced by local authority planning restrictions and building regulations. The Ministry of Housing, Communities and Local Government publishes guidance on how to comply with building regulations in the form of Approved Documents. These provide general advice on the performance expected of materials and building work, and practical solutions to some of the more common building situations.

Approved Documents are divided into the key areas listed below:
- Approved Document A: Structure
- Approved Document B: Fire safety
- Approved Document C: Site preparation and resistance to contaminants and moisture
- Approved Document D: Toxic substances
- Approved Document E: Resistance to the passage of sound
- Approved Document F: Ventilation
- Approved Document G: Sanitation, hot water safety and water efficiency
- Approved Document H: Drainage and waste disposal
- Approved Document J: Combustion appliances and fuel storage systems
- Approved Document K: Protection from falling, collision and impact
- Approved Document L: Conservation of fuel and power
- Approved Document M: Access to and use of buildings
- Approved Document N: Glazing – safety in relation to impact, opening and cleaning
- Approved Document O: Overheating
- Approved Document P: Electrical safety
- Approved Document Q: Security in dwellings
- Approved Document R: Infrastructure for high-speed electronic communications networks
- Approved Document S: Infrastructure for the charging of electric vehicles
- Approved Document 7: Materials and workmanship

Listed and heritage building regulations

Existing buildings may be protected by listed and heritage building regulations. If a building has special architectural or historical interest, not only will the structure of the building be protected, but also any other attached structures or features, including interiors.

Listed buildings are graded according to their national importance:
- Grade I (one) – buildings of exceptional interest
- Grade II* (two star) – particularly important buildings of more than special interest
- Grade II (two) – buildings of special interest warranting every effort to preserve them.

If a client wants to alter, extend or change the use of a listed building, they must apply to the local authority for 'listed building consent' and be granted permission before any work begins. The grade of listing usually determines what changes can be made and how they should be implemented, although any work will have to be carried out sensitively in order to protect the character and history of the building or structure.

▲ Figure 3.12 A listed building

Traditional versus modern building methods

It is important during the planning stage of any building project to consider carefully which construction method will be most appropriate.

Traditional methods of constructing walls, floors and roofs for **low-rise buildings** have evolved in recent years, in order to meet regulations and building standards on energy efficiency and sustainability. However, this type of construction can be labour-intensive and usually results in longer build times, because the entire superstructure is completed on site with low levels of **mechanisation**. During the initial stages of the construction work, before the building is watertight, the progress made is often weather dependent, for example, bricklayers cannot lay bricks at temperatures below 2°C or during periods of heavy rain. Traditional construction methods also tend to produce a lot of waste.

> ### Key terms
>
> **Low-rise buildings:** buildings with four storeys or fewer
>
> **Mechanisation:** the use of machines or automatic devices

Onsite Construction T Level: Core

Table 3.2 compares some of the most common traditional and modern construction methods.

▼ Table 3.2 Comparison of traditional and modern construction methods

Method of construction	Traditional/modern	Advantages	Disadvantages
Brick and block	Traditional	• Familiar • Good thermal performance • Good sound insulation • Can be used with concrete upper floors • Good structural performance • Durable • Good fire resistance • Good weather exclusion	• Thick walls due to insulation between inside and outside **skin** • Takes time to dry out • Cannot be carried out in heavy rain or freezing conditions • Production of cement creates high levels of carbon, which is harmful for the environment
Open panel – timber frame	Traditional	• Sustainable building material if the timber is from a managed forest • Factory-built framework reduces onsite build times • Good thermal performance and sound insulation (although this is reliant on the infill panels between the open frames)	• Can be more expensive than other building methods • Water can stain the exposed timber in the early stages of construction
Straw-bale construction (non-structural/infill system)	Traditional	• Sustainable • Low cost • Quick to build • Good thermal performance • Good sound insulation • No drying out required	• At risk of damage from vermin • Increased risk of fire • Uneven wall-surface finishes • Straw rots easily if exposed to moisture
Panelised – timber frame	Modern	• Manufactured off site in a factory, thereby reducing onsite build times, labour requirements/costs and health and safety risks (for example, working at height) • Quick to erect the shell of the building to make it **watertight** • Sustainable building material • Good thermal performance • Reduced waste (factory manufactured) • Accurate, easy to adjust and fix to • Thinner walls compared with other methods	• Unable to support concrete upper floors • Long **lead time** for panels to be made off site • Liable to rot if exposed to moisture

Chapter 3 Construction design principles

Method of construction	Traditional/modern	Advantages	Disadvantages
Insulated concrete formwork (ICF)	Modern	• Excellent thermal performance • Good sound insulation • Good weather exclusion • Durable • Excellent structural performance • Good fire resistance	• Inexperienced workers may need training to use the system • Poor preparation of the formwork (concrete mould) can lead to difficult and expensive repairs once the concrete has been poured and cured
Steel-frame construction	Traditional	• Quick to erect • Excellent weather resistance • Durable • Excellent structural performance • Could support concrete upper floors • Good fire resistance	• Expensive • Heavy • Requires a crane to lift the steel frame into position • Can distort if exposed to extreme heat (for example, in a fire)
Thin-joint blockwork/masonry	Modern	• No drying out of mortar joints needed • Quick to build • Excellent weather exclusion • Durable • Excellent structural performance • Fire resistant • Excellent thermal performance • Good sound insulation	• Slightly more expensive compared to traditional bricks and blocks • Accurate foundations needed to build on, because it is difficult to correct blockwork courses at a later stage as the walls are built
Structural insulated panels (SIPs)	Modern	• Good thermal performance • Manufactured off site in a factory, thereby reducing onsite build times, labour requirements/costs and health and safety risks (for example, working at height)	• Precise foundations required to align the SIPs; any deviation could result in costly delays • Requires a crane to lift the panels into position
Volumetric (pod/modular)	Modern	• Manufactured off site in a factory, therefore reducing onsite build times, labour requirements/costs and health and safety risks • Better quality control and accuracy maintained in a factory because of the level of automation used	• Initially more expensive • Requires a crane to move the pods into position, which can be expensive to acquire if not already on site

> **Key terms**
>
> **Skin:** a single thick masonry wall
>
> **Watertight:** when a building is at the stage of construction where the weather cannot penetrate it (for example, the roof is on and the windows and doors have been fitted)
>
> **Lead time:** the period of time between ordering and receiving goods or materials

> **Test yourself**
>
> What are the advantages of modern construction methods compared with traditional methods?

Buildability

Some modern construction methods involve parts or sections of a building being manufactured away from the site location. Large sections of walls and floors (referred to as cassettes), roof trusses and even whole rooms (known as pods) can be prefabricated more quickly in factories, with better quality control. Production is not delayed due to adverse weather conditions, and off-site factory manufacturing also reduces onsite build times, labour requirements/costs and some health and safety risks.

This method of combining factory-produced, pre-engineered units (or modules) to form major elements of a structure is referred to as modular construction. It results in improved '**buildability**', because potential assembly issues are usually designed out at an early stage for ease of construction onsite. However, the **feasibility** of using heavy lifting equipment and cranes on site to lift the large building sections into place, alongside the increased risks to the safety of workers and others, have to be factored into the planning of a project.

▲ Figure 3.13 A cassette being lifted into position on site

> **Key term**
>
> **Buildability:** the simplicity of manufacturing or constructing a building through its design

Choosing a method of construction

It can be difficult deciding whether to construct using traditional or modern building methods and materials.

Traditional construction generally uses less-sustainable technologies and materials that produce high levels of CO_2 during manufacture. However, the costs of maintaining modern methods are normally higher over the lifetime of the building. For more information on sustainable methods of construction, see Chapter 5.

Whichever method is chosen, it has to meet the minimum standards determined by building regulations. However, some buildings will naturally perform better than others with regards to heat retention and acoustics. The ability of a building to maintain an ambient temperature for its occupants is influenced not only by the U-value of the walls, floors and roof, but also by the construction of doors, windows and roof lights.

> **Key term**
>
> **Feasibility:** how easy or difficult it is to do something

Large windows, roof lanterns and bi-folding or sliding doors are popular features of modern homes. These allow natural daylight to illuminate large areas of the building, providing a feeling of more space and a connection between indoor and outdoor living.

The direction that a building faces on a plot and the carefully planned positioning of glazing will allow sunlight to warm a building through solar gain (the sun). This heat can also be stored in the fabric of the building during the day and released during the evening, which will reduce reliance on other sources of energy, decrease bills and maintain a comfortable living environment.

Windows and doors are important design features for access and egress (including for those with disabilities). They are also a means of escape in the event of a fire. In order to maintain the health and wellbeing of the occupants of a building, openings in windows and doors allow for ventilation, whereby stale air is removed from the building and replaced with natural fresh air.

Chapter 3 Construction design principles

▲ Figure 3.14 Natural light entering a building through a roof lantern

> **Improve your English**
>
> 'Luminaire' is a word used to describe a source of artificial light. Write a paragraph to explain the use of different types of luminaire in a building, and explain why natural light is always a better source of energy.

> **Test yourself**
>
> The installation and use of windows and doors are controlled by building regulations. Which Approved Documents provide information and guidance on this?

> **Research**
>
> Search online for the 'Building Research Establishment Environmental Assessment Method' (BREEAM) and explain the role it plays in the design of the built environment.

> **Test yourself**
>
> What factors need to be considered during the design of building services?

2.2 Stages and outcomes of the Royal Institute of British Architects Plan of Work

The Royal Institute of British Architects (RIBA) Plan of Work is a design and process management tool used by the UK building industry to bring greater clarity for the client at different stages of a project.

It organises the process of briefing, designing, constructing and operating building projects into eight stages. Each stage has intended outcomes, core tasks and information that should be exchanged with different parties.

In recent years, the Plan of Work has been updated to reflect changes in the industry and to support the government's target to be net zero carbon by 2050.

▼ Table 3.3 RIBA Plan of Work Stages and Stage Outcomes

		Stage	Outcome
Pre-design	0	Strategic definition	Determine the best way of achieving the client's requirements and the most appropriate solution.
	1	Preparation and briefing	Develop the client's concept and make sure it can be accommodated on site. Make sure everything needed for the next stage is in place.
Design	2	Concept design	Make sure the look and feel of the building are meeting the client's expectations and budget.
	3	Spatial co-ordination	Design the spaces within the structure of the building, before preparing detailed information about manufacturing and construction.
	4	Technical design	Develop information received from the design team and specialist subcontractors for the manufacture and construction of the building.
Construction	5	Manufacturing and construction	Manufacture and construct the building.
Handover	6	Handover	Complete the building works and address any defects that have been identified, to conclude the building contract between the client and the contractor.
In use	7	Use	The building should be used, operated and maintained efficiently until the end of its life. At this stage, the client may consider appointing professionals for aftercare activities such as servicing and maintenance.

Source: RIBA Plan of Work 2020 Stages and Stage Outcomes, reproduced courtesy of the Royal Institute of British Architects

> **Improve your English**
>
> The term 'spatial' is used in the RIBA Plan of Work to describe the space, position, area or size of things within a building. Write a sentence that includes the word 'spatial' to describe the arrangement of rooms on a level in a dwelling.

3 The role of different disciplines involved in design

3.1 Key job roles within construction design

There are many diverse disciplines that contribute to the design and execution of a building project. Each member of the project team is appointed by the client to provide specialist knowledge, skills and services.

The size of the development often determines how many people are involved, and many of them may have never worked together before. Complex information is communicated throughout the building and design team at various stages in the construction project; therefore it is important that roles and responsibilities are defined clearly at an early stage and recorded in the appointment documentation.

It is likely that some members of the project team will play only a brief role, and that the structure of the team will change throughout the development. In order to work effectively as a team and in the best interests of the client, communication needs to be clear and efficient, so that everyone is aware of what actions they need to take.

Let's look at some of the personnel who could be involved in the design and construction of a development.

Contractors

Contractors may be given responsibility by the client to design, plan, organise and control a construction project; this is commonly referred to as 'design and build'. Under the Construction (Design and Management) Regulations 2015, a principal contractor must be appointed when there is more than one contractor working on a project. The principal contractor then either appoints designers, or uses one of their own in-house designers, to manage this process as part of their project team.

The main benefit of this approach is that the client has fewer points of contact to communicate with at each phase of the development.

Many contractors start out as tradespeople to gain experience of the construction industry, before they progress into supervisory or principal contractor roles to manage building projects for clients.

Operatives

Operatives support different tradespeople at every stage of a building project by completing manual labour tasks, such as stacking and storing materials, mixing mortar and tidying the site. The availability of skilled labourers for a construction project and their experience of working with a particular building method may influence the design.

Skilled general operatives may progress into a trade, and with further experience into supervisory roles such as trade foreperson (there is further information on this role in Chapter 4).

Architects

Architects are normally appointed by the client to design new buildings or structures, or to conserve or redevelop old ones. Part of their role is to work closely with clients, the local authority and main contractors to prepare:
- detailed drawings
- specifications
- feasibility studies
- a project brief
- planning applications
- **tender documents**.

> **Key term**
>
> **Tender documents:** documents prepared to seek offers for the supply of goods or services; the client will use these documents to identify the most suitable contractor for a project

Architects may also be contracted to:
- prepare other documents, for example, a site waste management plan
- undertake site inspections
- offer advice to the client at various stages of the building work to resolve any technical difficulties.

Knowledgeable architects who work within a larger company may progress to a senior position as a lead architect or choose to work for themselves as a freelance consultant.

Chapter 3 Construction design principles

▲ Figure 3.15 An architect using computer-aided design (CAD)

Planners

The role of planners is to ensure that land in villages, towns, cities, the countryside and commercial sites is used effectively to meet economic, social and environmental needs. Their objective is to achieve a balance between encouraging innovation and growth in the development of housing, industry, agriculture, recreation and transport while trying to preserve the historical environment.

Planners can also work as consultants for clients and provide advice on local planning policy to architects and developers. They can consult with stakeholders to determine planning applications, contest appeals and enforce planning legislation.

Experienced planners may leave the local planning department to become independent, advising contractors and clients on planning and design to meet the local authority's requirements with regards to planning applications.

Building control officers/private building inspectors

Building control officers (BCOs) work for **Local Authority Building Control** (LABC). Private building inspectors work for government-approved building-inspection companies. Private building inspectors do not have the same powers of enforcement as BCOs, which means that their role may have to be transferred to LABC if there is a problem that cannot be resolved without its intervention.

> **Key term**
>
> *Local Authority Building Control:* local authority department responsible for inspecting building work against building regulations and signing-off completed projects

The role of the BCO is to inspect plans submitted for full planning approval, making sure they meet current building regulations. At intervals agreed between the BCO and the contractor, they will visit the construction site and check that it meets with regulatory standards of design and safety.

The progression route for a BCO might involve becoming a member of the local authority planning committee.

▲ Figure 3.16 A building control officer (BCO) on a construction site

> **Industry tip**
>
> Never deviate from approved working drawings during the construction phase without written consent from either Local Authority Building Control or a government-approved private building inspector. Changes to the design without permission can be expensive to put right if the work fails to meet building regulations approval.

Manufacturers

Manufacturers are responsible for the design, production and sale of building materials and goods. Before their products are sold and distributed to the end user, they are tested and certified to ensure they adhere to industry standards and regulations (for example, the Ecodesign for Energy-Related Products Regulations 2010).

Designers may choose certain materials based on their sustainability, eco-design and energy rating, which means that manufacturers can influence the design and use of buildings through their products.

Onsite Construction T Level: Core

Mechanical building services (design) engineers

Mechanical building services (design) engineers consult and advise clients on the design, installation, operation and maintenance of building services such as:
- heating systems
- ventilation systems
- air-conditioning systems
- renewable-energy systems
- sustainable technologies.

Their technical knowledge of building services may also be utilised by the client for the production of detailed drawings, specifications and calculations to meet with design standards and regulatory requirements. They may be involved in bids and tenders for work, and mentor the client from the start to the completion of a project.

Electrical building services (design) engineers

Electrical building services (design) engineers consult and advise clients on the design, installation, operation and maintenance of electrical systems for domestic, commercial and industrial projects, such as:
- lighting systems
- fire-safety systems
- security systems
- renewable-energy systems, for example, solar photovoltaic (PV) panels and electric-car charging points.

Mechanical design engineers (building services)

Mechanical design engineers use their specialist knowledge of complex mechanical systems to design, manage and supervise projects for the client, from concept to completion.

They often work closely with electrical engineers to ensure the design and installation work is carried out safely and in accordance with industry codes of practice.

Mechanical engineer design co-ordinators

Mechanical engineer design co-ordinators are usually appointed to assist the mechanical design engineers on larger projects, using their technical expertise to create innovative solutions to building-services problems. They may also organise tendering, project management, reporting and scheduling for the client when required.

Mechanical engineer CAD technicians

CAD technicians produce technical information and 2D/3D mechanical diagrams using computer-aided design (CAD). They create manufacturing drawings and work instructions for the building services team. This information is then used to improve the buildability of the project and reduce construction costs for the client.

BIM designers

BIM designers work with clients to implement **Building Information Modelling** (BIM). BIM is an intelligent, streamlined process of sharing **digital** information (for example, 3D models, drawings, specifications and schedules) between all parties and stakeholders. This cost-effective process allows a collaborative approach at every stage of planning, designing, building and managing a project, to improve efficiency and productivity throughout the life cycle of the building.

> **Key terms**
>
> **Building Information Modelling (BIM):** the use of digital technology to share construction documentation and provide a platform for collaboration
>
> **Digital:** in electronic form
>
> **Retrofit:** the process of adding new components to older structures

Retrofit assessors

Retrofit assessors assess of existing buildings in order to compile energy efficiency improvement plans for clients. They may also visit sites to:
- resolve issues (for example, maximising the use of PV panels on a building with a complex roof shape)
- review projects
- carry out audits to ensure health and safety and performance standards are being met.

Retrofit co-ordinators

Retrofit co-ordinators liaise with retrofit assessors, retrofit installers and clients about the installation of renewable-energy technologies in existing buildings, for example, solar thermal, biomass and air/ground source heat pumps.

An important part of their role is to gather data from the retrofit assessors, then check its accuracy and produce an improvement plan for the client. The client may also make arrangements for a co-ordinator to appoint an approved installer to carry out the work identified in the plan.

Chapter 3 Construction design principles

Retrofit installers

Retrofit installers (also known as retrofit installer technicians) are responsible for the installation, commissioning and final handover to the client of renewable energy systems and measures. Installers must be PAS 2030:2019 certified and meet these standards on all retrofit projects.

Competent and experienced retrofit installers may progress in their careers to become retrofit assessors.

> **Research**
>
> Search online for 'PAS 2030:2019 Specification for the installation of energy efficiency measures in existing dwellings and insulation in residential park homes'.
>
> Identify the standards of PAS 2030:2019 for retrofit installers and explain how they benefit the construction industry.

▲ Figure 3.17 Retrofit installer fitting an energy-efficient boiler

4 Design process from conception to completion

4.1 Key stages of the design process, and factors that may impact or influence design changes

House building can be a relatively straightforward process when good design principles are applied, the project has been well planned and there are no unforeseen problems encountered during the construction phase. However, when you are dealing with the natural environment no two sites are ever the same. It is therefore essential that the land is properly investigated to identify any potential issues early and reduce the risk of additional costs later in the build.

Individuals or organisations responsible for preparing or modifying designs for construction projects have legal duties under the Construction (Design and Management) Regulations 2015.

Designers must:
- ensure the client is aware of their duties under the CDM Regulations and help them to comply with those duties
- take into account any preconstruction information provided by the client
- eliminate foreseeable health and safety risks where possible
- seek to reduce or control any health and safety risk that cannot be eliminated
- provide design information for inclusion in the health and safety file for the project
- co-ordinate with other designers working on the project on matters of health and safety during the construction phase and beyond
- co-ordinate, communicate and co-operate with all contractors working on the project, taking into account their knowledge and experience of building design.

Further information on the CDM Regulations can be found in Chapter 1.

In this section, we will look at the importance of researching and analysing building plots, as well as the process of applying for planning permission, through to the final sign-off of the building work.

> **Test yourself**
>
> What legal duties do designers have under the Construction (Design and Management) Regulations 2015?

Research

Before construction work begins, a **desktop survey** is usually undertaken to identify and record details about previous and current uses of the site. The main objective of this survey is to support the design and construction processes, by allowing correct decisions to be made when planning the site layout.

> **Key term**
>
> **Desktop survey:** an investigation into a piece of land completed without visiting the site or taking physical samples of soil etc.

A desktop survey will look to establish the following information:
- site history
- waste records
- geology and hydrology
- contamination reports
- site boundaries
- position of existing services
- existing structures or buildings
- local roads
- access to the site
- topography of the land
- hedges, trees and fences
- wildlife and habitats.

Much of this information can be gathered from existing plans and records held by the Environment Agency and the Land Registry.

If a specialist or consultant is appointed, they will also produce an environmental risk assessment. This document is used to identify potential risks to the natural environment from the proposed building work.

A site of special scientific interest (SSSI) may be identified in the initial survey. This is an area that must be protected from construction activities due to its rare flora or fauna, or its physiographical or geological features.

▲ Figure 3.18 Some sites must be protected due to their features of special interest, such as their rare flora

▲ Figure 3.19 Some sites can become SSSIs due to their fauna

Key terms

Hydrology: the study of water in the earth and its relationship with the environment

Environmental risk assessment: a document used to identify and control potential risks to the natural environment from construction work

Flora: plants and trees in a particular region

Fauna: animals or insects in a particular region

Case study

A feasibility study for the construction of 75 new homes has identified the existence of fauna on the proposed building site.

Explain in detail what animals are protected and the potential outcomes now that these indigenous species have been discovered on the land.

Site analysis

Once the desktop survey has been completed, a walkover survey is usually undertaken. This is a physical inspection of the building site to identify any geological, ecological or topographical issues that may impact the project.

It involves drilling boreholes into the ground at various positions on the site to investigate the:
- composition of the soil
- load-bearing capacity of the ground
- position of the water table.

The water table is the point below the ground where the soil becomes saturated; this depth will vary depending on the location of the site.

The information gathered from the walkover survey is interpreted and used to determine the most appropriate form of foundation to suit the ground conditions.

Key term

Walkover survey: a physical inspection of a building site

Chapter 3 Construction design principles

▲ Figure 3.20 The water table

> **Research**
>
> Search online for the various methods of soil investigation and explain why they are used.

Planning

Under permitted development rights, you can extend or make certain alterations to an existing domestic property without the need for planning permission from the local authority. However, if the proposed changes are beyond the specific limitations of the rules, or new dwellings are to be constructed, then planning permission must be sought and granted.

> **Industry tip**
>
> Regardless of whether or not planning permission is needed, all building work must be completed in accordance with current building regulations.

The process of making an application for planning approval through to the final decision made by the planning committee can be lengthy and expensive, especially when you consider the investment in time, effort and money in designing the project.

To improve the efficiency of the application process and reduce any potential financial losses, a pre-planning application should be made to the local planning department. This type of application is usually processed much more quickly than a full application, where detailed drawings and specifications are often needed. Local planning departments encourage pre-applications so that they can offer support and guidance to resolve any issues in the proposal, improving the chance of the final application being successful.

If planning permission is required for a project, an application must be made to the local authority by the client or planning team, accompanied by a fee and supporting documentation. The design plans for the site will identify the boundary, **frontage line** and **building line**. These may be stipulated by the local planning department as a condition of planning approval.

At this stage, details of the planning application are advertised by the planning department in local newspapers and at the site of the proposal. Consultation letters are sent to neighbours, informing them about the application and advising how they can view the plans online or raise any valid objections to the project going ahead within a period of 21 days from the date of publishing.

> **Research**
>
> Find out what documents and types of drawing are needed to make a full planning application to a local planning department. Suggest the possible outcome if some of this detail is not provided.

> **Key terms**
>
> *Frontage line:* the front part of a building that faces a road
>
> *Building line:* a boundary line set by the local authority beyond which building work must not project

Onsite Construction T Level: Core

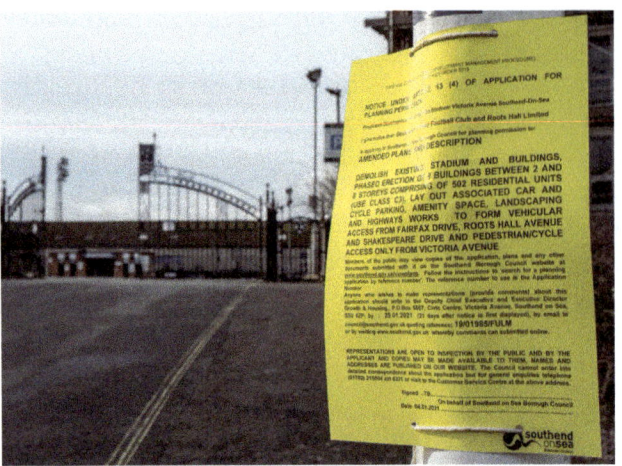

▲ Figure 3.21 A planning application notice

During the consultation period, a planning officer from the local planning department will visit the site, where they will take into account information provided in the application and any responses from the public or their representatives. They may also gather further site-specific information, such as measurements and photographs, in order to make their 'officer's report'. The planning officer is not responsible for making a planning decision at this stage, but will make recommendations for the authorised person or planning committee to base their final decision on.

The final outcome for the majority of domestic planning applications is decided by senior officers, under delegated powers from the planning committee.

There are three possible outcomes for a planning application:
- approved
- approved with conditions that must be complied with (for example, 'No further windows can be added to the proposal on the east elevation of the property')
- refused.

Applications that have been refused can be appealed; however this can be a lengthy and expensive process that is not always successful.

Construction work can be carried out on domestic buildings by submitting a 'building notice' to the local planning department, without the need to submit a full planning application. Although work can start immediately once a building notice has been submitted, lack of detail in the notice could result in the work not complying with building regulations and therefore having to be dismantled or corrected after it has been completed. It is therefore recommended that a building notice is used only for minor building works, or if the contractor has good knowledge of current building regulations.

▲ Figure 3.22 A building notice application

Research

Search online to find out how to submit a building notice to your local planning department. Download and complete (without submitting it!) a building notice application for a fictitious extension to your home.

Approval/review

Once building work has commenced on site, it must be inspected at regular intervals by either a building control officer (BCO) from Local Authority Building Control (LABC) or a private building inspector.

Chapter 3 Construction design principles

The quality and standards of work are checked against building regulations at the following stages:
- excavation of the foundation
- laying of foundation concrete
- installation of damp-proof course (DPC) and damp-proof membranes (DPM)
- laying of drains
- completion of the roof structure
- completion of first-fix installations (before plastering or dry lining)
- testing of drains
- completion of the project.

Where the standard of work falls below that expected, recommendations are made by the inspector to meet legislation.

Building regulations state that any person intending to carry out building work must notify LABC to determine stages of work and when they will be inspected. This is known as an 'inspection service plan'.

Project sign-off

As soon as possible after a building project has finished, the contractor or client must notify LABC, so that it can arrange a final visit to the site. During the inspection, the BCO will check that any outstanding actions from previous visits have been completed and verify that the dwelling meets with building regulations. Once satisfied with the building work, they will issue a completion certificate. Until this certificate is issued, a building is not 'signed-off' and therefore should not be occupied, and it may be difficult to insure it or secure a mortgage.

> **Key term**
>
> **Signed-off:** approved by a building control officer

4.2 Project planning

Part of successful planning for a construction project involves scheduling resources, materials and labour for various times throughout the building phase. If this is not given careful consideration at the planning stage, it could result in delays on site, missed completion deadlines and financial penalties for the contractor.

It is inevitable that certain events during construction work will impact on progress, for example, poor weather, equipment failure or accidents. However, these can be factored into a **programme of work** as and when they happen, so that adjustments can be made to reduce the impact further down the line.

▲ Figure 3.23 A site manager monitoring progress against the programme of work

> **Key term**
>
> **Programme of work:** a document used by construction managers to plan and organise resources for a building project

Construction scheduling software is often used to prepare programmes of work as part of BIM, because it produces documents that are clear, simple to amend and easy to share with all project stakeholders.

The most commonly used programmes of work are:
- Gantt charts
- critical path analysis (CPA).

Gantt charts

Gantt charts are a type of bar chart. They are used to record the project start and completion times, and the sequence in which construction activities are scheduled to take place in between. Different coloured references are used to:
- highlight planned activity durations
- plot the current status of the project
- flag any amendments that need to be made to complete the project on time and within budget.

105

Onsite Construction T Level: Core

	Task	Start date	End date	Planned duration (days)	March				April				May				June			
					1	8	15	22	29	5	12	19	26	3	10	17	24	31	7	14	21
1	Layout and preparation	02/03	06/03	5																	
2	Excavation	09/03	17/03	9																	
3	Reinforcement and formwork	16/03	27/03	12																	
4	Foundation construction	23/03	27/03	5																	
5	Structural steel	30/03	18/04	20																	
6	Masonry	14/04	01/05	18																	
7	Plumbing	21/04	25/04	5																	
8	Electrical	21/04	23/04	3																	
9	HVAC	21/04	25/04	5																	
10	Roofing	28/04	08/05	11																	
11	Plastering	08/05	12/05	5																	
12	Carpentry	11/05	22/05	12																	
13	Installation of windows and doors	18/05	22/05	5																	
14	Terrazzo	18/05	29/05	12																	
15	Glazing	01/06	12/06	12																	
16	Hardware	15/06	19/06	5																	
17	Painting	15/06	23/06	9																	
18	Exterior concrete	15/06	25/06	11																	

Key:
Red = planned activity durations
Blue = current status of the project
Green = amendments that need to be made to complete the project on time and within budget

▲ Figure 3.24 A Gantt chart

Critical path analysis (CPA)

Critical path analysis (CPA) is a decision-making tool used to plan complex building projects. The order and expected duration of activities are plotted using a networking diagram, connected by a series of **node points** containing critical information.

CPA is used by the project management team to identify when each activity can start and opportunities to relocate resources to improve efficiency; these are referred to as 'float times'.

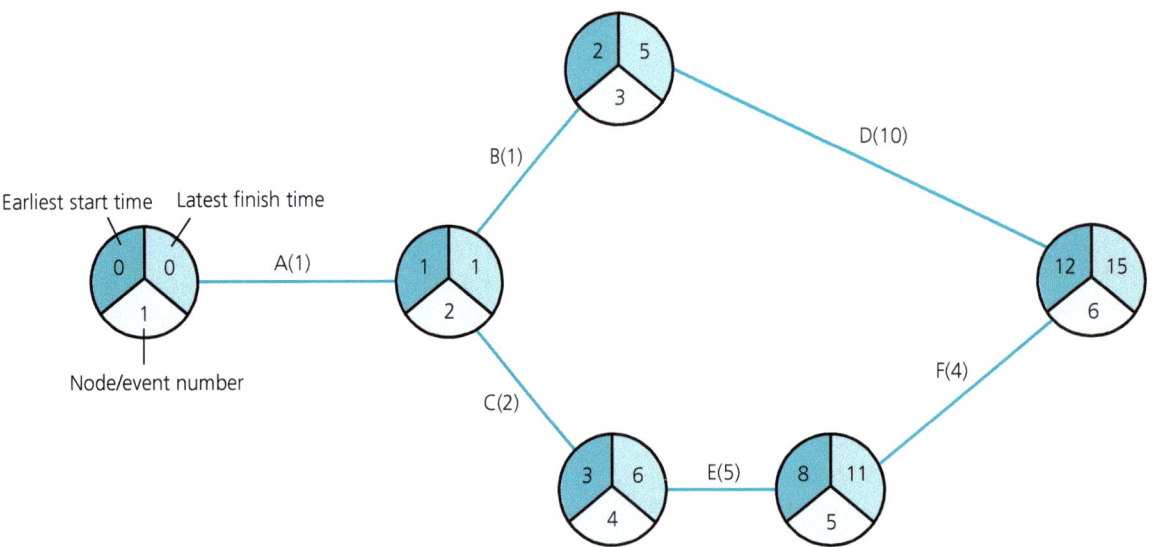

▲ Figure 3.25 Critical path analysis (CPA)

Key term
Node point: an intersection of lines or pathways in a diagram

Improve your maths
Search on YouTube for a tutorial on how to create a critical path analysis (CPA). Work with one of your peers to design a programme of work on a computer for a simple construction activity.

Chapter 3 Construction design principles

5 The concept of the 'whole building', including life cycle assessment

5.1 Life cycle assessment

Throughout this chapter, we have looked at how design and construction are influenced by different factors, such as legislation, and at the impact building works might have on the natural environment if measures are not taken to protect it. In order to design and construct '**green buildings**', we need to consider the '**whole building**' and understand how different construction systems work together and how this can determine project planning.

The term 'sustainable home' is sometimes broadly used when designing or planning; however, not all construction materials perform equally with regards to their carbon footprint. There is increasing demand among stakeholders for more sustainable and energy-efficient buildings, in order to reduce the effects on climate change. It is therefore important to understand the life cycle of construction materials, from their creation to their final use/disposal.

The impact that construction materials have on the environment can be calculated using a science-based tool known as 'life cycle assessment' (LCA). LCA accurately evaluates the effect that materials have at each stage of their life cycle, using data from Building Information Modelling (BIM) and other sources of information, and produces a report.

> **Key terms**
>
> **Green buildings:** buildings that have a low impact on the environment, during both their construction and use
>
> **Whole building:** the impact of a building in terms of resources and effect on the natural environment, from the initial sourcing of raw building materials to manufacturing and construction etc.

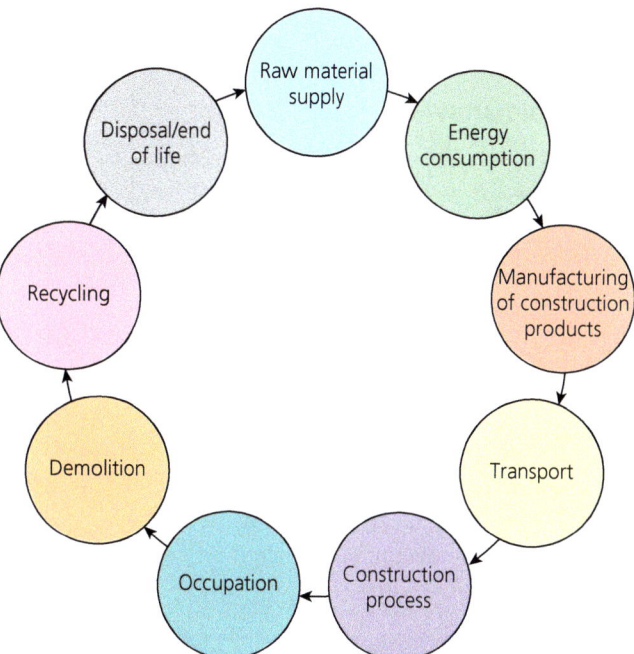

▲ Figure 3.26 Life cycle of building materials

LCA is a reliable source of data, controlled by international standards. Its findings identify areas in the building design that have the biggest impact on the environment. These areas can then be targeted by designers to alter the design or the materials used, to reduce the environmental impact.

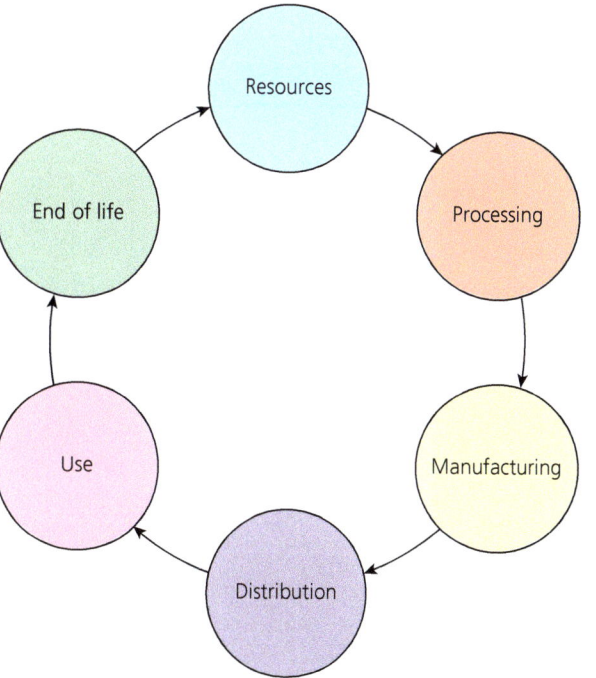

▲ Figure 3.27 Life cycle assessment (LCA)

Onsite Construction T Level: Core

Assessment practice

Short answer

1. Which Approved Document influences the design of a building's energy sources?
2. Under what rights are contractors and individuals allowed to carry out certain construction projects without planning permission?
3. Identify three stages of building work that are usually inspected by a building control officer.
4. What is the name of the design and process management tool used to bring greater clarity for the client at different stages of a building project?
5. Name one modern building method where components are constructed off site.

Long answer

6. Explain the difference between brownfield and greenfield sites.
7. Explain what is meant by vernacular construction and why it can impact on the profitability of projects.
8. Explain the role of a BIM designer.
9. What factors could influence changes to a building design?
10. List the steps for obtaining planning permission from the local planning department.

Project practice

A principal contractor has been appointed by a client for a 'design and build' project. A rural greenfield site has already been acquired by the client, but it does not have planning permission.

An application has been made to the local planning department, with plans for six new 3- and 4-bedroom low-rise dwellings. However, the plans have been initially refused by Local Authority Building Control.

Discuss in a group the potential grounds for LABC to oppose the planning application.

Bearing in mind the possible reasons you have identified for rejecting the planning application, prepare a new application to address each of the issues. To achieve this, you may need to:

- research construction materials to ascertain their properties and suitability
- consider sustainable construction solutions
- research corporate social responsibility towards the community.

Chapter 4 Construction and the built environment industry

Introduction

In this chapter, we will look at how the construction industry is structured and its contribution to the UK economy. We will identify the wide range of professionals and operatives who collaborate on various types of residential, commercial and industrial projects and examine the role of continuing professional development (CPD) in keeping their skills current.

We will then analyse the benefits of Building Information Modelling (BIM) for creating and managing information on a construction project, and look at the different types of documentation used.

We will also examine internal and external factors that may influence current and future building projects.

The chapter concludes by looking at handover procedures.

Learning outcomes

By the end of this chapter, you will understand:
1. the structure of the construction industry
2. how the construction industry serves the economy as a whole
3. integration of the supply chain through partnering and collaborative practices
4. procurement of projects within the construction sector
5. roles and responsibilities of construction professionals and operatives
6. the role of continuing professional development (CPD) in developing the knowledge and skills of those working in the sector
7. Building Information Modelling (BIM)
8. PESTLE factors

Onsite Construction T Level: Core

1 Structure of the construction industry

1.1 The construction industry

Construction is one of the biggest employment sectors in the UK, employing around 3.1 million people, with some of the largest contractors generating annual turnovers of several billion pounds. In 2018, the construction industry contributed £117 billion to the UK economy.

Construction work is broadly divided into three main categories:
- residential building (housing)
- non-residential building (for example, universities)
- infrastructure (**civil engineering**).

▲ Figure 4.1 Residential building

▲ Figure 4.2 Non-residential building

▲ Figure 4.3 Infrastructure

Work in the construction industry involves anything from building and maintaining homes, schools, offices, hospitals and factories to civil engineering projects, such as the creation of roads, bridges and airports. Demand for the construction sector is increasing, with the UK government expected to spend around £600 billion between 2021 and 2031, including £44 billion on residential housing.

> **Key term**
>
> **Civil engineering:** a profession involving the design, construction and maintenance of infrastructure that supports human activities, for example roads, bridges, airports and railways

> **Research**
>
> Define the following types of client and outline some examples of construction work they might be responsible for:
> - government
> - public limited company
> - commercial
> - private.

> **Improve your maths**
>
> Research the number of people working in each trade in the construction industry and produce a graph to illustrate this data. Analyse the information in your graph, and suggest reasons why some trade areas may be underrepresented.

110

Chapter 4 Construction and the built environment industry

> **Test yourself**
>
> What are the three main categories that construction work is divided into?

> **Improve your maths**
>
> If the VAT rate is 20 per cent, calculate the amount of VAT that a contractor would have to pay HMRC if they had a turnover of £137,000.

1.2 Business types

The size and scale of a project often determine which type of organisation is involved. For example, established and experienced contractors with proven track records of similar projects are usually on the preferred suppliers list for tenders. Private clients, who have a smaller amount of money to invest, may appoint subcontractors to complete their building work based on recommendations from previous customers/clients.

Outlined below are examples of different business types that may be involved in the construction sector. Further information on business types can be found in Chapter 11, section 1.1 (page 266).

Sole traders

A sole trader runs their own business as an individual and is **self-employed**. They are responsible for keeping financial business records of their sales and expenses. They also have to file a self-assessment tax return with HM Revenues and Customs (HMRC) each year, declare any profits and pay any tax due. Sole traders are personally **liable** for any debts or losses incurred by the business.

If annual turnover (sales) exceeds £85,000, the company must become **VAT** (value added tax) registered with HMRC. It must then charge VAT on all taxable sales it makes to its customers. As of August 2021, the VAT rate is 20 per cent, but this is subject to change.

> **Key terms**
>
> **Self-employed:** the state of working for oneself rather than an employer; a self-employed person is responsible for paying their own tax and National Insurance contributions on any earnings
>
> **Liable:** legally responsible
>
> **VAT:** value added tax – a sales duty that businesses have to charge their customers and pay to HMRC if their annual turnover exceeds £85,000

Contractors

Contractors work for a limited company registered with **Companies House**. Although they are responsible for the affairs and day-to-day running of the company, they have no personal liability for any of its financial losses if things go wrong. Limited companies are given more credibility than other types of business, and their owners also have greater control of their business with tax benefits.

There are several disadvantages of managing a limited company, including the number of **shareholders** cannot exceed 50. Setting up a limited company also involves a slightly more complicated process than setting up as a sole trader. To become a limited company, it must first have a suitable business name before paying an administration fee and registering with Companies House www.gov.uk/government/organisations/companies-house.

> **Key terms**
>
> **Companies House:** a government body that registers and stores information on all the limited companies in the UK and makes it available to the public
>
> **Shareholders:** investors in a company who have purchased at least one share of a company's stock

> **Research**
>
> Visit **www.gov.uk/topic/company-registration-filing/starting-company** and find out the process for registering a business as a limited company.

Subcontractors

Subcontractors are self-employed tradespeople who are hired by principal contractors to undertake specific work on building projects.

Self-employed workers are responsible for every part of running a business, for example estimating, invoicing, ordering and accounts, although they may appoint an accountant to manage some of their financial affairs. They do not receive some of the benefits that employed people enjoy, for example:
- holiday pay (including bank holidays)
- sick pay
- maternity or paternity leave or pay.

In order to cover the potential losses listed above, subcontractors have to charge their clients more. They also have to consider the days or even weeks that they may be absent through sickness. During these unforeseen periods, the business will cease trading and earning money while the self-employed person recovers.

Small, medium-sized and large organisations

An organisation can be privately owned, a partnership or a **corporation**. Its size can be measured in terms of the number of employees, annual **turnover**, profits or assets.

According to Companies House guidance:
- a small organisation employs on average no more than 50 people and has an annual turnover of £10.2 million or less
- a medium-sized organisation employs on average no more than 250 people and has an annual turnover of £36 million or less
- a large organisation employs at least 250 people and has an annual turnover in excess of £50 million.

Small construction businesses tend to have fewer resources and **overheads** than larger ones; therefore they often recruit local labour to work within a particular area or region of the country. The bigger an organisation becomes, the more resources it has at its disposal and the further it branches out nationally or internationally to compete for lucrative government or commercial construction contracts.

The **hierarchical structure** of personnel within medium-sized and large organisations is much more complex than in smaller businesses. This often causes delays, with information being processed at each level before being filtered down through the organisation. Smaller organisations have a simpler structure, resulting in decisions being made more easily. It is important to establish a hierarchy in an organisation, so that everyone understands lines of communication and areas of responsibility.

> **Key terms**
>
> **Corporation:** a business owned by its shareholders, who often appoint a board of directors to manage the day-to-day running of its activities; the business is a legal entity and the shareholders have no personal liability for its actions and finances
>
> **Turnover:** the amount of money that a business or individual has been paid by their clients for services they have provided
>
> **Overheads:** regular repeated costs associated with the day-to-day running of a business, for example rent and insurances
>
> **Hierarchical structure:** the arrangement of people and the positions they hold in an organisation according to their level of authority; there is a chain of command from those in the most senior positions to those in subordinate positions

> **Test yourself**
>
> Explain the term 'turnover' in regard to business finance.

1.3 The range of work undertaken in the construction industry

The impact of the construction industry is everywhere you look – from homes to public buildings to infrastructure. While structures may be diverse in their design, method of construction and intended purpose, they all illustrate just how much we depend on the construction industry to live, work and travel.

Construction work usually falls into one of a number of different categories, depending on the client and purpose. Table 4.1 outlines the wide range of work undertaken in the construction industry.

Chapter 4 Construction and the built environment industry

▼ Table 4.1 Range of work undertaken in the construction industry

Category of construction work	Examples of buildings/structures
Commercial	Offices, private hospitals, hotels and restaurants
Residential	Private and affordable housing, flats, apartment blocks, retirement villages and student accommodation
Industrial	Large buildings designed to manufacture goods, for example factories, agricultural buildings, warehouses and power plants
Health	Health centres, hospitals, clinics, treatment centres and GP surgeries
Retail	High-street shops, supermarkets, superstores, shopping centres, retail parks and distribution centres
Recreational	Stadiums, swimming pools, gymnasiums, golf and tennis facilities
Leisure	Public libraries, amusement parks, fitness centres, spas, cinemas and community centres
Utilities	Building services systems, for example water, electricity, gas, telecommunications and sewerage
Transport	Roads, railways, train stations, airports, bridges, footpaths and tunnels
New build	New housing developments
Retrofit	Modernisation or adaption of existing buildings to make them more energy efficient and sustainable

▲ Figure 4.4 Industrial construction

2 How the construction industry serves the economy as a whole

Earlier in this chapter, we highlighted the importance of the construction industry and how the wealth generated from construction developments contributes to the UK economy.

The industry is showing no signs of slowing down in terms of growth, which creates well-paid job opportunities for skilled workers. Where there are skills gaps or shortages of labour, local and national government often invests heavily in further and higher education, training and skills to meet the demand; however, this can be influenced by the government in power at the time and its priorities.

When people are employed and have realistic opportunities to progress their careers, they become financially stable and often spend or borrow money to invest in assets such as housing and cars. These investments create more demand for different types of buildings such as housing, transport and better infrastructure, leisure facilities, education establishments and hospitals, which in turn leads to greater employment and continues the cycle. Improvements of this nature in neglected urban areas or areas previously used for industry often lead to the regeneration of whole areas, which increases the value of property, creates employment and reduces crime in the local community.

The UK government is determined to continue investment in new technologies in the construction sector, to meet the national housing shortage and the need for more affordable homes. When the government and other large organisations are planning investments in housing, hospitals and infrastructure in the UK, market intelligence can forecast the labour needs in these particular areas. Forecasting may identify shortages of skilled workers in a specific region or at the planned time for an investment, for example if the number of people retiring from the industries is predicted to exceed the number of new workers. Market intelligence can also identify skills shortages, especially with the development of new technologies. If forecasting identifies future labour shortages, investments can be made in training and the recruitment of new skilled workers.

The UK government has invested in the construction industry by making a commitment to tackle climate change and the negative impact that construction has on the environment.

> **Key terms**
>
> **Carbon:** a chemical element that can be released into the atmosphere when fossil fuels are burned
>
> **Retrofitting:** the process of adding new features and technologies to existing buildings

▲ Figure 4.5 Gas hob using a finite fossil fuel

The UK government passed the Climate Change Act in November 2008, setting a target to reduce **carbon** emissions to net zero (carbon neutral) by 2050. To meet this target, the government proposed spending caps and total greenhouse gas emission targets for five-year periods, which should not be exceeded. So far, the country has met each target; however, if it were to miss the carbon budget for an agreed period, it could be challenged in a court of law.

Changes are being introduced in the construction industry to improve efficiencies and address the need for a cleaner environment, including:
- increased off-site manufacturing to reduce waste and speed up production
- installing smart technologies in homes
- using sustainable building materials
- creating energy-efficient buildings
- installing electric vehicle charging points
- installing alternative heating systems, for example ground and air source heat pumps
- **retrofitting** existing buildings to improve energy efficiency
- preparing infrastructure for the impact of climate change, for example flood defences
- adapting existing infrastructure to changes in technologies to protect the environment.

▲ Figure 4.6 Energy-efficient building

▲ Figure 4.7 Electric vehicle charging point

> **Test yourself**
>
> What are the objectives of the Climate Change Act 2008?

3 Integration of the supply chain through partnering and collaborative practices

Table 4.2 shows the integration and collaboration of partners in the supply chain for a building project.

▼ Table 4.2 Integration and collaboration of partners in the supply chain

Position	Role		Responsibilities
Customer	Client (individual or organisation)		• Develops the initial concept for a project • Appoints a design, management and building team • Finances the project • Complies with legal duties, for example the Construction (Design and Management) (CDM) Regulations 2015
Client's representative	Clerk of works (CoW)		Checks drawings and specifications, to ensure work is completed to the standards agreed in the building contract
Designers – planning team	Architect		• Interprets the client's design brief to produce detailed drawings and specifications • Complies with CDM duties • Leads the project team at the design stage • Oversees the construction phase • Makes the client aware of their legal responsibilities
	Surveyor	Quantity	• Studies building information and drawings/BIM to prepare tender packages • Controls budgets and costs • Provides professional advice to the client
		Land	• Plots, measures and gathers data on land for construction and civil-engineering projects • Advises the client on planning and construction
		Building	• Undertakes onsite property surveys • Identifies defects and makes recommendations for remedial work • Completes reports and advises the client on legal, planning or environmental issues
	Engineer	Structural	Collaborates with the rest of the design team to prepare structural calculations and designs for construction and civil-engineering projects
		Building services	Designs, plans and supervises the installation of building services, such as heating, water and electricity
		Civil	Designs, plans and supervises the construction and maintenance of public infrastructure, such as roads, railways, bridges and tunnels
	CAD operative/draughtsperson		• Creates and modifies technical drawings using computer-assisted design (CAD) • Visits sites to co-ordinate with architects, engineers and building services teams
	BIM manager/technician		• Creates management and planning documentation • Creates 2D and 3D BIM models
Building regulators	Local Authority Building Control (LABC)		• Consults and offers advice on planning applications • Grants or refuses planning applications
	Building control officer (BCO)		• Inspects building work against planning permission granted and building regulations • Monitors health and safety on construction sites • Signs off building work against building regulations
	Health and Safety Executive (HSE)		• Enforces the Health and Safety at Work etc. Act 1974 (HASAWA) in high-risk work environments • Carries out inspections, investigations and enforcement action where necessary
	Fire and Rescue Authority		• Has the same powers as the HSE to enforce HASAWA • Responsible for fire-safety measures in work environments
	Local Authority Environmental Health Department		Enforces HASAWA in low-risk work environments

▼ Table 4.2 Integration and collaboration of partners in the supply chain (continued)

Position	Role	Responsibilities
Contractor – construction management team	Managing director	• As the most senior person in the company, oversees the direction of the company • Meets with clients • Manages budgets, targets, business growth, personnel and performance • Makes sure the organisation's legal and contractual obligations are met
	Contracts/project manager	• Negotiates/completes tenders • Prepares contracts • Plans and co-ordinates projects • Prepares reports • Liaises with the client • Manages costs and timescales
	Site manager	• Manages progress and budgets on construction projects • Supervises workers and subcontractors on site • Manages health, safety and welfare • Collaborates with the project team and client to ensure work is completed to standard and on schedule • Conducts site inductions
	Assistant site manager	• Helps the site manager to fulfil their responsibilities on site • Monitors progress and supervises workers • Co-ordinates operations and the procurement of materials and other resources • Enforces the contractor's health and safety site rules
	Health and safety advisor	• Advises the contractor on health and safety matters • Prepares risk assessments and method statements • Monitors health and safety performance • Audits and investigates accidents
	Estimator	• Researches costs and sources materials, labour and equipment for contracts • Assists the contracts manager with tender bids • Supports buying activities
	Buyer	• Works closely with the company estimator • Procures resources for construction projects • Maintains accurate records of spending to maximise efficiencies
	Trade foreperson	• Supervises a small team of tradespeople • Liaises with the site manager on topics such as progress, materials and safety
	Fire marshal	• Carries out fire risk assessments • Makes sure everyone leaves the building safely in an emergency via the escape routes • Checks alarms, fire-fighting equipment and safety signage • Performs safety checks • Makes sure workers and visitors are aware of what to do in an emergency on site
	First aider	• Administers first-aid treatment • Looks after first-aid provisions and equipment, for example the first-aid box and eye-wash stations • Reports accidents
Operatives/ subcontractors	Craft operative (for example carpenter, bricklayer, gas fitter)	• Carries out their specific trade role on site • Produces risk assessments and method statements • Follows health and safety legislation and site rules • Attends toolbox talks • Reports unsafe conditions to the site manager
	General operative	• Assists tradespeople with moving and handling building materials • Carries out general housekeeping on site

Chapter 4 Construction and the built environment industry

Position	Role	Responsibilities
	Scaffolder	• Installs, maintains and inspects scaffolding systems to industry standards • Makes sure they do not put themselves and others at risk of harm due to their work activities
	Plant operator	• Inspects and operates heavy plant/equipment on site, for example telehandlers and dumpers • Follows safe working practices
Manufacturers	Manufacturer of building materials, resources and products for suppliers	Carries out manufacturing routinely through the supply chain or makes items to order, for example windows and doors
Suppliers	Supplier of standard building materials, equipment, machinery and tools	• Provides goods to contractors, subcontractors and the general public from a shop, warehouse or yard • Enters into credit agreements with contractors, with special rates and payment terms

▲ Figure 4.8 Land surveyors

▲ Figure 4.9 Building surveyor

All partners in the supply chain must have agreed roles and responsibilities at various stages of a building project, throughout the stages of design, planning and scheduling (**inventory management**). Integrated systems within an organisation should establish lines of communication for collaborative working between partners in the supply chain. Some of the benefits and consequences of the integration of the supply chain through partnering and collaborative practices are listed in Table 4.3.

▼ Table 4.3 Benefits and consequences of the integration of the supply chain

Benefits	Consequences
Project cost savings	Disruption/delays
Increased resources ensuing that the project is completed to standards	Increased cost, for example resource/over budget/penalty clauses
Project completed on time	Negative reputation
Project completed within the client's budget	

Key term

Inventory management: a process for recording and tracking stock levels, orders and sales

The management of an organisation may decide to change its management approaches with new goals or core values. This could involve new strategies for effecting change with the implementation of new technologies or processes that shareholders need to be made aware of. This means that effective communication and collaborative working within the supply chain are crucial in dealing with the changes throughout its transition period.

Onsite Construction T Level: Core

> **Test yourself**
>
> Explain the role of a clerk of works.

▲ Figure 4.10 Clerk of works

4 Procurement of projects within the construction sector

4.1 Need/demand

Construction work often requires huge financial investment by clients at each stage of a project, with an element of risk involved depending on the type of work undertaken. A client that has obtained a building plot without planning permission on it and drawn up proposals with an architect may risk the concept being refused by Local Authority Building Control. However, if planning permission has been granted and there is a demand for the project, it has every chance of being successful.

Before investing in construction work, it is important to establish the need for the project, in order to mitigate risk and avoid potential financial loss.

If the proposed project is speculative, with a view to selling or renting the building/s, the client must try to forecast these sales and potential income. It is not uncommon for building work on housing developments to stop if homes are not selling as well as expected, until sales pick up.

If the work is a government contract for a new school or hospital, then it already has a purpose and does not have to be sold upon completion of the building work.

4.2 Tendering and bidding processes

Once the design for a project has been approved by Local Authority Building Control, the process of finding a suitable contractor begins. There are several methods the client can use to establish the right contractor for a project; however the following factors need to be considered when deciding on the most suitable procurement method:
- financing and budgets
- time
- quality
- risk
- project limitations.

The client may already know which contractor they would like to use for their project, based on recommendations from others or having previously appointed them on another project. However, while the client may have total confidence in the contractor's ability to complete the job to the standards expected, they will not have any comparable quotes from other contractors, and therefore may not be getting the best value for money.

Open tendering

The client may invite offers for the contract from anyone, based on the information provided. This is referred to as open **tendering**.

The advantage of open tendering is that the process is fair and without bias. However, the client may have to choose a contractor they are not familiar with or one that may not have experience of completing similar projects. The client may have no alternative but to appoint a contractor based on their quotation and references from previous clients.

> **Key term**
>
> **Tendering:** the process of inviting bids from contractors to carry out specific projects

Two-stage tender

If the client would like to secure the services of a contractor early, without having all the relevant design information in place, they can appoint them using a two-stage tender. This involves an agreement with the contractor for work that they are able to price for based on the information provided, so that they are able to start work. At the second stage, when all the design information has been supplied, the contractor is able to negotiate a fixed price for completion of the contract.

Preferred supplier (selective)

Clients with previous experience working on similar projects may have developed successful relationships with a range of contractors. Some of these contractors may be included on the client's list of preferred suppliers. From this list, the client may select a number of contractors and invite them to tender for a contract, although the contractor is not obliged to submit a bid.

Negotiated tendering

Where a client has invited just one contractor to price for a project, this is referred to as negotiated tendering. Before building contracts are signed, the client usually negotiates the best price for the work.

> **Test yourself**
>
> List the different methods of tendering, and explain why they may be used by a client.

> **Industry tip**
>
> Selecting the contractor that has provided the cheapest tender is not always the best option. Clients should also consider:
> - the size of the contractor's company
> - the contractor's experience of completing similar projects
> - references from the contractor's previous clients
> - the contractor's health and safety records, including possible HSE enforcement actions.

4.3 Tender documentation

The client or client's representative is responsible for gathering all the documentation needed for the tender package. The same tender package is usually sent to three or four contractors, depending on the tendering process used, so that they can calculate their costs fairly based on exactly the same information, before returning it to the client with their final estimate.

The information contained in a tender package will be specific to each project. However, it may contain some of the following documents:
- Letter inviting the contractor to submit a bid for the work
- Outline of the proposal
- Form of tender and timeline to return the completed bid
- Form of contract and conditions (including the process for payments and interim valuations)
- Programme of work
- Design drawings
- Specifications
- Site-specific information or issues
- **Preliminaries**
- Special planning-permission requirements
- **Bill of quantities** (cost framework)
- Tender return document.

> **Key terms**
>
> **Preliminaries:** pre-construction information that outlines items that are necessary for a contractor to complete the works but are not actually part of the works, for example general plant, welfare facilities and site security
>
> **Bill of quantities:** a document that breaks down a construction project into an itemised list of work to be carried out, including quality, quantities and costs

Specification	
102 External cavity walling	
• Walling below ground:	
- Type:	Cavity wall, concrete filled.
- Masonry units:	Common bricks.
- Mortar:	Class M6 mortar.
• DPC at ground floor:	Flexible cavity trays.
• Walling above ground:	
- External leaf above ground:	
Masonry units:	Facing bricks.
Bond or coursing:	Flemish bond.
- Internal leaf above ground:	
Masonry units:	Aerated concrete blocks.
- Mortar:	
Type:	Class M4 mortar.
Joint profile to external faces:	Bucket handle.
- Wall ties:	Insulation retaining wall ties.
- Cavity insulation:	Full fill cavity insulation.
- Ventilation components:	Air bricks and sub-floor ventilation ducts.
• Openings:	
- Lintels:	
Type:	Manufactured stone lintels.
Cavity tray cover:	Flexible cavity trays.
- Cavity closers:	Flexible insulated DPCs.
- Sills:	
Type:	
DPC below:	Manufactured stone sills.
• Abutments:	Natural stone sills.
Cavity trays and DPCs:	Precast concrete sills.
Flashings built into masonry:	As drawings.

▲ Figure 4.11 Specification

Item No.	Description	Unit	Rate	Amount
1	Setting out works	Lump sum		
2	Temporary works, e.g. welfare facilities, site office and canteen	Lump sum		
3	Scaffolding	Lump sum		
4	Power, heating and lighting	Lump sum		
5	Clear construction site on completion	Lump sum		

▲ Figure 4.12 Preliminaries

Improve your English

Draft a formal letter inviting a contractor to submit a tender bid for a construction project.

The contractor is often asked to provide information about their business and previous work they have completed. This will include proof of employer's insurance and public liability insurance, information about the key personnel that will represent their business on site and references from previous clients.

Chapter 4 Construction and the built environment industry

> ### Research
>
> The most common procurement routes for construction projects are listed below. Research each type and explain why it may be selected by the client:
> - Contractor led
> - Design and build
> - Fast track
> - Lump sum
> - Single stage
> - Two stage.

5 Roles and responsibilities of construction professionals and operatives

5.1 Construction professionals

A construction professional is someone who is trained and qualified in their chosen field. Training may involve many years of study in order to gain a recognised qualification. They may also be a member of a professional organisation or body.

For example, an architect will have completed a degree in architecture and be a member of the Architects Registration Board (ARB) or Royal Institute of British Architects (RIBA). The importance of professional membership is explained later in this chapter.

Some of the key professional job roles in the construction industry are outlined below.

Architects

When a client has a concept for a construction project, they usually appoint a professional architect, also known as a designer, to discuss their ideas, needs and budget. A client's project could be a new housing development, an extension or alterations to an existing building. It could also involve the **restoration** or conservation of older buildings of historical interest.

Architects initially translate information provided by clients and surveys into 2D drawings and 3D digital models, using **computer-aided design (CAD)**. While planning a project, they have a responsibility to protect the environment through their designs, choice of materials and construction methods.

Initial designs often evolve and include further building documentation, until the client is satisfied that the proposal is ready to be submitted for planning approval if necessary. Design information provided by an architect also informs the BIM process shared with the project team. For example, information from the architect might allow structural engineers to calculate loads and stresses on different parts of a building to determine the size of structural components such as steel beams and foundations.

An architect could be an individual or organisation; however, where there is more than one architect working on a single project, the client must appoint a principal designer under the Construction (Design and Management) Regulations 2015. The role of the principal designer is to lead the project by planning, managing, monitoring and co-ordinating health and safety during the pre-construction phase.

While there is no requirement to use an architect, their expert knowledge of the construction industry, building regulations and materials makes the design and build process much easier for the client. The client could even extend the services of an architect beyond the design stages to advise them during the building phase and oversee the whole construction project.

> ### Key terms
>
> **Restoration:** the process of returning a building to its original condition
>
> **Computer-aided design (CAD):** using computer software to develop designs for buildings and structures

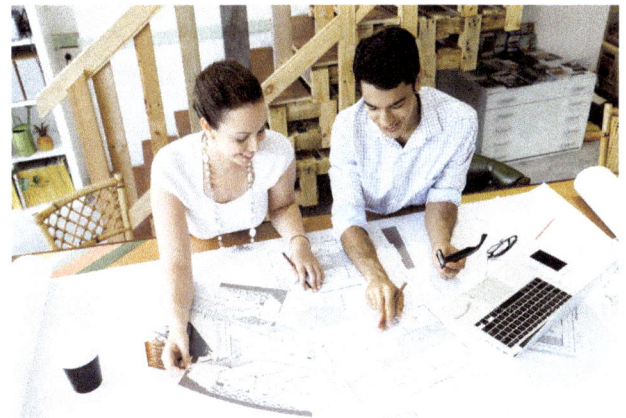

▲ Figure 4.13 Architect

Onsite Construction T Level: Core

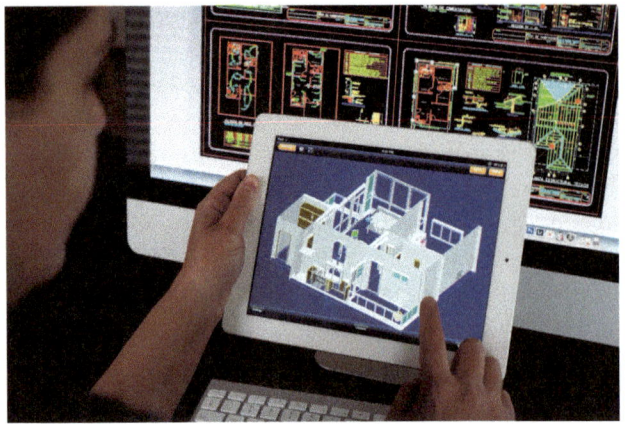

▲ Figure 4.14 Computer-aided design (CAD)

Civil engineers

Civil engineers are also classified as designers, although they manage the design, construction and maintenance of infrastructure rather than buildings. Typically, they are involved in projects such as:
- roads
- bridges
- tunnels
- airports
- harbours
- railways
- water and sewage systems
- power plants.

▲ Figure 4.15 Civil engineering

There are many different subdisciplines in civil engineering. People usually specialise in one of the following fields:
- transportation engineering
- structural engineering
- environmental engineering
- **geotechnical engineering**
- hydraulic engineering
- construction engineering.

As civil engineers are responsible for the design and implementation of infrastructure in construction projects, their collaboration with other professionals in the design and construction team is vital to the success of any project.

> **Key term**
>
> **Geotechnical engineering:** a field of civil engineering that deals with the behaviour of earth materials such as soil and rock

▲ Figure 4.16 Geotechnical civil engineering

Building services design engineers

Building services design engineers consult and advise clients on the concepts and possible approaches that could be integrated into their buildings to supply the following services:
- heating
- ventilation
- air conditioning
- **renewable energy**, for example **solar photovoltaic (PV) panels** and electric car charging points
- sustainable technologies
- lighting
- fire and security systems.

> **Key terms**
>
> **Renewable energy:** energy that comes from natural sources or processes that are replenished or replaced, such as water, sun and wind
>
> **Solar photovoltaic (PV) panels:** panels that use solar cells to make clean, renewable energy by converting sunlight directly into electricity

Chapter 4 Construction and the built environment industry

▲ Figure 4.17 Renewable energy

▲ Figure 4.18 Building services engineer

Their technical knowledge of building services may also be utilised by the client for the production of detailed drawings, specifications and calculations to meet with design standards and regulatory requirements. They may also be involved in bids and tenders for work and advise the client from start to completion of a project.

Building services engineer technicians

Building services engineer technicians are responsible for:
- assisting the project team with design solutions, specifications and planning for building services engineering systems (for example, water and drainage, lighting and power)
- supervising specialist contractors during installations
- taking the lead on health and safety
- **commissioning** systems while minimising the impact construction work has on the environment
- monitoring quality control, recording progress and reporting to the project team during installations.

> **Key term**
>
> **Commissioning:** the process of ensuring that a building system is performing or working as it has been designed

Building services engineering site managers

Building services engineering site managers oversee the installation of complex environmental systems in construction projects, such as heating, lighting and electrical power. During the construction phase, they work closely with the project team to make sure the design information is understood, in order to maintain quality and productivity during installation on site.

Facilities managers

Facilities managers take responsibility for the operation, servicing and maintenance of building services once the building work has been completed, signed-off and handed over to the client. When contractors are needed for work on the systems, the facilities manager will prepare tender documents, schedule repairs and monitor work under their control to minimise disruption and protect people from work activities.

Client representatives

It is not always possible or practical for the client to be on site all of the time during the construction phase of a project, and their technical knowledge about the building industry may be limited. Consequently, either the principal designer or the client will appoint a clerk of works (CoW).

The CoW acts as the client's representative on site to oversee the quality and safety of work on the project. In the interests of the client, they work closely with the construction staff, surveyors and engineers to make sure plans and specifications are followed properly. In doing so, the CoW will refer to working drawings, building regulations and health and safety legislation. Where standards have not been achieved in terms of the quality of work, materials or safety, the CoW will report to the site manager and make suggestions for improvement.

As progress is made throughout the construction phase, the CoW will keep the client informed with accurate reports/records from the site, at intervals agreed at the start of the project.

Contract managers

Contract managers are employed by contractors to assist in preparing tenders for clients and securing future business. When tender bids have been accepted by a client, a legal contract is drawn up by the contract manager. The terms and conditions of the contract are negotiated with the client and other stakeholders with regards to budget, the service that will be provided and project timescales. The contract manager is usually the main point of contact for the client and the site/project managers for the duration of the project.

Once construction work has started on site, the contract manager will monitor progress against the agreed schedule and technical standards by attending regular site meetings with the management team. Where unexpected costs arise or the terms of the contract change during the project, the contract manager will have to resolve these issues as quickly and effectively as possible.

Site managers

Site managers, also known as site supervisors, are responsible for organising work on construction sites to ensure it is completed safely, on time and within the client's budget. They are usually based in a temporary site office and will remain on site throughout the construction phase of the project to organise labour, equipment and materials for each stage of the build.

Site managers:
- supervise workers, subcontractors and visitors, for example delivery drivers
- monitor the quality of work and the progress made against the programme of work
- take responsibility for the health, safety and welfare of workers and others that may be affected by work activities; this involves making arrangements to protect people from harm and ensuring workers are suitably trained and informed about risks and control measures, usually through a site induction.

Health and safety standards are closely monitored by the site manager against risk assessments, method statements and health and safety legislation, and where necessary they will take action to maintain or improve safety on site.

Throughout the construction phase, the site manager will liaise with other professionals, such as the quantity surveyor to discuss budgets and the architect or surveyors regarding design issues or amendments. They may also have to deal with accidents, near misses or other emergency situations.

While key information on progress and budgets can quickly be communicated through the project team using collaborative methods, regular site/progress meetings are often chaired by the site manager.

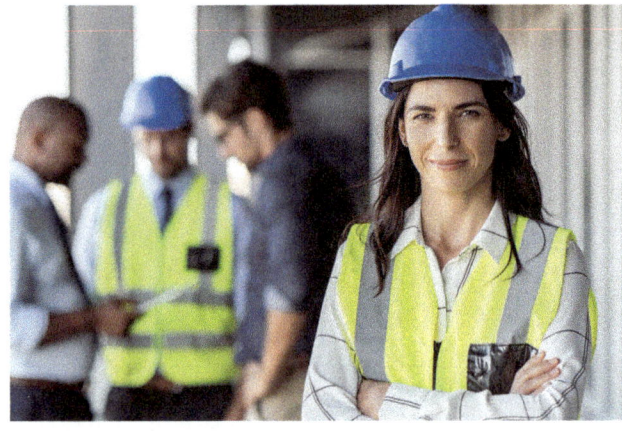

▲ Figure 4.19 Site manager

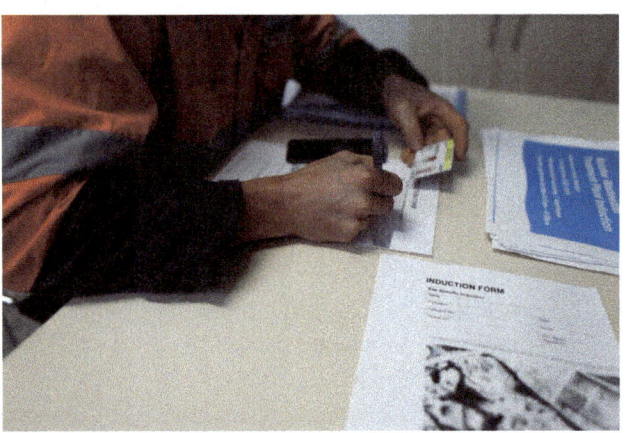

▲ Figure 4.20 Site induction

5.2 Construction operatives

Most construction operatives are experienced and skilled workers that specialise in a particular area of the building industry. The level of qualifications and registrations needed to become a tradesperson are often determined by the occupation; however the most widely recognised are National Vocational Qualifications (NVQs, SNVQs in Scotland) and Apprenticeship Standards.

In addition to these qualifications, construction operatives may also need a current CSCS card to demonstrate to their employer that they have the minimum level of health and safety knowledge to work safely on site; the CSCS card scheme is covered in more detail in Chapter 1.

Chapter 4 Construction and the built environment industry

Carpenters and joiners

Carpenters and joiners both work with wood:
- Joiners are usually based in a workshop, where they manufacture purpose-made items to order, for example doors, windows, frames and staircases, using a range of machinery, power tools and hand tools.
- Carpenters usually work on construction sites. **First-fix** carpentry includes the installation of floors, roofs, walls and stairs, and **second-fix** carpentry involves the fitting of skirting, architraves, doors and kitchens.

Carpenters and joiners both work from drawings and specifications provided by the design team; however, they may need to gather further information from the management team during a site visit before they begin any manufacturing off site.

> **Key terms**
>
> **First fix:** a phase of construction work completed before plastering
>
> **Second fix:** a phase of construction work completed after plastering, for example installing kitchens, doors and tiling

Plasterers

Plasterers are tradespeople used by contractors to apply smooth and textured finishes to internal walls and ceilings, using a range of materials such as **gypsum**, cement or lime. Highly skilled plasterers also have the ability to carry out repairs or cast ornamental mouldings in plaster to match existing period features in a property. Plasterers can also apply smooth or textured finishes to external walls, using lime, cement or flexible modern materials such as polymer or acrylic renders.

Plasterers can usually start to apply finishes to the internal walls and ceilings as soon as tradespeople have completed the installation of their respective first-fix items. Where work is needed externally, this is often weather dependent, unless the work is under cover or protected; therefore plasterers will often work closely with the site manager to schedule this type of work when the weather permits.

> **Key term**
>
> **Gypsum:** a natural mineral, often used in building products such as plaster and plasterboard

▲ Figure 4.21 Carpenter

▲ Figure 4.22 Joiner

▲ Figure 4.23 Gypsum plaster

Tilers

Tilers may be subcontracted by the principal contractor to lay decorative and protective wall and floor tiles. Tiles can be ceramic, clay, marble, slate or glass and are available in a range of different sizes, shapes, designs and textures.

A tiler will often calculate, cut and lay tiles in kitchens and bathrooms, as areas most exposed to high levels of moisture. A tiler could also be appointed to work outdoors to lay tiles on patios, terraces or swimming pools. These tasks will be completed in the final stages of a project, to protect the areas from damage from other building work.

Bricklayers

Bricklayers set out, build and repair walls, piers and archways for domestic and commercial projects. They work from plans and design specifications to calculate costs for clients and quantities of materials needed.

Bricklayers can build single as well as double **cavity walls** using a range of different bricks, blocks or stone, for internal or external projects. Their job is very physical and often involves working at height using a range of access equipment; therefore, it is essential that they have a strong understanding of health and safety practices.

The stages at which bricklayers are scheduled into a project are determined by the build method chosen. However, the majority of their work is often structural, which would result in them being on site once the foundations have been poured.

> **Key term**
>
> **Cavity walls:** external, load-bearing, structural walls consisting of two individual leaves (skins) of masonry with a gap (cavity) between them

▲ Figure 4.24 Bricklayer

Plumbers

Plumbers install, maintain and repair water, heating and drainage systems in new and existing buildings. This includes cutting, shaping and fixing pipes and fixtures for the installation, testing and commissioning of the following:
- hot and cold running water
- heating systems
- baths
- sinks
- toilets
- showers
- dishwashers and washing machines.

During the first-fix stage of work, plumbers prepare for the installation of services on site by fitting pipework. In domestic projects, pipework is run through floors, walls and ceiling voids where possible, so that it can be hidden from view. On other projects, such as industrial or commercial, the specification may state that pipework has to be located in surface-mounted ducting, so that it can easily be maintained or adapted without causing damage.

At the second-fix stage, plumbers must work closely with other trades to co-ordinate the installation of fixtures and fittings, to protect them from damage and to avoid complications for work that follows.

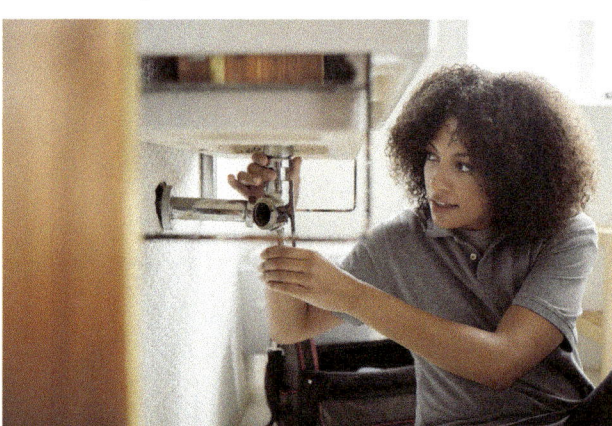

▲ Figure 4.25 Plumber

Electricians

Electricians are responsible for designing, installing, servicing and repairing electrical systems in new and existing buildings. At the first-fix stage, electricians interpret wiring diagrams and the client's specification to run the wiring system throughout the building to power lighting, sockets and alarm systems. Later in the project they will be scheduled by the project manager to return to site to install/connect the second-fix electrical fixtures and fittings, for example the fuse board, light fittings and switches.

Once the system has been installed, the electrician will test and commission it to make sure it is safe to put into service. If a new electrical system has been fitted, the electrician will issue the client with an Electrical Installation Certificate (EIC) to show that it has been installed to a satisfactory standard. If the electrician has carried out work on an existing system, they will issue the client with a Minor Electrical Installation Works Certificate (MEIWC) to demonstrate that the work they have completed is safe.

Painters and decorators

Painters and decorators are usually one of the last trades to be scheduled during a construction project. Part of their role is to prepare surfaces such as walls, ceilings, metal and woodwork for decorative and protective finishes. Some of these finishes can be applied internally or externally with brushes, rags, sponges, rollers or spray systems. Skilled decorators can also apply other finishes, such as hanging wallpaper, stencilling, **graining** and creating marbling effects.

> **Key term**
>
> *Graining:* a method used by decorators to create woodgrain effects on different surfaces

Trade supervisors

Trade supervisors are often experienced construction workers who have progressed from their chosen profession to a supervisory role, where they are responsible for a small team of tradespeople working on a particular project. In most situations, they remain 'on the tools' throughout the building phase, while managing the standard of work, resources and health and safety of operatives under their control.

Trade supervisors may be required to attend scheduled site meetings with the construction management team to represent their trade area, report their progress and raise any concerns that they are unable to resolve themselves.

Non-skilled operatives

Non-skilled operatives usually undertake manual-labouring tasks to support tradespeople and other skilled workers on site, such as preparing building materials, moving equipment and general housekeeping to keep people safe. They play an important role in the success of any construction project.

▲ Figure 4.26 Non-skilled operative

> **Research**
>
> There are other roles that are necessary on a construction project. Research the roles and responsibilities of a plant operator and ground worker, and explain during which stages they would be involved.

▲ Figure 4.27 Plant operator

> **Research**
>
> Find out the qualifications needed to become a tradesperson for each of the occupations discussed in section 5.2 and explain why they are so important.

6 The role of continuing professional development (CPD) in developing the knowledge and skills of those working in the sector

- health, safety and welfare
- digital technology
- conservation and refurbishment
- sustainability
- maintenance
- tools and equipment
- industry standards and best practice.

> **Key term**
>
> *Continuing professional development (CPD):* the process of maintaining, improving and developing knowledge and skills related to one's profession in order to demonstrate competence

6.1 Role of CPD

Construction is a fast-moving industry, and the people working within it must keep pace with any new developments.

Professionals and skilled operatives beginning a career in the sector must undertake training and demonstrate competence in order to achieve formal qualifications. These achievements often create employment opportunities, allowing those new to the industry to gain valuable work experience and develop a greater understanding of their job role.

Some professionals and operatives have legal responsibilities to belong to professional, accredited or certified organisations in order to actively continue with their job roles; for example gas fitters have to keep a yearly registration with Gas Safe. This often involves having to maintain professional standards and keep records of **continuing professional development (CPD)**.

Many businesses insist on employees maintaining professional CPD and outline workforce planning in contracts of employment or codes of practice. There is further information on chartered, professional, accreditation and certification bodies in Chapter 7.

Often people working in the construction industry have to undertake CPD to demonstrate the currency of their professional competence. For this reason, they may focus on a particular area of weakness that they have identified or that has been brought to their attention, such as:
- legislation and regulations
- management and supervision

6.2 Types of CPD

There are many different ways to undertake continuing professional development (CPD), from formal qualifications to less informal training. The list below provides some examples of how CPD can be maintained while working in the construction industry:
- work experience
- qualifications
- informal training/updating
- formal training (structured and systematic)
- self-learning or self-directed
- chartered (mandatory CPD as part of a membership or affiliation to an organisation).

6.3 Benefits of CPD

CPD is a legal requirement for some professionals. However, there are also many benefits for employers and individuals in upskilling, including:
- protecting clients, customers and the public
- keeping up to date with industry best practice, for example changes in regulations, product developments and technological advancements
- developing product knowledge
- working more efficiently
- improving knowledge and skills
- enhancing the company image
- career progression.

Chapter 4 Construction and the built environment industry

▲ Figure 4.28 Construction workers updating their CPD

Businesses that actively encourage employee CPD will not only have a highly skilled and professional team, but also members of staff that are committed and remain loyal. The more businesses invest time, money and effort in the development of their staff, the safer and more skilled the building industry will become as a whole.

> **Research**
>
> Research the following organisations and suggest which professionals may belong to them:
> ▶ Chartered Institute of Building (CIOB)
> ▶ Royal Institution of Chartered Surveyors (RICS)
> ▶ Royal Institute of British Architects (RIBA).

> **Improve your English**
>
> Create a template on your computer and input the CPD you have completed over the past year. Reflect on each area and how it may have benefited you in your job role. Ask your tutor to check your spelling, grammar and punctuation.

7 Building Information Modelling

Building Information Modelling (BIM) uses **smart technology** to allow effective and efficient collaboration between designers and the construction team at every stage of a building project. It is adaptable to suit the size and complexity of each project and allows technical information to be shared throughout the management and construction teams.

> **Key term**
>
> **Smart technology:** computers, smartphones and tablets with software that connects to the internet in order to control, report, monitor and analyse devices and appliances remotely

▲ Figure 4.29 Building Information Modelling (BIM)

7.1 BIM process

At the start of a project, the client usually meets with the designer and contractor to discuss the information they want to receive at each stage up until handover, and how this will be shared with them through BIM. This detail is recorded in a document known as the Employer's Information Requirements (EIR), so that the construction team understands the service it needs to provide to the client.

Designers usually start the BIM process by translating information captured from the construction site and the client's drawings into digital 3D models of the building and the infrastructure around it. Clients and other stakeholders can use BIM for virtual-reality tours throughout the building before it has been constructed, helping them to gain a better understanding of the structure.

BIM illustrates every detail of the project in graphical form (drawings) and non-graphical form (written information), recording the relationships between components and how they all fit together. This information can be shared easily with all members of the project team, so that they can analyse every aspect of the design, and where necessary make changes to the model in real time.

BIM brings different professionals together to show how their work fits into the overall project by analysing data and exploring **visualisations** to help understand how a building is to be constructed. Working collaboratively on complex building projects using BIM can eliminate mistakes before they happen, save costs and achieve better project outcomes.

At the design stage, the intelligent model process is used to generate documentation for construction specifications and schedules for the building phase of the project, referred to as the Digital Plan of Work (DPoW). During the building phase, this information is used by the contractor and subcontractors to inform planning and organise resources for an efficient site, and ensure the smooth running of the project.

Another advantage of BIM is that it can be accessed by any member of the project team on the construction site using a mobile device, and it is often used with augmented reality (AR) technology. In AR, digitally generated images are superimposed over real-world images in real time. This technology is useful for building services engineers when planning the arrangement of services, such as air-conditioning ducts and vents.

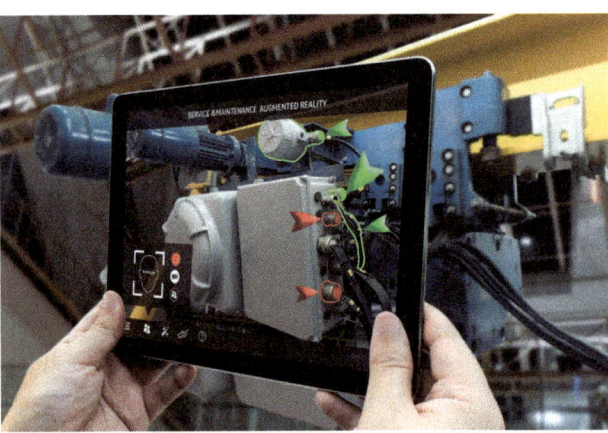

▲ Figure 4.30 Augmented reality (AR)

Key term

Visualisations: digital or virtual representations of a structure

Research

Find out about the use of augmented reality (AR) in construction. Explain how it can be used by the project team to manage building work.

7.2 BIM levels

The extent to which BIM is used will vary, depending on the nature and size of a project. The government recognises the positive steps the construction industry has taken towards the adoption of BIM and how it has contributed to substantial savings. It also understands the complexity of BIM and how it could be used broadly to describe many different systems.

To bring clarity to how BIM is used in a project, the government has defined a number of levels:

▶ Level 0 – Designers are using computer-aided design (CAD) to produce 2D drawings and plans; however this information has to be printed to be shared with the project team. There is no digital collaboration between stakeholders using the same platform.
▶ Level 1 – A Common Data Environment (CDE) is established, usually by the contractor, allowing graphical and non-graphical information to be stored centrally and accessed by the whole project team. CAD is used in the production of 2D drawings, 3D BIM models and other construction documentation. Clear roles and responsibilities are established in the project team, and common forms of data are used to share information between different parties.
▶ Level 2 – Any CAD software can be used at this level, but it must be capable of being exported to common file formats. The project team will also be able to work collaboratively using different systems throughout a construction project. This is the minimum level required by the government for all public construction projects.
▶ Level 3 – This level is in its preliminary stages of development; however it is intended to improve on level 2 BIM with the use of a single server. This will allow an 'open data' standard, so that all stakeholders are able to work simultaneously on the same project, from anywhere in the world.

▲ Figure 4.31 BIM: 2D computer-aided design (CAD)

8 PESTLE factors

There are many internal and external factors that can affect the way a business is controlled. Some of these factors can be planned for and managed; however there are some that are beyond the control of the business and will inevitably happen.

It is important that businesses develop strategies to prepare for these factors before they happen, by analysing their market from many different angles and the environment as a whole.

Factors that are beyond the control of a business are known by the acronym PESTLE:

- **P**olitical – political situations could affect local and national government spending in the construction sector, for example a change of government.
- **E**conomic –
  - If people are not spending money, the UK economy will slow down. When this happens, the building industry is one of the first to be impacted.
  - Changes in gross domestic product (GDP) value and exchange rates will affect the market value of construction goods and services.
  - When interest rates increase, people are more likely to save rather than spend their money. If interest rates decrease, people will often borrow money to spend on new homes or invest in improvements to their existing properties.
  - Higher taxation also has an impact on businesses and the self-employed, because it results in increased costs of building materials and labour.
- **S**ocial – population demographics and the movement of social and community groups can influence the type of buildings and structures constructed in particular areas to meet the needs of the community.
- **T**echnology – technological innovations could affect how a building or structure is designed and constructed, for example the use of green energy as a source of power.
- **L**egal – legislation made by Parliament can influence the planning, design and construction of buildings. For example, following the Grenfell Tower fire in 2017 there was a review of the Building Regulations 2010 and Approved Document B (Fire safety), effectively banning the use of combustible cladding in external walls of relevant high-rise buildings.
- **E**nvironment – companies have corporate social responsibility (CSR) and protect the environment by lowering carbon emissions produced during manufacturing and construction.

▲ Figure 4.32 PESTLE

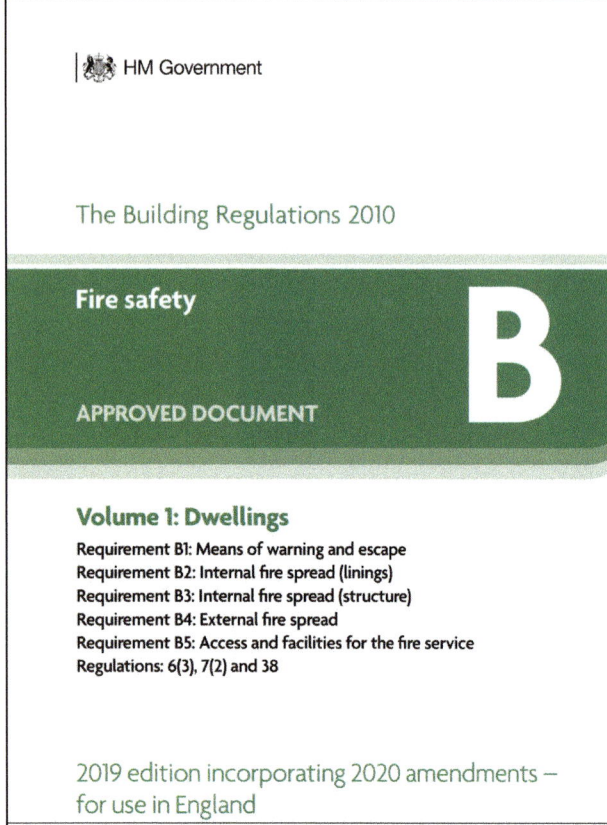

▲ Figure 4.33 A PESTLE factor that can affect construction businesses

> **Test yourself**
>
> List the internal and external factors that can affect a business over which a contractor may have no control.

9 Documentation used in construction projects

At every stage of a construction project, people have to collaborate to share different types of information. If these lines of communication are not efficient, or the information is not clear, it could be detrimental to the project in terms of mistakes and financial implications. While verbal communication is quick and effective in many situations, it can be misunderstood or forgotten, and there is nothing to refer back to at a later date or if a dispute occurs.

In the construction industry, important information is recorded in different types of written documentation.

It is advantageous to record information in this way because it is clear, understood by everyone and can easily be duplicated and distributed at various levels in the project team.

Other than health and safety documentation (which is covered in Chapter 1), the following documents are commonly used in construction projects:

▶ take-off sheets
▶ contracts
▶ schedule of rates
▶ estimates
▶ quotations
▶ delivery notes
▶ purchase orders
▶ bills of quantities
▶ wiring diagrams.

9.1 Take-off sheets

To calculate quantities of building materials and labour costs for a construction project, estimators complete take-off sheets (also referred to as material take-off) using information contained in working drawings and specifications.

While this estimation process may not be necessary for small construction jobs, it is always used for major projects to determine their feasibility, as well as being a financial planning tool. Take-off sheets are also used by estimators to produce bills of quantities for the tendering process, and to support bids for future contracts.

Take-off sheets can be completed manually, referring to hard copies of drawings to determine measurements and quantities using a scale rule. However, this is reliant on the methodical approach of an estimator and is at risk from human error. Accurate take-off sheets can be created using computer software, providing estimates that can easily be shared between the office and construction sites.

Requisitioning materials necessary for a construction project from take-off sheets can result in workers getting the resources they need on time, for smoother running of the job.

> **Test yourself**
>
> Explain the purpose of take-off sheets.

DOOR & WINDOW TAKE-OFF											
Project:						Sheet # 1 of 1				
Take-off by:						Date:				
DOORS											
How many	No.	Width	Height	Thickness	Type	Material	Glass	Frame type	Frame material	Jamb size	Notes
	1	3'0	6'8	1 ¾"	S/C	Birch	(2) 1'6" sidelights		Ponderosa pine	4 ¼"	6'0×6'8 opening
	2	6'0	6'8		Sliding	Glass	¼" insulated		Aluminium	4 9/16"	Includes screen
	3	2'8	6'8	1 ⅜"	H/C	Birch			Ponderosa pine	"	
	4	3'0	6'8	1 ¾"	S/C	-			"	"	4-1/4" jamb for door from garage to garden
	5	2'4	6'8	1 ¾"	H/C	-			"	"	
	6	7'0	8'0		Overhead	Wood			"	4 ¼"	Sectional garage door
											** Allow £150 for exterior doors
											*** £190 allowance for all hardware
WINDOWS											
How many	No.	Width	Height	Thickness	Type	Material	Glass	Frame type	Frame material	Jamb size	Notes
	1	6'0	5'0		Casement	Wood					
	2	4'0	3'0		"	"					
	3	3'0	4'0		"	"					
	4	4'0	4'0		"	"					

▲ Figure 4.34 Take-off sheets

9.2 Contracts

Lump-sum and measurement contracts

Once a client has decided which contractor they are going to appoint for a project, either through the tendering process or otherwise, they can formally secure their services with a lump-sum contract. A lump sum is a fixed price for the project based on drawings, specifications and the bill of quantities (see 9.8).

If a lump sum cannot be accurately determined by the contractor before work starts on site, a measurement contract may be used. This type of contract is based on agreed values or rates (prices) for work between the client and the contractor, where the total cost is not finalised until the end of the project.

When neither of these options can be used, a contract can be agreed for work to be completed on the understanding that the client will be charged the **prime cost** of the materials, labour and plant for the project, plus an additional fee to cover the contractor's profit margin and overheads. This is known as a cost-reimbursement contract. The extra charges added to the prime cost can be based on a percentage of the overall project value, or a fixed sum of money decided before the contract is agreed.

Where the services of a contractor may be required for design and build, management only or construction management purposes, other forms of contract can be designed to suit these requirements.

> **Key term**
>
> **Prime cost:** the actual value of goods and services without any additional costs added, for example profit margins

When both parties have agreed to the terms and conditions of a contract, they will sign it in the presence of a witness to complete the legally binding document. This is known as an agreement.

> **Research**
>
> Research the Joint Contracts Tribunal (JCT) for further information on standard forms of building contracts. Explain the factors that may influence the choice of contract agreed between a client and their contractor.

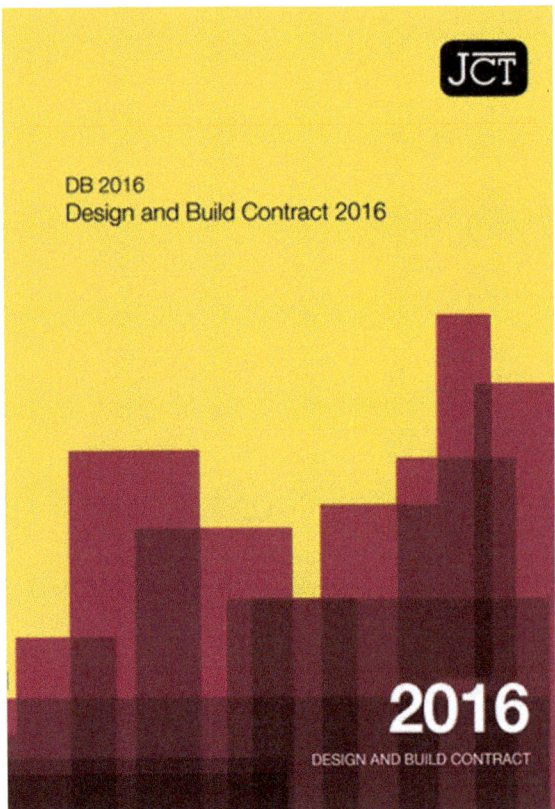

▲ Figure 4.35 Building contract

Payment schedule

An important part of the terms and conditions of any building contract is the payment schedule. This determines how and when the client will pay the contractor for the agreed work. Where minor works are undertaken for a short duration, one payment could be agreed to be made at the end of the project. However, for bigger contracts payments are usually made regularly throughout the project until completion.

Without interim payments, the contractor will have to finance the client's project by paying labour, materials and plant for the duration of the contract, which could cause cash-flow problems. Payments can be agreed to be made in advance of each significant stage of building work, for example:
- the first payment when the contracts have been signed
- the second payment when the foundations are built and inspected
- and so on for each stage of the project.

Most clients prefer to pay contractors interim payments for work completed. This reduces the risk for the client if the contractor is unable to complete the project after payments have been made.

Retention

Clients sometimes withhold a percentage of money due to the contractor at each stage of the building work; this is known as a retention. The exact percentage of money to be withheld has to be agreed between the client and contractor before work starts and is usually 3–5 per cent. The retention acts as financial security for the client, to make sure the contractor finishes the building work and any **snagging** within a reasonable amount of time after completion of the work; this is known as the defects liability period.

If the contractor does not return to complete the work within the period agreed in the contract, the client has reasonable grounds to use the money to instruct other contractors to undertake the outstanding work. On the other hand, if the contractor does complete the project and resolve all of the defects identified within the defects liability period, the client must release the outstanding retention payment without delay.

▲ Figure 4.36 Snagging

> **Key term**
>
> **Snagging:** corrective work undertaken by a contractor or their subcontractors that has been identified by the client or their representative

Penalty clauses

To ensure a building project is completed by the date stipulated in the contract, a penalty clause is usually included. In the event that the contractor does not finish the agreed work by the completion date, a sum of money will be owed to the client by the contractor. This amount will increase with every day past the original deadline, until the project is completed.

Chapter 4 Construction and the built environment industry

Other documents

There are various other documents included in a contract between a client and a contractor, which may include:
- Building Information Modelling (BIM)
- working drawings
- specifications
- a schedule of work.

9.3 Schedule of rates

The cost of building materials will naturally increase over a long period of time due to inflation. However, it can also quickly rise above this level if there is an increase in transport costs or demand and a shortage of materials.

When prices for materials are not stable, it can cause difficulties for contractors and subcontractors in the period between pricing or tendering for contracts and actually starting the projects when purchase orders are placed.

So that contractors are not caught out by price increases, suppliers often agree a schedule of rates for a set period of time. This allows contractors to determine more accurate quotations for clients, without having to overcompensate for unforeseen price increases after they have submitted their fixed price for work.

9.4 Estimates

An estimate is a prediction of costs for building work provided by contractors for clients, based on all the information provided for a job. Although there has to be a degree of accuracy in an estimate, there are factors that are liable to change after the costs have been finally calculated, which could lead to additional charges once the work has been completed.

When a client is considering the affordability of a project, they may not be able to provide all the information needed by the contractor, for example the final details of finishes. A percentage of money is usually added to the estimate as a **contingency**, to cover any unknown financial risk until more information becomes available.

> **Key term**
>
> **Contingency:** provision for an unforeseen circumstance; financial contingencies are often considered when planning for a construction project

▲ Figure 4.37 Written estimate

> **Industry tip**
>
> Never agree to verbal estimates; they often lead to disputes with the client once the work has been completed.

> **Improve your maths**
>
> Research the average hourly rate for workers employed in your area in the same trade as you. Consider a routine task that you would undertake and calculate the total cost. This should include labour, materials, consumables, overheads and profit. You are VAT registered; therefore you will also have to include a further 20 per cent tax.

> **Improve your English**
>
> Create a template for an estimate on your computer and include the costs calculated in the 'Improve your maths' task. You could use your parent, carer or guardian as the customer to complete all the details needed for the estimate.
>
> Once you have completed the task, ask your parent, carer or guardian for feedback.

> **Test yourself**
>
> Explain the difference between an estimate and a quotation.

9.5 Quotations

A quotation is different to an estimate; this is a fixed price for goods and services offered by contractors or subcontractors to potential clients. Once a quotation has been agreed in writing between both parties and work has started on site, the price cannot be amended, even if this means that the contractor has underestimated their work. To avoid disputes over money between the client and contractor at the end of the project, it is important that the quotation details exactly what services and goods are being provided.

Most written quotations have an expiry time, for example 30 days from the date provided, after which time the quote is no longer valid. If this date is not stipulated on the document, then the cost of construction materials and overheads may increase due to increased transport costs or demand, which could reduce profit margins for the contractor. Where the client is responsible for a sizable project, they will often request quotations from different contractors as part of the tendering process.

9.6 Delivery notes

A delivery note is a document that accompanies a shipment from a supplier and describes the goods and quantities being delivered.

When goods are delivered to site, the delivery driver will ask for a dated signature on the delivery note from a responsible person, confirming that the correct items have been received in good condition and that they match the original purchase order. Any damaged or missing goods must be recorded on the delivery note before signing it or allowing the driver to leave. Defects or missing goods unreported when a delivery is made are much more difficult to justify to the supplier or put right at a later date.

While the format for delivery notes varies between different suppliers, the information contained on them is basically the same and includes:
▶ purchase order number or reference
▶ purchaser's name or company
▶ delivery address
▶ order date
▶ item reference/description
▶ quantities
▶ sizes
▶ signatures of the delivery driver and recipient
▶ delivery date.

While paper delivery notes are still used by many suppliers of construction materials, some use electronic pads for signatures, with receipts emailed afterwards. This avoids the need to store the documents on site where they could get lost or damaged, as well as providing a digital copy of the delivery note that can easily be stored.

▲ Figure 4.38 Quotation

Chapter 4 Construction and the built environment industry

Construction Supplies Ltd
Delivery note

Customer name and address:	Delivery Date: 16/12/14
CPF Building Co	Delivery time: 9am
Penburthy House	
Falmouth	Order number: 26213263CPF
Cornwall	

Item number	Quantity	Description	Unit Price	Total
X22433	100	75 mm 4 mm gauge countersunk brass screws slotted	30p	£30
YK7334	100	Brass cups to suit	5p	£5
V23879	1 litre	Sadikkens water based clear varnish	£20	£20

Subtotal	£55.00
VAT	20%
Total	£66.00

Discrepancies: ....
Customer Signature: ....
Print name: ....
Date: ....

▲ Figure 4.39 Delivery note

Neither version of delivery notes contains individual prices, nor the total cost for the goods; those details are sent to the purchaser's business address on an invoice. Contractors that have credit accounts with suppliers will receive an invoice after each delivery or collection of goods. At the end of each month, the contractor will receive a statement listing each invoice number and cost, amounting to the total money due for that period.

9.7 Purchase orders

To keep control of spending and budgets, contractors usually set up a **credit agreement** with their suppliers. Only then can goods be ordered with an official purchase order (PO) from the contractor or their nominated employees, for example a buyer. Hard-copy purchase orders can be used, although these are often much slower to process than electronic versions, which can be emailed directly to the supplier without delay.

Before issuing a purchase order, the contractor or their buyer will usually negotiate the best prices for the goods or services with different suppliers, until they are satisfied that they have value for money.

All purchase orders have a unique reference number or code that links them to a particular job or contract. The same references are then used by suppliers on their delivery notes and invoices. This makes it easier for the contractor to assign money actually spent on resources for a job against estimates and quotations; it also prevents unauthorised purchases from suppliers.

> **Key term**
>
> **Credit agreement:** a legal contract made between a person or party borrowing money and a lender; it states the terms and conditions of the acceptance of credit, including how the debt will be repaid

9.8 Bill of quantities

A bill of quantities (BoQ) is a document usually produced for the client by a quantity surveyor at the planning stage of a building project. It contains a list of all the quantities of materials and resources needed to complete the work, measured in number, length, area, volume or time.

The bill of quantities is part of tender documentation, allowing potential contractors to provide itemised costs for work based on the same specified quantities (rather than taking off quantities from drawings and specifications). The client is then able to compare easily both the overall cost and individual item costs between tenderers.

The bill of quantities is not just used for cost planning during the tendering process, but also later for construction planning, material scheduling and the valuation of interim payments for the contractor.

BILL OF QUANTITIES						
Number	Item description		Unit	Quantity	Rate	Amount
						£	p
	CLASS A: GENERAL ITEMS					
	Specified requirements					
	Testing of materials					
A250	Testing of recycled and secondary aggregates		sum			
	Information to be provided by the contractor					
A290	Production of materials management plan		sum			
	Method related charges					
	Recycling plant/equipment					
A339.01	Mobilise; fixed		sum			
A339.02	Operate; time-related		sum			
A339.03	De-mobilise; fixed		sum			
	CLASS D: DEMOLITION AND SITE CLEARANCE					
	Other structures					
D522.01	Other structures; concrete		sum			
D522.02	Grading/processing of demolition material to produce recycled and secondary aggregates		m³	70		
D522.03	Disposal of demolition material offsite		m³	30		
	CLASS E: EARTHWORKS					
	Excavation ancillaries					
E542	Double handling of recycled and secondary aggregates produced from demolition material		m³	70		
	Filling					
E615	Importing primary aggregates for filling to structures		m³	15		
E619.1	Importing recycled and secondary aggregates for filling to structures		m³	15		

▲ Figure 4.40 Bill of quantities (BoQ)

> **Test yourself**
>
> What is the difference between a bill of quantities and a take-off sheet?

9.9 Wiring diagrams

Wiring diagrams are technical drawings provided by the client to building services engineers, such as electricians, for the installation of electrical systems and circuits. They are simplified visual representations of the layout of electrical systems or circuits. They show how electrical wires are interconnected and where fixtures such as switches, sockets and lighting components are positioned, both internally and externally on every level of the building. Symbols, abbreviations and notes are used on the diagrams to indicate the incoming power source, explain how components relate to one another and provide other important information such as voltages and ratings, as well as the size and capacity of the system.

At the end of the project, the wiring diagrams are handed over to the occupants in the health and safety file, and they should be kept for the lifetime of the building. They can be referred to at a later date during any further building work, or during the repair or maintenance of the electrical system.

> **Assessment practice**
>
> **Short answer**
>
> 1. When must a business become VAT registered?
> 2. Identify one type of client that may have construction work undertaken by a contractor.
> 3. There are many different sub-disciplines in civil engineering. Name one specialised field of civil engineering.
> 4. Which document is used to record building faults and defects discovered in work or materials in the construction industry?
> 5. Identify one factor that can have an impact on a business that it is unable to control.
>
> **Long answer**
>
> 6. List the disadvantages of being a self-employed subcontractor.
> 7. Explain the purpose of Building Information Modelling (BIM).
> 8. Explain the term 'retention' often found in a building contract.
> 9. Explain the benefits of continuing professional development (CPD).
> 10. List the documents that should be included in a tender package.

> **Project practice**
>
> Your employer has noticed that the construction team has not been collaborating well recently, which has reduced its efficiency and caused inaccuracies in information that has been shared.
>
> Your task is to:
> ▶ outline the roles and responsibilities of each member of the construction team
> ▶ create a hierarchy of job roles within the company on a computer, illustrating the lines of communication throughout the team
> ▶ present your work to the team, explaining how it will improve collaborative working.

Chapter 5 Sustainability principles

Introduction

This chapter explores principles of sustainability in the construction industry, which play an important role when planning and delivering projects. It identifies a range of sustainable solutions, including different materials currently available.

We will then look at responsibilities under environmental legislation and the key measures for environmental performance. In particular, we will focus on the importance of waste management, as well as on how renewable energy and energy conservation systems can be incorporated into buildings.

Learning outcomes

By the end of this chapter, you will understand:
1. sustainability when planning and delivering a construction project
2. types of sustainable solutions
3. environmental legislation
4. environmental policies and initiatives
5. environmental performance measures
6. principles of heritage and conservation
7. lean construction
8. waste management legislation
9. waste management
10. energy production and energy use

1 Sustainability when planning and delivering a construction project

1.1 Sustainability

When planning and delivering a construction project, sustainability is achieved by:
- using renewable and recyclable resources
- sourcing materials locally
- protecting resources
- reusing and refurbishing materials
- reducing energy consumption and waste
- creating a healthy and eco-friendly environment
- protecting the natural environment.

> **Research**
>
> Choose one of the following stages of project development:
> - design
> - planning
> - delivery.
>
> Find out about the importance of sustainability at this stage of the construction process. Present your findings to a classmate: why is it important to consider sustainability at this point?

1.2 Assessment methods

Various assessment methods are used to determine how well a building performs against environmental, social and economic standards. Buildings are evaluated according to regional, national and global criteria, to assist architects and planners during the design stages of a project.

Building Research Establishment Environmental Assessment Method (BREEAM)

The Building Research Establishment (BRE) has created best practice standards for the environmental performance of buildings through their design, specification, construction and operation. BREEAM sets out benchmarks for standard categories of development and also offers a scheme for non-standard buildings. The assessment method can be applied to both new developments and refurbishment projects.

> **Research**
>
> Look up the BREEAM rating benchmarks and list each one.

BREEAM is an example of **life cycle assessment**, which covers all stages in the life of a construction project, including the transportation of goods, the extraction and manufacture of raw building materials, **demolition** of the building, and how much of the building is reused or repurposed.

> **Key terms**
>
> **Life cycle assessment:** assessing the total environmental impact of a building, considering all stages of the life of the products and processes used in it
>
> **Demolition:** when something (such as a building) is torn down and destroyed

Leadership in Energy and Environmental Design (LEED)

Leadership in Energy and Environmental Design (LEED) is an environmental certification scheme, developed by the US Green Building Council. It offers a set of rating systems covering the design, construction, operation and maintenance of green buildings, homes and neighbourhoods, and provides **third-party verification** that a project meets high standards of sustainability. It aims to help building owners and operators be environmentally responsible and use resources efficiently.

LEED-certified buildings are healthier, more productive places that reduce stress on the environment. They are energy and resource efficient, and enjoy increased building value and decreased utility costs.

> **Key term**
>
> **Third-party verification:** confirmation from an independent party that a project meets standards

Chapter 5 Sustainability principles

> **Test yourself**
>
> List five ways sustainability can be achieved when planning and delivering a construction project.

> **Research**
>
> Visit the following websites:
>
> www.breeam.com/
>
> http://leed.usgbc.org/
>
> Research and compare the strengths and weaknesses of BREEAM and LEED. Present your findings in a PowerPoint presentation.

Timber Research and Development Association (TRADA)

The Timber Research and Development Association (TRADA) is an international membership organisation dedicated to inspiring and informing best practice design, specification and use of wood in the built environment and related fields.

It has a comprehensive online library, which is free to access for its members. Among its resources are wood information sheets, case studies and technical guidance relating to healthy buildings.

TRADA also provides training and events throughout the country to educate designers and engineers.

▲ Figure 5.1 Use of timber in construction

> **Research**
>
> Research the different types of wood information provided on the TRADA website and produce a written report.

WELL Building Standard

The WELL Building Standard measures, certifies and monitors aspects of the built environment that affect human health and wellbeing. It looks at seven different factors:

- Air: achieve optimum air quality by preventing and removing contaminants and purifying the air.
- Water: achieve optimum water quality by using filtration and treatment, and ensure water is accessible by placing water points in optimum positions.
- Nourishment: ensure the availability of and access to healthier food choices, and encourage better eating habits by providing information to users of the building.
- Light: ensure lighting levels are appropriate for the tasks carried out in the building through the design of windows and lighting.
- Fitness: provide opportunities for physical activity for users of the building, to maintain a fitness regime.
- Comfort: create a soothing and comforting environment, for example by ensuring that thermal and acoustic parameters that might cause discomfort can be controlled.
- Mind: support mental and emotional health by providing building users with relaxation spaces and knowledge about their environment.

> **Test yourself**
>
> List the seven factors that the WELL Building Standard considers.

1.3 Carbon footprint

Carbon footprint describes the amount of CO_2 emissions that are created by construction projects. It is estimated that the built environment contributes to around 40 per cent of the UK's total carbon footprint. Almost half of this is from the energy used in buildings in construction, operation and maintenance.

Approximately 10 per cent of CO_2 emissions are directly associated with construction. This includes:
- the extraction of raw materials
- manufacturing of building products
- transporting the materials to site
- the construction process
- disposal of waste materials.

CO_2 emissions associated with a construction project can be reduced by implementing a number of strategies.
- Use of sustainable materials – cutting CO_2 emissions can be achieved by using less polluting materials such as timber, compressed earth and other natural materials instead of cement-based products.
- Transportation – the use of electric vehicles, sourcing materials locally and reducing the number of deliveries required can all contribute to reducing CO_2 levels.
- Waste management – this plays a vital part in reducing the CO_2 levels of a construction project. Accurate measurement and adopting the three Rs (reduce, reuse and recycle) throughout the project can ensure reductions are made.

The design of a building also plays a big part in reducing the CO_2 emissions associated with the built environment. Aspects to consider include:
- energy use and energy source, including renewable energy systems such as heat pumps and solar PV
- water management – the use of water conservation techniques and reduced terminal flow fittings
- ecology and landscape of the surrounding area
- flexibility and durability, ensuring the building will last and can be repurposed
- consideration of how building users will travel to the building, such as links to local transport and cycle ways.

Other strategies that can be undertaken to reduce CO_2 on construction sites include educating staff on their own energy use, ensuring efficient energy management in site offices such intelligent electrical installations, avoiding leaving plant and vehicles idling and recycling materials.

> **Research**
>
> Research and give two examples of considerations to be taken regarding carbon footprint when planning a construction project.

2 Types of sustainable solutions

Sustainability involves a commitment to environmental, economic and social objectives:
- Environmental sustainability is about acting responsibly to avoid the depletion of or damage to natural resources. It involves protecting the environment from the impact of emissions, sewage and waste.
- Economic sustainability is about supporting long-term economic growth and ensuring profitability. It involves making more efficient use of resources such as labour, materials, energy and water.
- Social sustainability focuses on wellbeing and quality of life. It involves recognising the needs of everyone impacted by construction projects, from design to demolition. This includes construction workers, local communities, project supply chains and users of the building.

Sustainable construction is guided by these objectives during all stages of a project.

2.1 Sustainable solutions

Prefabricated construction

Prefabricated construction combines pre-engineered units to form major elements of a building. They are manufactured in factories and then transported to site where they are assembled.

Benefits of prefabricated construction include:
- reduced waste (a huge environmental benefit)
- lower build costs
- design and build flexibility
- consistent accuracy and quality of components
- reduced construction time.

See Chapter 3, section 2.1 and Chapter 7, section 1.2 for more details.

Self-healing concrete

Self-healing concrete contains the spores of limestone-producing bacteria and a food source. When cracks occur, moisture in the air causes the spores to germinate. The reactivated bacteria then eat the food source and excrete calcite to heal the crack.

Green roofs

A green roof, also known as a living roof, is an attractive and sustainable roof system that involves installing additional waterproof membranes and

Chapter 5 Sustainability principles

drainage mediums, onto which soil is added to allow growth of vegetation.

The benefits of green roofs include:
- reduction of water run-off from roof areas
- extended roof life
- insulation of the building (keeping it warm in winter and cool in summer)
- sound insulation
- providing a habitat for wildlife.

▲ Figure 5.2 Green roof

Smart glass

Smart glass (also known as switchable glass) changes from transparent to translucent (and vice versa) when exposed to specific levels of voltage, heat or light.

Benefits of smart glass include:
- reduced energy costs
- increased privacy
- controlled room temperatures
- improved security.

▲ Figure 5.3 Smart glass

Grey water

Grey water is waste water that has been used for washing and is generated from hand basins, washing machines, showers and baths. Rather than sending it down the drain, it can be reused for watering plants and flushing toilets, thereby reducing mains water usage.

There are several different types of grey-water system available in the UK, but they all have similar features, including some form of treatment, a storage tank for the treated water, a pump and a distribution system.

Grey water should not be used for:
- drinking
- personal washing, showering and bathing
- food preparation
- dishwashing (by hand or machine).

If grey water is filtered, stored correctly and treated where necessary, it is suitable for:
- flushing toilets
- washing clothes with a washing machine (requires treatment)
- watering garden lawns, plants and flowers
- washing cars.

▼ Table 5.1 Advantages and disadvantages of grey-water recycling

Advantages	• Conserves water • Indirectly reduces energy consumption and carbon emissions • Wide range of system options available • Annual cost of water reduces where a water meter is fitted
Disadvantages	• Potentially long payback periods • Not straightforward to install in existing buildings • Risk of contamination or cross-connection • Only certain types of outlet and appliance can be supplied

Key term

Grey water: water that has not been purified for the purpose of drinking, for example recycled water from a sink

143

Onsite Construction T Level: Core

▲ Figure 5.4 Grey-water recycling

Reed beds

Reed beds are artificially constructed wetlands that use natural filtration and biological processes to break down organic matter in waste water and sewage **effluent**.

There are two different types:
- horizontal flow
- vertical flow.

> **Key term**
>
> *Effluent:* liquid waste or wastewater

> **Case study**
>
> John and his family live in a remote detached property in the Scottish Highlands. As there is no access to mains drainage, they would like to explore the use of reed beds as a form of sustainable drainage.
> - Research the types of reed bed suitable for the property.
> - Provide an overview of the working principles of each type of reed bed.
> - List the building regulations that would have to be complied with for the project.

Soakaways

Soakaways are an effective way of dealing with surface water. They are essentially large underground holes, filled with coarse stones or purpose-made plastic crates, which allow water to filter through and soak into the ground.

Smart cement

Smart cement contains potassium ions, which allow it to store electricity for long periods of time. If a building was constructed using this cement, and connected to a power source such as photovoltaic panels, it could store power during the day and release it at night.

> **Research**
>
> Find out how the following materials are used in the construction of buildings and roofs:
> - recycled bricks
> - recycled tiles/slates
> - sustainable timber.

> **Test yourself**
>
> Describe two benefits of a green roof.

3 Environmental legislation

3.1 Environmental Protection Act 1990

The Environmental Protection Act 1990 defines legal responsibilities for the management of waste and pollution, and places a duty on local authorities for collecting waste. Under this legislation, businesses have a duty to handle waste safely.

3.2 Climate Change Act 2008

The Climate Change Act 2008 set a target for the UK to reduce its greenhouse gas emissions by 80 per cent by 2050, compared to 1990 levels. This target was updated in 2019 to 100 per cent – net zero.

The main elements of the act are:
- setting emissions reduction targets and carbon budgeting
- establishing a system for annual government reporting on emissions
- creating an independent advisory body (the Committee on Climate Change)
- giving powers to enable the government to introduce trading schemes to lower emissions
- providing a procedure for assessing the risks of climate change
- requiring the government to develop an adaptation programme for sustainable development
- supporting emissions reductions through various policy measures, such as amendments to improve renewable transport fuel obligations, powers to introduce charges for single-use plastic bags and powers to trial incentive schemes for household waste minimisation and recycling.

3.3 Clean Air Act 1993

The Clean Air Act was first enacted in 1956, in response to the Great Smog of London of 1952 – a dense yellow smog that descended on London and contributed to an estimated 4000 deaths. It was modified by the Clean Air Act 1968, and repealed by the Clean Air Act 1993.

The act covers a range of topics, including:
- prohibition of dark smoke from chimneys
- prohibition of dark smoke from industrial or trade premises
- the requirement for new furnaces to be smokeless (so far as is reasonably practicable)
- the emission of grit and dust from furnaces
- the height of chimneys for furnaces
- declarations of smoke control areas by local authorities.

Under this legislation, local authorities may declare the whole or part of the district of the authority to be a smoke control area, where it is an offence to emit smoke from the chimney of any building, furnace or fixed boiler.

▲ Figure 5.5 It is an offence to emit smoke in a smoke control area

Improve your English

Your client lives in a smoke control area and therefore cannot emit smoke from their chimney unless they use authorised fuels or exempt appliances.

Research either authorised fuels or exempt appliances and write an email to your client explaining the options available to them. What do you recommend?

3.4 Water Act 2014

The main aims of the Water Act 2014 were to:
- reform the water industry to make it more innovative and responsive to customers
- increase the resilience of water supplies to natural hazards such as droughts and floods
- address the availability and affordability of insurance for households at high risk of flooding.

Source: www.gov.uk/government/publications/2010-to-2015-government-policy-water-industry

It includes information on topics such as:
- the water industry:
  - water supply licences and sewerage licences
  - **water and sewerage undertakers**
  - regulation of the water industry
- water resources:
  - water abstraction reform
  - main rivers in England and Wales
  - maps of waterworks
- environmental regulation
- flood insurance:
  - the flood reinsurance scheme
  - flood insurance obligations.

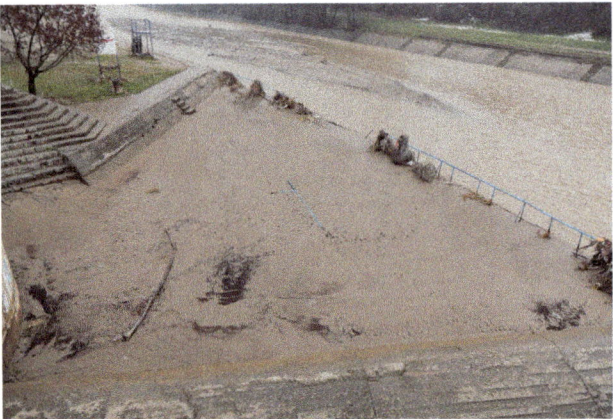

▲ Figure 5.6 Some areas are at higher risk of flooding

> **Key term**
>
> **Water and sewerage undertakers:** suppliers of wholesome water and sewerage services to properties

3.5 Building Regulations 2010

Nearly all new construction work and alterations to existing structures have to comply with the Building Regulations 2010. The requirements are set out in Schedule 1 and cover a range of topics (fire safety, sanitation, security in dwellings, infrastructure for the charging of electric vehicles, etc.).

To help people comply with the regulations, the Ministry of Housing, Communities and Local Government publishes approved documents that offer general guidance on each part of the law (more on these in Chapter 7, section 3.1).

Sustainability is a common theme throughout the approved documents. For example, Approved Document G states that water consumption in a new building should not be greater than 125 litres per person per day. Approved Document L sets minimum appliance efficiencies and control requirements for gas, oil and solid-fuel heating equipment.

The Domestic Building Services Compliance Guide supports Approved Document L and provides guidance for the installation of fixed building services in new and existing dwellings to help compliance with the energy-efficiency requirements of the Building Regulations.

The Domestic Ventilation Compliance Guide supports Approved Document F. It helps architects, planners and installers to comply with the Building Regulations by ensuring the provision of adequate ventilation while minimising energy use and environmental issues.

3.6 Control of Substances Hazardous to Health Regulations 2002

The Control of Substances Hazardous to Health (COSHH) Regulations 2002 are intended to protect people from ill health caused by exposure to hazardous substances. Details of the COSHH regulations can be found in Chapter 1.

Manufacturers and suppliers of hazardous substances produce safety data sheets that contain important information about how products should be transported, used, stored and safely disposed of after use, any special conditions you should be aware of and how to deal with the substance in an emergency.

The COSHH Regulations detail measures that must be taken to ensure substances do not pose a hazard to the environment, for example chemicals being discharged or leaked into water sources.

> **Industry tip**
>
> Manufacturers produce material data sheets, which provide all the required information relating to each chemical.

Chapter 5 Sustainability principles

> **Test yourself**
>
> What is the purpose of the Water Act 2014?
>
> Which of the 2010 Building Regulations provides detail on conservation of fuel and power?
>
> What was the first Clean Air Act (1956) a response to?

3.7 Hazardous Waste (England and Wales) Regulations 2005

Under environmental legislation, waste is considered hazardous if it contains substances or has properties that might make it harmful to either human health or the environment.

The Hazardous Waste (England and Wales) Regulations 2005 restrict the production, movement, receiving and disposal of hazardous waste, such as fluorescent tubes, refrigerators and asbestos. They introduced a registration process for producers of hazardous waste and a new system for controlling, tracking and recording the movement of hazardous waste.

> **Research**
>
> Visit: www.gov.uk/dispose-hazardous-waste
>
> Find out the responsibilities of the following with regard to hazardous waste:
> - a producer or holder (who produces or stores waste)
> - a carrier (who collects and transports waste)
> - a consignee (who receives waste for recycling or disposal).

3.8 Control of Pollution (Oil Storage) (England) Regulations 2001

Also known as the Oil Storage Regulations or simply OSR, these regulations were designed to reduce incidents of oil escaping into the environment. They require anyone in England who stores more than 200 litres of oil to provide a more secure containment facility for tanks, drums, bulk containers and **mobile bowsers**.

> **Key term**
>
> **Mobile bowser:** a wheeled trailer fitted with a tank for carrying oil

All types of oil are covered by the regulations, except for waste mineral oil.

▲ Figure 5.7 Oil storage

3.9 Energy Performance of Buildings Directive

The Energy Performance of Buildings Directive is an EU legislative instrument that aims to reduce the carbon emissions produced by buildings. It requires:
- the production of an energy performance certificate whenever a building is sold, rented out or constructed
- the production of a display energy certificate for large public buildings, which must be displayed in a prominent place
- the regular inspection of air-conditioning systems and boilers.

The requirements of the directive were implemented on a phased basis by the Energy Performance of Buildings (Certificates and Inspections) (England and Wales) Regulations 2007. Later, the Energy Performance (England and Wales) Regulations 2012 consolidated and revoked all previous regulations.

> **Research**
>
> Visit: www.legislation.gov.uk/uksi/2012/3118/contents/made
>
> Who is required to display energy certificates? What should the display certificate include?

3.10 Waste Electrical and Electronic Equipment Directive

Details of the Waste Electrical and Electronic Equipment (WEEE) Directive can be found in section 8.1 of this chapter.

4 Environmental policies and initiatives

There is a range of environmental policies and procedures that are applicable to design and construction. These have been implemented to ensure buildings are designed and constructed with sustainability and energy efficiency in mind.

4.1 Conservation of fuel and power approved document Part L

This document provides guidance on the energy efficiency of different structures. It is updated regularly to reflect developments in building materials and new technologies; therefore people must ensure they are using the most up-to-date version. Approved Document L is divided into four parts, with each providing specific details on a particular building type:
- L1A – new dwellings
- L1B – existing dwellings
- L2A – new buildings other than dwellings
- L2B – existing buildings other than dwellings.

The documents cover the energy efficiency of heating, ventilation and air-conditioning systems, insulation, boiler productivity, lighting and hot water storage.

4.2 BREEAM

The Building Research Establishment (BRE) has created best practice standards for the environmental performance of buildings through their design, specification, construction and operation. More information can be found in section 1.2 of this chapter.

4.3 Energy performance certificate

Energy performance certificates (EPCs) are a legal requirement for domestic and commercial properties in the UK that are available to buy or rent. More information can be found in section 1 of Chapter 3.

4.4 Hazardous Waste (England and Wales) Regulations 2005

More information can be found in section 3.7 of Chapter 5.

4.5 Home Quality Mark

The Home Quality Mark is a voluntary assessment and certification scheme. It was developed by BRE and it is used to assess the impacts of new build homes. The mark provides a rating out of 5 stars and other indicators which are assessed against particular criteria, including the home's surroundings, the comfort of the home and other considerations such as construction energy use, water use, site waste and considerate construction methods.

> **Research**
>
> Visit the following website and produce a leaflet detailing each aim of the quality mark:
> www.ukbuildingcompliance.co.uk/wp-content/uploads/2018/02/home-quailty-pdf.pdf

5 Environmental performance measures

5.1 Performance measures

Environmental performance can be measured by looking at a range of factors during the design, construction, use and demolition of a building.

Source and use of materials

Sourcing materials locally has a lower environmental impact, as it uses less transportation and therefore causes fewer carbon emissions.

When choosing materials to construct a building, it is also important to consider their:
- recyclability
- maintenance requirements
- thermal properties.

Chapter 5 Sustainability principles

▲ Figure 5.8 Using locally sourced materials will cause fewer carbon emissions

Energy source and consumption

The energy efficiency of a building depends on its use, design, orientation, location and the materials used in its construction.

Choosing a form of heating that is based on renewable energy, such as ground or air source heat pumps, will improve a building's environmental performance. Similarly, environmentally friendly electrical systems can generate power using solar panels or **wind turbines**.

It is also important to ensure a thermally efficient building, that reduces heat losses/gains and so requires minimal heating and cooling. Materials with a lower U-value lose less heat and are more efficient.

Maximising daylight in a building can help to reduce lighting costs, as well as energy consumption due to solar heat gain.

> **Key term**
>
> **Wind turbine:** a vaned wheel that is rotated by wind to generate electricity

Water source and usage

In order to improve a building's environmental impact, it is important to minimise water usage. This can be achieved by installing:
- grey-water recycling systems
- rainwater harvesting systems
- flow-limiting valves
- dual-flush toilets
- water-saving showers
- infrared taps and non-concussive percussion taps.

Flexibility

The more easily a change of use can be accommodated by a building, the better its sustainability. A building should therefore be designed to allow adaptation, conversion and extension in the future, for example using building components that can be easily disassembled and repositioned for flexibility.

Adapting an established structure for a new purpose can support sustainability, since a substantial part of the existing building fabric will be reused. Depending on the repurposed design, fewer new materials will be required, which translates into less energy being consumed during manufacture. It also avoids the negative impact of demolition on the environment.

Durability and resilience

The environmental impact of a building depends on its:
- resistance to degradation over time (durability)
- ability to adapt and respond to changing conditions while maintaining functionality (resilience).

A durable and resilient building will have a lower environmental impact as it will require less maintenance and repair throughout its life cycle.

Pollution and waste processing

Construction that generates large amounts of waste or causes pollution to land, air or water will have a negative environmental impact. It is important to design waste out of all the stages of a building's life cycle, from construction to use to demolition.

For example, waste can be reduced by:
- using standardised components (for example, bricks and blocks)
- storing materials carefully to avoid damage or deterioration (for example, cement or lengths of timber)
- recycling, reusing and repurposing materials (for example, demolition materials)
- installing energy-efficient and non-polluting systems (for example, heating systems).

Transport

Transporting construction materials over long distances will have a negative effect on the environment due to vehicle emissions. Therefore, buildings that use locally sourced materials are considered more sustainable.

Buildings that are easily accessible using local public transport systems will have a lower impact on the environment than those that are only accessible by car.

Demolition

Well-planned management of demolition means that building materials used during construction can be recycled, thereby reducing environmental impact. The more of a building that can be recycled, the better. For example:

- Concrete can be broken down as aggregate.
- Drywall's paper covering can be recycled like most paper products, and the gypsum core is ideal for recycling as garden fertiliser.
- Glass can either be recycled into new glass or broken down as aggregate.
- Steel can easily be melted down and recycled.

Energy-efficient lighting

Energy-efficient lighting not only lowers electricity bills but it also benefits the environment by reducing carbon emissions.

Incandescent bulbs are inefficient, converting around five per cent of the electricity they use into visible light. Halogen bulbs use the same filament technology as traditional bulbs but run at a higher temperature, making them slightly more efficient. These types of lighting have now been phased out and replaced with more efficient types:

- Compact fluorescent lamps (CFLs) comprise a mercury-filled glass tube with an inner coating of phosphor powder which glows when charged with electricity. They use around 60–80 per cent less electricity than equivalent traditional bulbs and last around ten times longer.
- Light emitting diodes (LEDs) use 90 per cent less energy than traditional bulbs and have begun to replace CFLs. They turn on instantly at full brightness and are available to fit almost any light fitting. There may be a large number of LEDs within a single bulb to create sufficient brightness.

▲ Figure 5.9 Energy-efficient lightbulbs

Appliance efficiency ratings

The Energy-related Products (ErP) Directive was introduced in 2009 to help EU countries reduce energy consumption and emissions.

▲ Figure 5.10 Energy efficiency ratings

Energy labels are a common sight on many appliances, such as white goods, grading them according to their energy efficiency with a letter-based rating. These labels were first introduced in 1995 to drive innovation and help consumers make informed decisions about the products they purchase. However, in 2021 a new generation of labels was introduced for refrigerators, freezers, washing machines, washer-dryers, dishwashers, televisions and electronic displays, and light sources.

The energy efficiency rating system for these appliances now runs from A (most efficient) to G (least efficient). The energy labels also include QR codes, clearer guidance on energy consumption and extra information specific to product types, such as noise levels and water usage for washing machines.

Energy labels on other appliances, such as tumble dryers, ovens and air conditioners, still use the old rating system of A+++ to G, although there are plans to update these in 2022. For now, the UK is following EU legislation changes, although post-Brexit it also has the freedom to make its own changes.

Manufacturers are responsible for ensuring their products are accurately labelled.

Since September 2015, boilers, water heaters and heating products have been labelled in the same way as refrigerators and freezers to show how energy efficient they are. Under the ErP Directive, boilers must be labelled with their efficiency level from G (lowest efficiency) to A+++ (highest efficiency). Boiler manufacturers and retailers are responsible for ensuring their products are accurately labelled.

Insulation for pipework and ductwork

All pipes installed in vulnerable or exposed locations inside and outside a building, such as unheated cellars, roof spaces, garages and outbuildings, must be insulated to improve energy efficiency, prevent freezing and retain heat energy.

Types of pipework insulation include:
- PVC foam
- expanded polystyrene
- extruded polystyrene
- cross-linked polyethylene foam
- expanded nitrile rubber
- expanded synthetic rubber.

Ductwork should be insulated in order to:
- keep air within the ducts at the required temperature
- prevent ducts from leaking air.

Ductwork carrying warm air should be insulated to conserve energy. Ductwork carrying cold air should be insulated to conserve energy and control condensation on the outer surface.

Various thermal insulation products are available for ducting, including rigid preformed slabs or boards and flexible rolls, blankets, mats and mattresses.

Building insulation

Insulation is needed to reduce heat loss from a building during cold weather (see Figure 5.11). This in turn reduces the energy required to maintain a comfortable temperature for the occupants.

Insulation should also reduce the heat entering a building in warm weather. This saves energy by reducing the need for air-conditioning systems.

There are many different ways to insulate a building, making it more energy efficient and lowering running costs.

▲ Figure 5.11 Heat loss from a building

Cavity wall insulation

Cavity walls are constructed from two leaves of masonry, separated by a 50+ mm gap (most cavities are now likely to be over 75 mm). The outer leaf is usually facing brickwork, and the inner leaf is blockwork. Insulating an existing cavity wall involves drilling a series of holes through the outer leaf and blowing or injecting the insulation material into the cavity.

Cavity wall insulation materials include:
- blown mineral fibre
- polystyrene beads (EPS)
- rigid board insulation (installed during construction)
- urea formaldehyde foam.

External and internal wall insulation

External wall insulation involves fixing an insulating layer over the existing wall, either mechanically or adhesively, and covering it with a **render**.

Where external insulation is not an option, for example in conservation areas, internal wall insulation can be installed. This consists of thermally engineered insulated studs and insulation slabs. The use of internal wall insulation will reduce the internal space.

> **Key term**
>
> **Render:** a pre-mixed layer of sand and cement, similar to mortar, used to make masonry walls flat and prepare them for top coats of finishing plaster

Loft insulation

Installing loft insulation, or upgrading existing loft insulation, is the quickest and most efficient method of improving the thermal performance of an existing building. Loft insulation materials include:

- blown insulation material
- insulation boards
- roll loft insulation.

Current building regulations state that insulation materials should be laid to a minimum thickness of 270 mm.

Floor insulation

Around 15 per cent of heat is lost through an uninsulated floor. Insulating a ground floor will keep the building warmer and reduce energy bills. This can be done using rigid board insulation.

Upstairs floors do not need to be insulated, as they are within the thermal envelope of the building.

> **Research**
>
> Research the purpose, limitations and benefits of insulation materials. Copy and complete the following table:
>
> | Insulation method | Purpose | Benefits | Limitations |
> |---|---|---|---|
> | Foam board | | | |
> | Loose fill | | | |
> | Insulation roll | | | |
> | Sprayed foam | | | |
> | Reflective system | | | |
> | Rigid fibrous | | | |

> **Test yourself**
>
> Describe three methods that can be used to insulate a property.

5.2 Schemes

Several schemes can be used to certify levels of environmental performance in construction, for example:
- Building Research Establishment Environmental Assessment Method (BREEAM) (see section 1.2 of this chapter)
- Leadership in Energy and Environmental Design (LEED) (see section 1.2 of this chapter)
- **Passivhaus**.

> **Key term**
>
> **Passivhaus:** ('Passive house' in English) an energy performance standard intended primarily for new buildings, which ensures that buildings are so well constructed, insulated and ventilated that they require little energy for heating or cooling

> **Test yourself**
>
> Explain what a U-value is and why it is important to consider when designing buildings.

6 Principles of heritage and conservation

6.1 Listed buildings

There are three types of listed status for buildings in England and Wales:
- Grade I – buildings of exceptional interest
- Grade II* – particularly important buildings of more than special interest
- Grade II – buildings of special interest, warranting every effort to preserve them.

Buildings are listed in their entirety, even though some parts may be more important than others. The designation regime is set out in the Planning (Listed Buildings and Conservation Areas) Act 1990.

The government document *Principles of Selection for Listed Buildings* sets out the criteria that the Secretary of State applies when assessing whether a building is of special architectural or historic interest:
- To be of special architectural interest, a building must be of importance in its architectural design, decoration or craftsmanship.
- To be of special historic interest, a building must illustrate important aspects of the nation's social, economic, cultural or military history.

Listed building consent is required for all works of demolition, alteration or extension to a listed building that affect its character. This is to ensure the impact of any proposed changes is reviewed before they are approved. This is usually carried out by a conservation officer within the local authority planning department.

Local planning authorities may serve a building preservation notice (BPN) on the owner and occupier

Chapter 5 Sustainability principles

of a building which is not listed, but which they consider is of special architectural or historic interest and which is in danger of demolition or of alteration in such a way as to affect its character.

When carrying out work on listed buildings, it is important that you do not use modern repair methods on traditional construction. For example:
- Using cement on older buildings made from materials such as lime mortar can cause irreparable damage.
- Traditional cast-iron soil stacks should be replaced like for like, and not substituted with modern plastic pipework.
- Any original architectural features such as doors, decorative stonework, fireplaces or windows should not be altered.

Any changes and the materials used must be agreed with the conservation officer.

On a listed building, you cannot paint or render stonework, demolish chimney stacks, or add new pipework, flues or alarm boxes on principal elevations of the property.

> **Research**
>
> Review this page: https://historicengland.org.uk/advice/technical-advice/buildings/building-materials-for-historic-buildings
>
> What can be used for consolidating weathered limestone? How is this substance applied? Produce a short materials guide.

The Heritage Protection Bill is a legislative and policy framework which protects the historic environment. It requires consents and permissions to protect England's heritage via a balanced, democratic and informed approach to managing changes in historic places. The different heritage assets are protected in different ways and various consents are required when carrying out works to existing buildings and for new development.

6.2 Conservation areas

Local planning authorities have the power to designate any area of special architectural or historic interest as a conservation area, where the character or appearance should be preserved. The special character of these areas is not just made up of buildings but can also be defined by features which contribute to particular scenic views, such as woodland or open spaces.

6.3 Town and Country Planning Act 1990

Under the Town and Country Planning Act 1990, planning permission needs to be sought when carrying out work to:
- build a new property
- increase the size of an existing property
- make significant alterations to an existing property
- change the use of an existing property.

The process of applying for planning permission is straightforward and completed with the local authority planning department.

> **Test yourself**
>
> List and describe the three types of listed status for buildings in England and Wales.

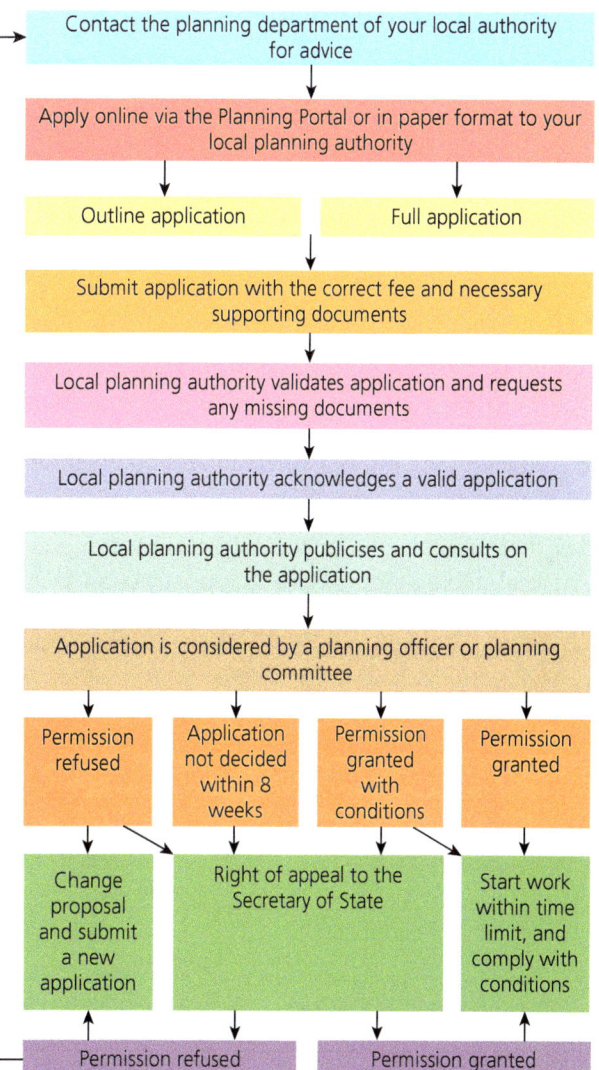

▲ Figure 5.12 Steps in planning permission

> **Research**
>
> Research the local planning permission requirements and application process in your area.

7 Lean construction

7.1 Principles

Lean construction is a construction methodology that aims to minimise waste in terms of costs, materials, time and effort, while maximising productivity and value.

The principles of lean construction are:
- efficiency
- best value
- ensuring the work environment is clean and safe
- improving planning
- continuous review and improvement.

7.2 Techniques

A common example of a lean construction technique is just-in-time deliveries. This is a method of providing the required materials for a project in precisely the correct order and quantity at exactly the right time for installation.

The benefits of just-in-time deliveries include:
- elimination of waste
- less storage space needed
- a stabilised work schedule
- reduction in the cost of inventories and inventory requirements.

Another example of a lean construction technique is artificial intelligence (AI). In the construction sector, AI is commonly used to develop safety systems for construction worksites. This includes tracking the interactions of tasks, workers, machinery and objects on a construction site and alerting supervisors to potential safety issues, construction errors and productivity issues, thereby reducing errors and health and safety risks during the construction of buildings.

Recycling is another commonly used method of lean construction. Recycling materials reduces the requirements for the use of new raw materials during the construction stage, eliminating waste removal from site and avoiding over ordering of materials, which can result in additional waste and an increase in the carbon footprint of the building.

> **Research**
>
> Research other methods of lean construction used in the industry today. Explain two to a classmate using a PowerPoint presentation.

8 Waste management legislation

8.1 Waste Electrical and Electronic Equipment Regulations 2013

The Waste Electrical and Electronic Equipment (WEEE) Regulations 2013 implemented in UK law the provisions of the EU WEEE Directive. They aim to support sustainable production and consumption through the collection, reuse, recycling, recovery and treatment of end-of-life electrical and electronic equipment (EEE). This both reduces the need to produce new equipment and minimises the amount of waste sent to landfill, reducing associated environmental impacts.

The WEEE Regulations cover the categories of appliance shown in Table 5.2.

▼ Table 5.2 Examples of appliances covered by the WEEE Regulations

Type of appliance	Examples
Large household appliances	Refrigerators, cookers, microwaves, washing machines, dishwashers, air conditioners
Small household appliances	Vacuum cleaners, irons, toasters, clocks, kettles, fryers, electric knives, body care appliances
IT and telecommunications equipment	Personal computers, copying equipment, telephones, pocket calculators, screens, printers
Consumer equipment and photovoltaic panels	Radios, televisions, hi-fi equipment, camcorders, musical instruments, photovoltaic panels, radios
Lighting equipment	Straight and compact fluorescent tubes, high-intensity discharge lamps

Chapter 5 Sustainability principles

Type of appliance	Examples
Electrical and electronic tools	Drills, saws, sewing machines, electric lawnmowers, jigsaws and other gardening equipment
Toys, leisure and sports equipment	Electric trains, games consoles, running machines
Medical devices	(Non-infected) dialysis machines, analysers, medical freezers, cardiology equipment
Monitoring and control instruments	Smoke detectors, thermostats, heating regulators, heating controls, alarm components
Automatic dispensers	Hot drinks dispensers, money dispensers, snack dispensers

Under the regulations, producers of EEE have obligations with regards to:
- the EEE they sell
- financing the collection, treatment, recovery and environmentally sound disposal of WEEE.

There are specific obligations on producers of non-household equipment when it is discarded as waste by non-household end users. In practice, this means that electrical tools used in construction by contractors must be disposed of in the correct way.

The regulations also require producers to prioritise, where appropriate, the reuse of whole appliances.

Regulations state that if you are a large producer (i.e. have more than 5 tonnes of EEE on the market), you must join a producer compliance scheme (PCS). The PCS takes on your obligations to finance the collection, treatment, recovery and environmentally sound disposal of household WEEE collected in the UK.

Distributors of EEE also have specific responsibilities. They must offer to take back waste of the same (or similar) type as the item a customer is purchasing from them, regardless of the method of purchase (in store, online or via mail order) and irrespective of the brand of the item.

They can discharge these take-back obligations by:
- joining the Distributor Takeback Scheme (DTS)
- offering in-store take-back
- providing an alternative free take-back service.

> **Research**
>
> Find out what the current UK WEEE targets are. Does data show that the UK is meeting these targets?

8.2 Waste carrier licences

Under section 34 of the Environmental Protection Act 1990, there is a duty of care to ensure that waste is managed and disposed of properly.

If a business transports waste, either for itself or for someone else, it needs to register with the Environment Agency as a waste carrier. If a business' waste is being collected by someone else, it must ensure that the carrier is registered.

There are two types of licence, depending on the type of waste to be transported:
- An upper-tier waste carrier transports other people's waste on a professional basis (for example, a waste management company) or its own construction or demolition waste (for example, a builder).
- A lower-tier waste carrier was either previously exempt from registration or carries its own (non-construction/non-demolition) waste on a regular and normal basis. This might include carpet fitters taking away offcuts, businesses carrying waste from maintenance work or businesses collecting confidential waste from different locations and transporting it to a centralised office.

▲ Figure 5.13 Waste carriers

8.3 The Fluorinated Greenhouse Gases Regulations (2018)

Fluorinated greenhouse gases (known as F-gases) are synthetic and originate from human activities. The most common type is hydrofluorocarbons, which are used in applications such as refrigeration, air conditioning, fire extinguishers and aerosols.

While these gases are ozone friendly, and therefore often used in place of ozone-depleting substances, they have a high global warming potential (GWP) and therefore require regulation.

Under the Fluorinated Greenhouse Gases Regulations:
- the installation, leak testing, maintenance, repair and disposal of equipment containing F-gases must be performed by trained and certified personnel
- equipment containing F-gases must be tested regularly to check for leaks and any leaks must be repaired as soon as possible
- equipment containing F-gases must be clearly labelled with the type and quantity of the gas (refrigeration and air-conditioning equipment should have an information plate affixed to it)
- records must be kept of:
  - types and quantities of gases in use
  - leaks detected and repairs carried out
  - maintenance history of equipment, including details of the maintenance company or operatives
- F-gases must be recovered at the end of equipment life and managed appropriately.

Further information on the requirements for working with F-gases can be found by searching on www.gov.uk.

> **Industry tip**
>
> Operatives who work on systems that contain F-gases, for example heat pumps or air conditioning, must be F-gas qualified.

9 Waste management

The waste hierarchy sets out the order in which actions should be taken to manage waste, from the most to least preferable in terms of environmental impact. There are five actions:

1. Reduce: this is about producing less waste and can be achieved by using fewer materials during design and manufacture, keeping products for longer, choosing products with less packaging and using fewer hazardous materials.
2. Reuse: this involves either using a product multiple times, rather than single use, or repurposing products or their parts at the end of their life by checking, cleaning, repairing or refurbishing them.
3. Recycle: this involves turning waste into new substances or products and, if it meets quality protocols, composting the material.
4. Recover: by using the waste products as fuel to provide heat and power; other methods include **anaerobic digestion**.
5. Landfill: this is the last resort and involves disposal and incineration without energy recovery.

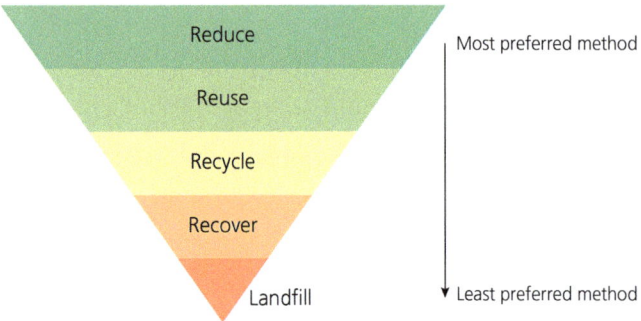

▲ Figure 5.14 Waste hierarchy

> **Key term**
>
> *Anaerobic digestion:* a sequence of processes by which microorganisms break down biodegradable material in the absence of oxygen

9.1 Site waste management plans

While no longer a legal requirement, a site waste management plan (SWMP) is an important document for setting out how waste will be managed and disposed of during a construction project. It should be compiled by the design team, contractor and subcontractors and refer to the waste hierarchy to reduce the volume of waste to landfill and increase the use of sustainable materials. This ensures compliance with environment legislation, improves resource efficiency and increases profitability.

A typical SWMP should contain information on:
- who has overall responsibility for the management of waste on site
- which contractors will be involved in the waste management process and in what capacity
- which types of waste will be generated, including hazardous waste

Chapter 5 Sustainability principles

- the expected quantity of waste that will be produced
- how waste will be managed both on site and off site.

Managing waste responsibly can contribute to an improved business reputation, leading to future contracts.

9.2 Waste segregation

Waste segregation means dividing waste into different categories for efficient disposal. Not only does this support the waste hierarchy and protect the environment, by offering opportunities to reuse and recycle waste before resorting to landfill, but it also offers cost savings to construction businesses. Sorted waste is cheaper to dispose of, and some types of waste can even be sold as a source of income.

Hazardous waste in particular should be segregated and disposed of carefully, due to the risk of harm to human health and the environment. There are specific regulations that cover this, namely the Hazardous Waste Regulations 2005 (see section 3.7 in this chapter).

Waste on a construction site is usually segregated into the following streams:

- general (for example, insulating materials that do not contain asbestos)
- hazardous (for example, asbestos)
- clean fill (material that can be recycled or reused in future construction projects)
- hard fill (for example, soil, concrete, bricks and blocks)
- plastic
- metal
- wood
- plasterboard.

It is important that all staff on a construction site are trained on the site's waste policy and the waste hierarchy, to ensure waste segregation is maximised.

9.3 Recycling

Prior to the adoption of environmentally friendly methods of managing waste, such as recycling, reusing and repurposing, most waste was disposed of by burying it in landfill sites or burning it at extremely high temperatures (incineration), depending on the type of material. Some waste is still disposed of in this way.

Recycling or reusing materials can reduce the overall carbon emissions that are produced during manufacturing.

Metals

Often, a building may appear to consist mostly of masonry or concrete. However, surprising quantities of different metals may be present in the structure. Scrap or waste metal can be recovered and reprocessed into new products, for example:

- structural steel columns, beams and lintels
- pipework made from lead (in older buildings) or copper
- cables with copper or aluminium cores.

Metal to be recycled is usually separated into ferrous (containing iron) and non-ferrous before collection.

Plastic

Plastic waste can be very harmful to the environment. When it is disposed of in landfill, it can take hundreds of years to break down, and if it ends up in our oceans, it can damage ecosystems and kill marine life.

It is therefore important to segregate and recycle as much plastic waste as possible. Waste plastic can be reprocessed into many different products, from drinks bottles to car components to building materials.

Timber

Offcuts of timber can be reprocessed into new resources, such as:

- chipboard and medium density fibreboard (MDF)
- paper and cardboard
- mulch
- bedding for animals
- biomass fuel.

Bricks and blocks

Undamaged bricks and blocks can easily be reused in new building projects. Brick and block waste from cutting operations can be crushed and used as aggregate in some types of concrete.

> **Test yourself**
>
> Which new products can offcuts of timber be reprocessed into?

Onsite Construction T Level: Core

10 Energy production and energy use

10.1 Energy production

In 1831, Michael Faraday discovered that electricity could be generated by moving a bar magnet through a wire coil. This is the principle upon which modern generators work.

In a power station generator, a huge magnet is mounted on a central rotating shaft; this is known as the rotor. Around the rotor is a series of coils known as the stator, which is where the electricity is generated by the rotating magnetic field. As the rotor and stator may weigh several hundred tons, an energy source is required to turn the rotor, for example a turbine.

In the UK, electricity is generated in several different ways, for example by:
- burning fossil fuels, such as coal, gas and oil
- using nuclear fission
- using renewable energy, such as wind, wave, hydro, biomass and solar.

Most power stations are steam powered. The heat created by burning fossil fuels or through nuclear fission is used to produce superheated steam to drive a turbine. This causes the rotor to turn, creating an **alternator** as in Faraday's prototype. The rotor turns at 3000 revolutions per minute (50 revolutions per second) and produces an alternating current (AC) with a frequency of 50 hertz (cycles per second). The electromotive force causes current to flow.

See Chapter 2, section 3, for more information on electricity generation and transmission.

> **Key term**
>
> **Alternator:** an electrical generator that converts mechanical energy into electrical energy

Non-renewable energy sources

Fossil fuel is a term used to describe an energy source formed by the decomposition of organic matter beneath the Earth's surface over millions of years:
- coal is formed from dead trees and plants
- crude oil and gas are formed from dead marine plants and animals.

Fossil fuels are predominantly composed of carbon and hydrogen. They are burned in power stations to heat water and provide steam for turbines. However, this process releases carbon dioxide, which can contribute to the greenhouse effect and climate change.

Fossil fuels are a finite resource, which means that they are being used up faster than they can be replaced.

Nuclear power plants use reactors to split atoms, causing a large amount of thermal energy to be released. This process is called nuclear fission. Uranium is the main fuel for nuclear reactors.
- Nuclear reactions generate thermal energy.
- Thermal energy generates water vapour (steam).
- Steam powers a turbine, producing mechanical energy (kinetic energy).
- Mechanical energy actuates an electrical generator, producing electricity.

▲ Figure 5.15 How electricity is produced

Chapter 5 Sustainability principles

▼ Table 5.3 Advantages and disadvantages of nuclear power

Advantages	• Does not produce polluting gases • Does not contribute to global warming • Low fuel cost • Low fuel quantity required • Power station has a long life
Disadvantages	• Costs are high for building and decommissioning • Waste is radioactive • Accidents can be catastrophic

Renewable energy sources

Renewable energy sources are obtained from the environment around us and are naturally replenished. As the availability of non-renewable energy diminishes and the effects of global warming intensify, it is important that society considers the use of these technologies as alternatives.

Bioenergy

Bioenergy is created using biomass – organic material such as wood, plants, agricultural crops, food waste or even sewage.

Biomass can be burned in power stations to heat water and produce high-pressure steam. The steam flows over turbine blades, causing them to rotate, which in turn drives a generator and produces electricity.

Biomass can also be converted into liquid fuels, for example biodiesel for use in vehicles.

Wind energy

The natural **kinetic energy** of the wind can be used to drive an aerodynamic bladed turbine, rotating a generator and producing electricity. Often many wind turbines are grouped together to form a wind farm and provide bulk power to the National Grid.

Turbines can be installed on land or offshore where there is a reliable source of wind.

▲ Figure 5.16 How wind turbines generate electricity

> **Industry tip**
>
> The greater the wind speed, the faster the rotation of a wind turbine's blades, generating more electricity.

> **Key term**
>
> *Kinetic energy:* energy derived from motion

Solar energy

Solar energy utilises photovoltaic (PV) panels to convert the sun's radiation into electricity. The panels can be installed on the roofs of buildings or in larger-scale solar farms. Solar farms are capable of delivering bulk power to the National Grid.

In order to maximise the energy generated by PV panels, it is important to install them carefully, for example where there is enough sunlight and facing South.

Solar photovoltaics

As mentioned earlier, solar energy uses photovoltaic (PV) panels to convert the sun's radiation into electricity.

PV systems can be divided into two categories:
- off-grid systems, where the PV modules are used to charge batteries
- on-grid systems, where the PV modules are connected to the National Grid via an inverter.

▲ Figure 5.17 Solar photovoltaics

▼ Table 5.4 Key components of off-grid and on-grid PV systems

Off-grid systems	On-grid systems
PV modules	PV modules
PV module mounting system	PV module mounting system
DC cabling	DC cabling
Charge controller	Inverter
Deep-discharge battery bank	AC cabling
Inverter	Metering
	Connection to the National Grid

▲ Figure 5.18 Off-grid system components

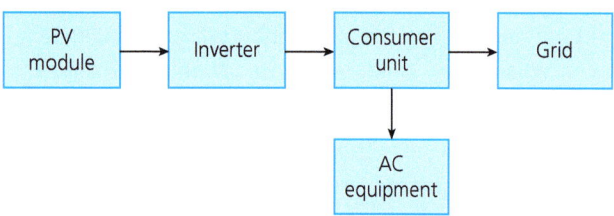

▲ Figure 5.19 On-grid system components

Health and safety

When working on solar PV systems it is important to follow industry safe isolation procedures.

PV modules are available in different efficiencies:
- Monocrystalline modules range in efficiency from 15 to 20 per cent.
- Polycrystalline modules range in efficiency from 13 to 16 per cent.
- Amorphous film ranges in efficiency from 5 to 7 per cent.

Chapter 5 Sustainability principles

Photovoltaic modules can be fitted in different ways:
- On-roof systems comprise aluminium rails, which are fixed to the roof structure by means of roof hooks.
- In-roof systems replace roof tiles with the PV modules.
- Ground-mount and pole-mount systems provide free-standing PV arrays, which can also be installed as computer-controlled motorised mounting systems that rotate and track the sun as it moves across the sky.

An inverter converts the DC input to a 230 V AC 50 Hz output and synchronises it with the frequency of the mains supply.

> **Research**
>
> Research the operating principles of solar photovoltaic (PV) panels and provide an overview of how they convert sunlight to electricity.

Hydroelectric energy

Hydroelectric power plants produce energy by channelling running water through a turbine connected to an electrical generator. The electricity is then fed into the National Grid for distribution.

They can use the natural flow of a river as it falls from a greater to a lesser height, or artificial reservoirs and dams, which hold back water and release it as required.

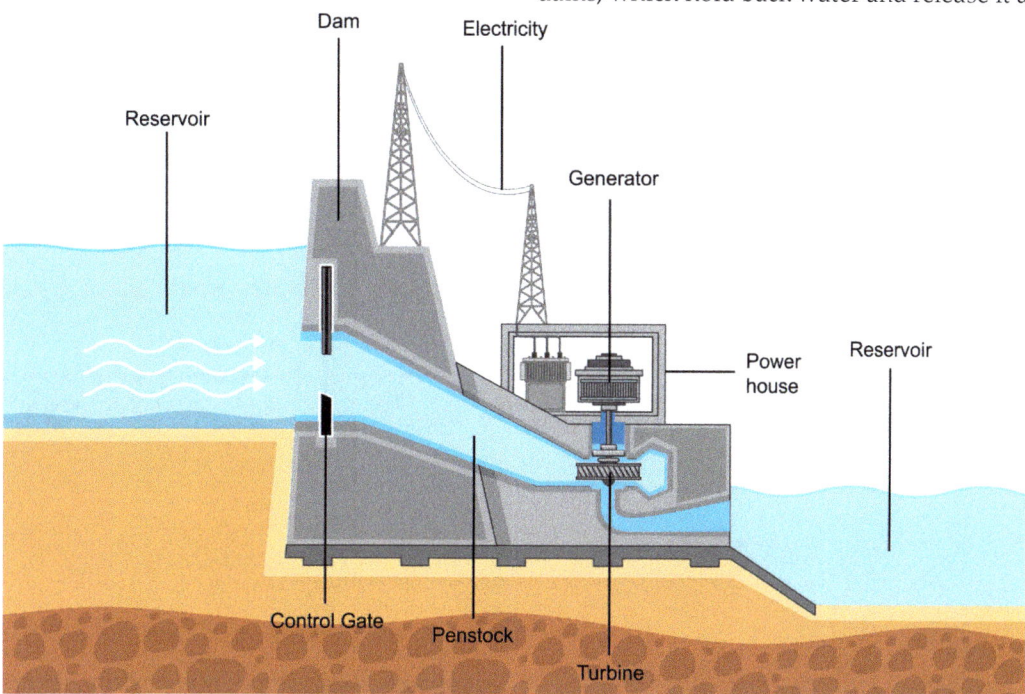

▲ Figure 5.20 Generating electricity through hydroelectric power

Wave and tidal energy

Energy from the movement of waves and tidal flows can be used to generate electricity. Wave and tidal generators are similar to wind generators, except that they use the ocean's current to rotate the turbine instead of wind. This type of energy production is ever-changing, with a wide range of options available to harness the power of waves and tides. These options include absorbers, attenuators, water columns and pendulum devices.

The amount of energy created is determined by the wave's height, speed, wavelength and water density. Wave power is much more predictable than wind power and it increases during the winter, when electricity demand is at its highest. Tidal energy is also predictable and consistent.

> **Research**
>
> Research the advantages and disadvantages of using non-renewable and renewable energy sources and create a table to show your findings.

10.2 Energy use

The **Department for Business, Energy and Industrial Strategy (BEIS)** publishes an annual statistical publication titled 'Energy Consumption in the United Kingdom' (ECUK). This collates data from a variety of sources to provide a comprehensive review of UK energy consumption and changes in intensity and output since the 1970s.

Onsite Construction T Level: Core

> **Key term**
>
> **Department for Business, Energy and Industrial Strategy (BEIS):** the department of the UK government responsible for the UK's business, energy and industrial strategy

Energy consumption is reported annually for the following sectors:

- domestic
- transport
- industry
- services.

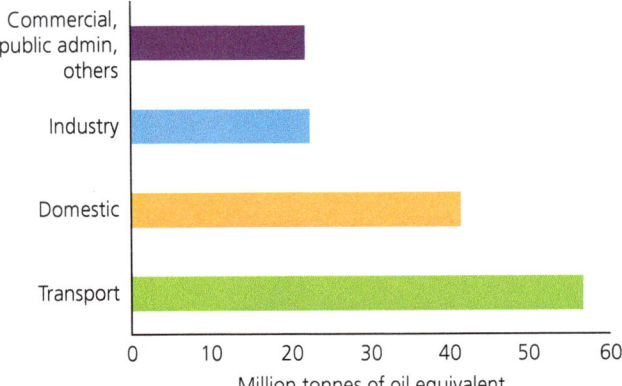

▲ Figure 5.21 Energy consumption in the UK 2019

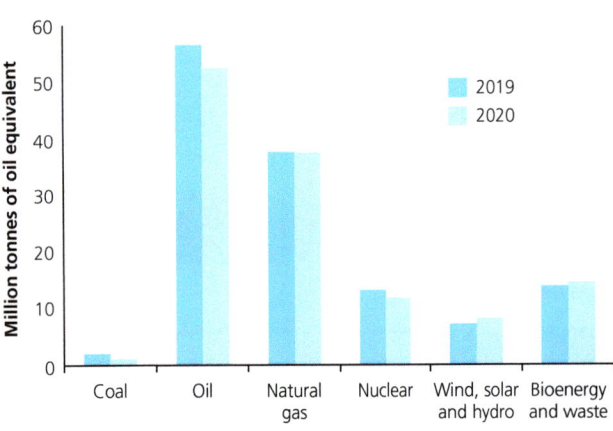

▲ Figure 5.22 Primary energy consumption in the UK in 2019 and 2020 by fuel/technology type

> **Test yourself**
>
> Describe how energy can be harnessed from wave and tidal power.

Uses of renewable energy sources

As mentioned earlier, renewable energy sources are naturally replenished, such as solar, wind, tidal and geothermal. They can be used in a wide range of applications, from small-scale systems such as solar thermal providing hot water for one property to large-scale wind farms producing energy for national use.

Non-renewable energy sources are finite and cannot be replenished as quickly as they are used, for example coal, oil and gas.

Energy sources fall into one of three categories according to their level of carbon emissions: high carbon, low carbon and zero carbon.

▼ Table 5.5 Energy sources categorised according to their carbon emissions

Category	Energy sources
High carbon	• Coal • Oil • Natural gas • Liquified petroleum gas (LPG)
Low carbon	• Solar thermal • Biomass • Heat pumps • Combined heat and power (CHP) • Combined cooling, heat and power (CCHP) • Hydrogen fuel cells
Zero carbon	• Wind • Wave/tidal • Hydroelectric • Solar photovoltaic • Nuclear

> **Research**
>
> Find out how hydrogen fuel cells work and what they are used for.

Solar thermal

Solar thermal systems use either a roof-mounted or ground-mounted collector to harness the sun's energy and heat domestic water. A conventional boiler or immersion heater (auxiliary heat source) is used to top up the water temperature to 60°C.

Chapter 5 Sustainability principles

▲ Figure 5.23 Solar thermal system components

Key components of a solar thermal system are as follows:
- A solar collector absorbs heat radiation from the sun. There are two types:
  - a flat plate collector, consisting of a series of flat glass or plastic pipes with a black metal surface, used to absorb the heat
  - an evacuated tube collector, consisting of a set of tubes connected to a header, used to absorb the heat and transfer it to the water contained within it.
- A differential temperature controller (DTC) has sensors connected to the solar collector and the hot water cylinder. Its purpose is to monitor the temperature at both points of the system and switch the circulating pump on and off.
- A circulating pump is controlled by the DTC and circulates the system's heat transfer liquid around the solar hot water circuit.
- A hot water storage cylinder enables the transfer of heat from the solar collector circuit to the stored water for use at terminal fittings.
- An auxiliary heat source is used to provide hot water when solar energy is unavailable.

> **Industry tip**
>
> The best direction for a solar collector to face is due south.

▼ Table 5.6 Advantages and disadvantages of solar thermal systems

Advantages	• Reduced carbon emissions • Lower energy costs • Low maintenance • Improved energy performance certificate rating
Disadvantages	• Not compatible with all existing hot water systems • Less solar energy available in the winter months • High initial installation costs • Require an auxiliary heat source

> **Research**
>
> Research the different types of solar thermal systems in the UK and their uses.

Ground source heat pumps

Ground source heat pumps (GSHPs) extract heat from the ground using a system of underground pipes filled with a mixture of water and antifreeze known as glycol. A heat exchanger then transfers the heat to the pump itself, ready for distribution to:
- warm-air heating systems (ground to air pumps)
- underfloor heating systems, radiators and hot water cylinders (ground to water pumps).

Key components of a GSHP include:
- a heat collection ground loop, set out in one of two arrays:
  - a horizontal array has flattened, overlapping coils known as slinkies that are spread out in shallow trenches about 1.5–2 m deep
  - a vertical array has pipes installed in holes bored to a depth of 15–60 m, depending on soil conditions
- a circulating pump, which circulates a mixture of water and antifreeze around the pipes
- a heat pump, which extracts heat from the ground and transfers it for domestic use by means of an electric compressor
- a heating system, which is used to heat the property.

Onsite Construction T Level: Core

▲ Figure 5.24 Ground source heat pump

▲ Figure 5.25 Slinkies

> **Research**
>
> Research how a GSHP works in greater detail and produce a poster to showcase this to your classmates.

> **Industry tip**
>
> A 10 m slinky trench will produce around 1 kW of heating load.

> **Research**
>
> Select a GSHP manufacturer and research the installation requirements. Produce a pocket guide for a classmate.

> **Health and safety**
>
> Working in excavations can be dangerous. Ensure that adequate control measures have been implemented such as trench supports and correct access.

Air source heat pumps

Air source heat pumps (ASHPs) extract heat from outside air. Using a heat exchanger, compressor and expansion valve, they distribute the heat to:

- warm-air convectors (air to air pumps, usually used in commercial buildings as a reverse cycle heat pump that can be used for both heating and cooling)
- underfloor heating systems, radiators, hot water cylinders and even swimming pools (air to water pumps).

ASHPs can get heat from the air even when the temperature is as low as –15°C.

▲ Figure 5.26 Air source heat pump

Water source heat pumps (WSHPs)

Water source heat pumps (WSHPs) operate in a similar way to GSHPs, with the collector located in a body of water (such as a lake) to extract heat.

Heat pumps are not able to provide instant heat, so they work better when they run continuously. Start-stop operations will shorten their lifespan. A buffer tank is incorporated into the system so that when heat is not required, the heat pump can 'dump' the heat into the vessel and keep running. When heat is needed, it can be drawn from the buffer tank.

▲ Figure 5.27 Buffer tank

164

▼ Table 5.7 Advantages and disadvantages of heat pumps

Advantages	• Reduced carbon emissions • Typical efficiencies between 300 and 500 per cent • Low maintenance • Improved energy performance certificate rating
Disadvantages	• Not usually suitable for connection to existing heating systems using panel radiators • High initial installation costs • Air source installations can present a noise issue • Ground source installations require a large ground area or a borehole

Biomass

Biomass is plant or animal material that can be burned to create heat. Unlike fossil fuels, which have taken millions of years to form, biomass material has been sourced recently.

When burned, both fossil fuels and biomass produce carbon dioxide, which is a greenhouse gas linked to global warming. However, biomass material absorbs carbon dioxide as it grows, reducing the amount in the atmosphere, so burning biomass does not lead to a net increase in levels.

Biomass can be used to heat:
▶ individual rooms using stoves (these may be fitted with a back boiler to heat water too)
▶ whole properties, using boilers that supply central heating and hot water systems.

▼ Table 5.8 Advantages and disadvantages of biomass systems

Advantages	• Carbon-neutral and inexhaustible energy source • Lower fuel costs • Do not rely on building orientation or weather conditions to operate effectively
Disadvantages	• Require a suitable flue/chimney • High initial installation costs • Require a large space to store fuel • Not suitable for smaller properties

Micro-wind

Micro-wind turbines harness energy from the wind and turn it into electricity. The UK is an ideal location for the installation of wind turbines, as about 40 per cent of Europe's wind energy passes over the UK.

The working principles of a wind turbine are as follows:
▶ The wind passes over the rotor blades of a turbine, causing it to turn.
▶ The hub is connected by a low-speed shaft to a gearbox.
▶ The gearbox output is connected to a high-speed shaft that drives a generator which, in turn, produces electricity.

Turbines are available as either horizontal-axis wind turbines (HAWTs) or vertical-axis wind turbines (VAWTs).

▲ Figure 5.28 Horizontal-axis wind turbine

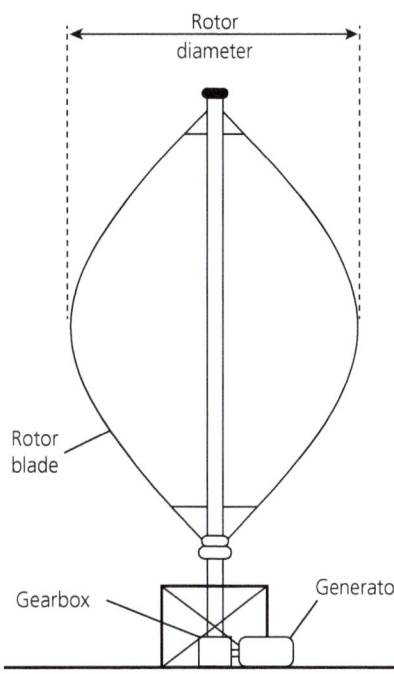

▲ Figure 5.29 Vertical-axis wind turbine

There are two types of micro-wind turbine suitable for domestic installation:
- pole-mounted (free-standing)
- building-mounted (fitted directly to the building, generally smaller than pole-mounted turbines).

As with solar PV, there are several connection options:
- off-grid systems charge batteries in order to store electricity for later use
- on-grid (grid-tied) systems are connected in parallel with the grid supply via an inverter.

▼ Table 5.9 Advantages and disadvantages of micro-wind systems

Advantages	• Good electricity output levels can be achieved • Zero-carbon technology • Effective where no mains electricity is available
Disadvantages	• Require a mounting site away from buildings and obstructions • High initial installation costs • Variable performance according to the availability of wind and location • Turbines can cause noise, vibration and **shadow flicker**

Key term

Shadow flicker: when the rotating blades of a wind turbine create moving shadows

Improve your English

Referring to your local planning requirements, research the location and building requirements for the installation of a micro-wind turbine. Draft an email to a potential customer who is considering one.

Micro-hydro

Micro-hydro is a type of hydroelectric power that harnesses kinetic energy as water moves from a higher to a lower level. The water passes across or through a turbine, which turns a generator and produces electricity.

As with the other micro-generation technologies, there are two possible system arrangements for micro-hydro: on-grid and off-grid systems.

The main components of the watercourse construction are:
- intake (the point where some of the river's water is diverted from the main stream)
- the canal (connecting the intake to the forebay)
- the forebay (a reservoir of water ensuring that the penstock is always pressurised and allowing surges in demand to be catered for)
- the penstock (pipework taking water from the forebay to the turbine)
- the powerhouse (the building housing the turbine and generator)
- the tailrace (the outlet that takes the water exiting the turbine and returns it to the mainstream of the river).

▲ Figure 5.30 Watercourse for a micro-hydro system

To ascertain the suitability of a water source for hydroelectric generation, it is necessary to consider the head (vertical height difference between the proposed inlet position and the proposed outlet) and the amount of water flowing through the watercourse.

Chapter 5 Sustainability principles

▼ Table 5.10 Advantages and disadvantages of micro-hydro systems

Advantages	• Excellent payback potential • Zero-carbon technology • Effective where no mains electricity is available
Disadvantages	• Require a watercourse with suitable head and flow • High initial installation costs • Require planning permission from the Environment Agency

> **Industry tip**
>
> A survey is required prior to the installation of any renewable technology.

Micro combined heat and power (micro CHP)

Micro combined heat and power (micro CHP) systems generate useable heat and electricity for properties at the same time. The fuel source is usually natural gas or liquefied petroleum gas (LPG), but it could also be biomass.

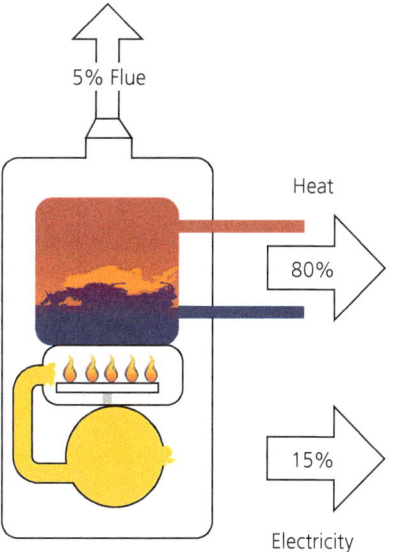

▲ Figure 5.31 Micro CHP boiler

The key components of a micro CHP boiler are the:
▶ engine burner (1)
▶ Stirling generator (2)
▶ supplementary burner (3)
▶ heat exchanger (4).

▲ Figure 5.32 Components of a micro CHP boiler

The boiler system works as follows:
▶ When there is a call for heat, the engine burner fires and starts the Stirling generator. The engine burner produces about 25 per cent of the full heat output. The burner preheats the heating-system return water before it passes to the main heat exchanger.
▶ Hot flue gases from the engine burner are passed across the heat exchanger to heat the heating-system water even more. If there is more demand than supply, the supplementary burner operates.
▶ The Stirling generator uses the expansion and contraction of internal gas (often helium) to operate the piston. When the engine burner fires, the gas expands, forcing the piston downwards.
▶ The return water from the heating system passes across the engine, causing the gas to contract. A spring mechanism in the engine returns the piston to the stop, and the process starts again.

Onsite Construction T Level: Core

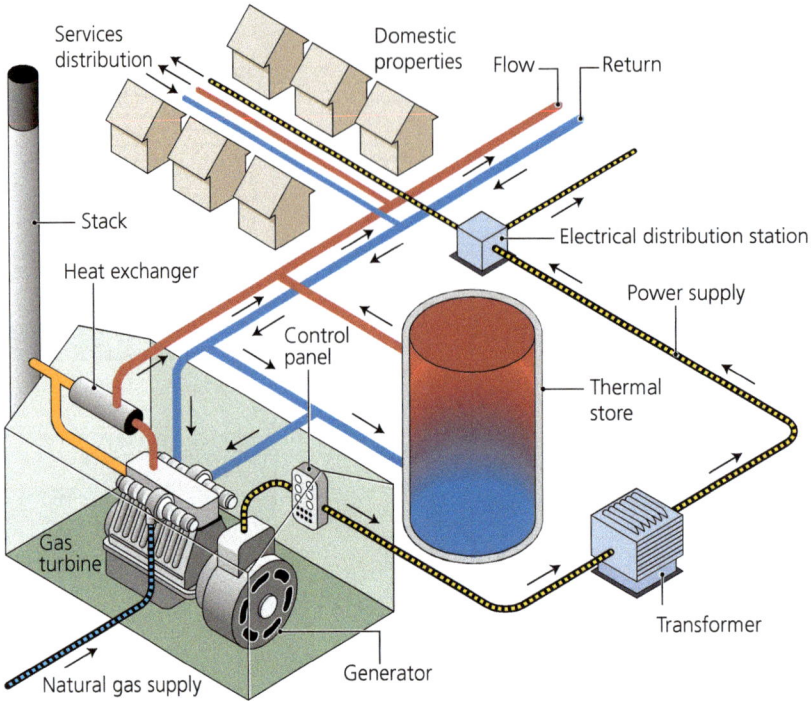

▲ Figure 5.33 Combined heat and power system

▼ Table 5.11 Advantages and disadvantages of micro CHP systems

Advantages	• Domestic units now similar in size to central-heating boilers • Produce free electricity while generating heat • Do not rely on building orientation or weather conditions to operate effectively
Disadvantages	• More expensive than central-heating boilers • Not suitable for properties with low heat demand • Limited electrical generation capacity • Low-carbon rather than zero-carbon

Combined heat and power (CHP)

Combined heat and power (CHP) systems (also known as cogeneration systems) use waste thermal energy from the generation of electricity for district or industrial heating purposes. The heat is produced centrally at one or more locations and supplied to an unrestricted number of residential and industrial customers, using either steam or high-temperature water.

▲ Figure 5.34 Rainwater harvesting

Chapter 5 Sustainability principles

▼ Table 5.12 Advantages and disadvantages of combined heat and power systems

Advantages	• Reduce energy costs • Reduce carbon emissions • Wide choice of fuels available, including coal, biogas, heating oil and bioethanol
Disadvantages	• Not suitable for all sites because of constant and consistent power demands • Can be expensive: initial cost is high due to the technology and payback periods can be very long

Heat recovery ventilation (HRV) systems

Heat recovery ventilation (HRV) systems reduce the heating and cooling demands of a building. Installed at the top of a building, they extract warm, stale air and pass it through a heat exchanger. The heat is then drawn to incoming fresh air, which is transferred back into the building. They recover between 60 and 95 per cent of the heat in exhaust air and markedly improve a building's energy efficiency.

▲ Figure 5.35 Heat recovery

> **Research**
>
> Research one of the following types of heat recovery system:
> ▶ counterflow heat exchangers
> ▶ crossflow heat exchangers
> ▶ rotary wheel heat exchangers.
>
> Explain this system to a classmate.

> **Test yourself**
>
> List three domestic uses for recycled water.

Electric vehicle charging points

New cars and vans powered wholly by petrol and diesel will be banned in the UK from 2030, resulting in a move towards electric vehicles. These are powered by rechargeable batteries and there are currently three types of public charging point:
▶ Rapid chargers have outputs of 43 kW and use a Type 2 connector. They are the fastest way to charge an electric vehicle and take between 30 and 80 minutes to reach 80 per cent charge.
▶ Fast chargers have outputs between 7 and 22 kW and use Type 1 or Type 2 sockets. It typically takes between three and four hours to charge a vehicle.
▶ Slow chargers have a three-pin plug, like a home charger. It can take up to ten hours to fully recharge a vehicle.

▲ Figure 5.36 Electric car charging point

> **Research**
>
> Visit the following website, which introduces electric vehicle charging: www.voltimum.co.uk/articles/installation-electric-vehicle-charging
>
> Find out:
> ▶ What type of circuit is required?
> ▶ What type of protection is required?
> ▶ How is electric shock prevented?

Onsite Construction T Level: Core

Project practice

Services plc is an established building contractor, with branches throughout the UK. Each branch works on different types of development, including new-build residential and commercial retail.

Services plc is preparing a tender submission for the building of a new school. You are part of the team working together to respond to the tender.
- Suggest a range of renewable technologies that could be included in the build.
- Summarise the purpose of each system, including advantages and disadvantages.
- For each system, explain the components and their purpose.
- Give examples of sustainable solutions that could be incorporated into the building, for example automatic lighting.
- Give examples of how waste will be managed on the construction project, including completion of a basic site waste management plan.

The information should be included in a project plan, which will be discussed in a group as part of the tender process.

Assessment practice

Short answer

1. State two common sustainability assessment methods.
2. List three control and monitoring systems for BSE systems.
3. List three sustainable solutions for construction.
4. Which document provides guidance for the installation of fixed building services in new and existing dwellings to help compliance with the energy-efficiency requirements of the Building Regulations?
5. State the three types of listed status for buildings in England and Wales.

Long answer

6. Explain how sustainability is achieved when planning and delivering a construction project.
7. Explain the term 'social sustainability'.
8. Explain what a passive house is.
9. Describe the operating principles of grey-water recycling.

Chapter 6 Measurement principles

Introduction

This chapter looks at the importance of accurate and appropriate measurement, identifying the standard units of measurement and appropriate measurement techniques used in construction.

By applying tried and tested construction measurement principles, it is possible to ensure accuracy both in calculating the quantities of materials needed for a project and in setting out and building a structure. We will also look at the possible consequences of inaccurate measurements for both client and contractor.

The chapter concludes by looking at scales and tolerances used in construction.

Learning outcomes

By the end of this chapter, you will understand:
1. accurate and appropriate measurement
2. standard units of measurement and measurement techniques
3. measurement standards, guidance and practice.

1 Accurate and appropriate measurement

The methods of measurement used for construction activities vary depending on the work task and the operational phase of the construction project.

Well before construction work starts on site, a range of personnel will have taken many accurate measurements in order to design, plan and prepare for a project. Measurements will continue to be taken with accuracy during the subsequent construction phase.

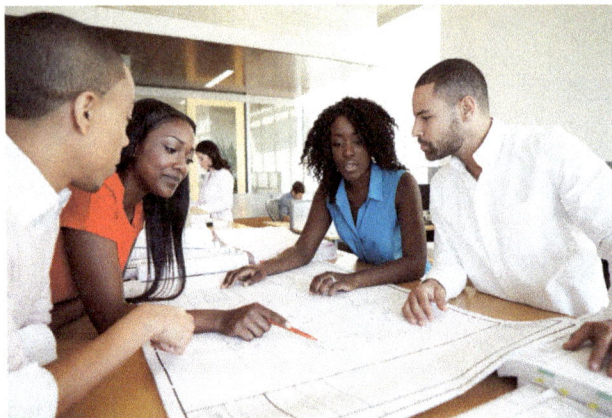

▲ Figure 6.1 Appropriate measurements are taken when planning a project

> **Improve your English**
>
> Search online for definitions of the word 'measurement'. Select three definitions that you think could be applied to construction activities and combine them into one definition in your own words.

1.1 The benefits of accurate measurement

Accurate measurement ensures a project meets the requirements of the client and runs smoothly through to completion for the contractor.

Benefits to the client

When a client commissions a project, they have certain expectations:
- The work should be completed within an agreed budget and timescale.
- The **design brief** should be wholly fulfilled.

> **Key term**
>
> **Design brief:** a working document which specifies what a client wants; it makes clear all the design requirements of a project

These outcomes can only be achieved if accurate and appropriate methods of measurement are used during all phases of construction.

During the planning stage, accurate measurement of quantities of materials and components required for the job will mean the right amounts are delivered to site. This avoids a shortage of materials, which can cause production delays, or a surplus of materials, which can cause waste. Resulting cost savings can be passed on to the client.

▲ Figure 6.2 Accurate calculation of materials for delivery to site avoids production delays and waste

> **Industry tip**
>
> Costing a project of any size is a balance between making an acceptable profit for the contractor and offering a fair price to the client. Accurate costings at the beginning of a project may allow a contractor to pass savings on to the client while still maintaining healthy profit margins.

Using established principles when calculating required quantities of materials allows for assessment of the time needed to use and install those materials, contributing to the creation of a realistic work programme. Completion targets can then be set to support efficiency and productivity and satisfy the client's requirements for timely handover of the building.

Chapter 6 Measurement principles

Creation of a work programme founded on accurate calculations will also allow workforce requirements for the project to be established, which will sustain productivity and avoid potential delays due to shortfalls in workforce numbers. The client benefits because the project is more likely to be delivered on time.

▲ Figure 6.3 Accurate calculation of quantities of materials contributes to creating a realistic work programme

> ### Research
>
> Once all the quantities of materials for a project have been accurately calculated, an evaluation of costs can be carried out, and the amount of time needed to complete specific tasks can be estimated.
>
> Examine the room you are in carefully. Make a list of all the materials you think are contained in the room, for example timber for the floor and cables for electrics. Then for each material in your list, assign a tradesperson or operative who might use it. The list will vary depending on the use of the room, so repeat the exercise with a partner in a second room and discuss your findings. How many materials have you identified?

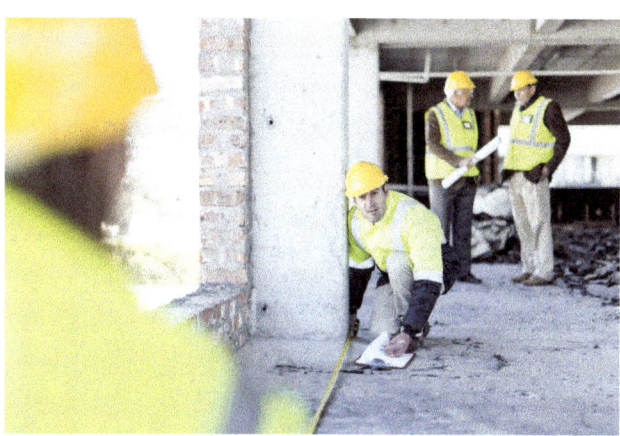

▲ Figure 6.4 A new building must be constructed using accurate measurements

In order to construct a building exactly as shown in the drawings and in accordance with specifications, it is vital to use accurate and appropriate methods of measurement.

If measurement principles are not applied correctly, it can cause adverse consequences:
- ▶ If a building that is not dimensionally accurate or not **built to square** is handed over to the client, significant problems can emerge when the installation of items such as floor tiles is attempted, or fittings and **modular components** such as kitchen units are positioned.
- ▶ Using unspecified materials or components may mean the building is not suitable for its intended purpose, for example reducing the width of timbers used in floor joists.

173

> **Key terms**
>
> **Built to square:** where all the corners of a square or rectangular building are accurately set at 90°
>
> **Modular components:** standardised units that fit together in a number of ways

> **Health and safety**
>
> If a building is not constructed in accordance with drawings and specifications because of carelessness in measurement, the structural integrity of the design could be compromised, leading to dangerous structural failure.

> **Test yourself**
>
> State two benefits to the client of accurate measurement.

Benefits to the contractor

Accurate measurement allows for accurate costing of a project, which in turn can:

- help a contractor to manage profitability and cash flow, to maintain business activity and ensure company growth
- allow a contractor to negotiate with suppliers for favourable prices on bulk materials and specified components, to improve company profit margins.

▲ Figure 6.5 Costing a project must be done carefully

Careful use of accurate and appropriate methods of measurement during the construction of a building prevents errors, which may be costly to remedy. It would be especially problematic if errors were discovered once the building was occupied after handover to the client. Avoiding errors means the job runs smoothly and is more likely to be completed on time.

If a company consistently completes projects on time and to a high standard, by accurately constructing buildings to specification, it will over time build up a good reputation and ensure continuity of work. Carelessness and inaccurate work can quickly destroy a hard-earned and valuable reputation.

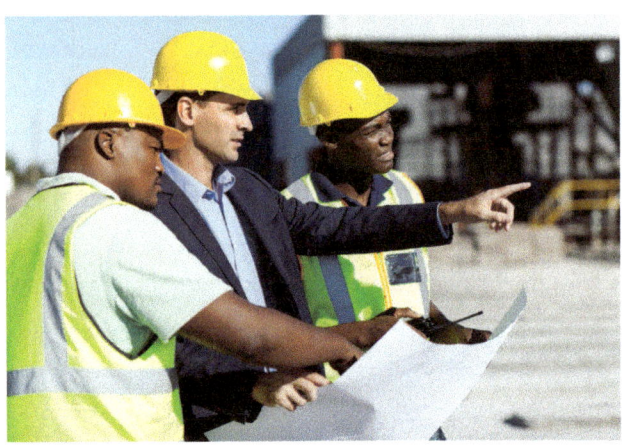

▲ Figure 6.6 Accurate work can result in a good company reputation

> **Test yourself**
>
> State one benefit to a contractor of accurate measurement.

1.2 Choosing appropriate methods of measurement

Throughout the stages of a construction project, different work activities require different methods of measurement. These should always be applied accurately.

Chapter 6 Measurement principles

▼ Table 6.1 Appropriate methods of measurement and the benefits of accuracy

Construction activity	Method of measurement	How the method is used	Benefits of accuracy
Site survey ▲ A survey of a new site must be undertaken	Calculation of area	A building must fit on a proposed site, with allowances for roadways, paths, parking and landscaped areas. There should be consideration for adequate space between the new building and existing structures.	Accurate calculation of the area of the site is needed to confirm that the proposed project is feasible. Early confirmation that the site is suitable allows for an appropriate design concept to be created.
Locating services on a site ▲ Existing services must be accurately located	Linear measurement	One of the first tasks on a new construction project is to locate the position of any existing services on the project site. Records regarding the position and route of services are used to accurately measure from stated positions on site. Once located, the services will be removed or rerouted.	Finding the precise location of services by accurate measurement is essential so that subsequent site work can proceed safely. Accidental damage to existing services could also cause delays and create added costs.
Groundworks including removal of topsoil and excavations ▲ Large volumes of material may need to be moved	Calculation of volume	Before a building can be constructed, the soft topsoil must be removed from the area on which the structure will be positioned. To measure the amount to be moved, the volume of material is calculated in cubic metres (m^3). Excavations for foundations and underground drainage systems also require calculation of the volume of material to be moved.	Moving large volumes of soil or other similar material involves the use of heavy machinery, which is expensive to hire or buy. Accurate measurement of the quantity of material to be moved means the number of machines and the time needed can be established and the cost confirmed.
Correctly locating the new structure on the plot ▲ Accurate measurements are needed when positioning a building on a plot	Linear measurement	Project drawings show where a building will be located on a site. Accurate measurement from given reference points is needed to position the building in the correct location.	If a structure is wrongly positioned, it may interfere with other elements of the development, or its position may contravene planning requirements. In extreme cases, a wrongly located building has had to be demolished and rebuilt in the correct position – at great expense.

175

▼ Table 6.1 Appropriate methods of measurement and the benefits of accuracy (continued)

Construction activity	Method of measurement	How the method is used	Benefits of accuracy
Building the new structure ▲ Accurate measurements are needed throughout the course of construction	Linear measurement, calculation of area, calculation of volume	Accurate linear measurement is needed to set out the outline of a building, the position of internal walls, doors and windows, vertical dimensions to establish floor and roof heights and much more. Accurate calculation of area will confirm quantities of materials required for activities such as plastering, bricklaying and roofing. Calculation of volume identifies requirements in cubic metres for materials such as concrete or screed required for solid floors.	Accurate measurement of the outline and features of a building is essential if the building is to be constructed exactly as designed. For example, wrongly positioned internal walls may interfere with other design elements. Measurements to calculate quantities of materials must be accurate to avoid shortfalls or excess materials, affecting productivity or adding to costs.

Key terms

Linear measurement: the distance between two given points along a line

Screed: a levelled layer of material (often sand and cement) applied to a floor or other surface

Health and safety

It is essential to accurately locate existing services in order to allow subsequent work to be conducted safely:
- Damage to water services can cause flooding and disrupt supply to others in the locality.
- Accidental contact with electricity services can cause death or injury to workers or damage to machinery.
- Gas services carry a risk of explosion if pipes are damaged or breached.

Test yourself

For what type of work task would linear measurement be used? Give two examples.

Improve your English

Write a clear explanation of the difference between area and volume.

1.3 Costing techniques for different projects

Construction projects vary greatly in size and complexity. A project may be relatively simple, such as building a boundary wall, or highly complex, such as building a multi-storey office block.

Calculating the cost of such a wide variety of project types requires the use of different techniques to arrive at figures that are as accurate as possible. Let us look at some frequently used costing techniques.

▲ Figure 6.7 A complex construction project in London

Job costing

This technique is usually applied to specific client requirements for a distinct project or 'job'. It involves analysing the job in detail, breaking down the costs of each element and tracking and recording them as the job proceeds.

The cost elements are broken down into labour, materials and overheads. This costing technique is sometimes confused with process costing, where the steps to complete a job are identified and an average cost is applied to each step based on past experience.

> **Industry tip**
>
> When a contractor prices a job, the calculated labour and materials are sometimes referred to as 'direct labour' and 'direct materials', since they are sourced and paid for by the contractor.

Job costing sometimes includes a 'cost reimbursement' condition, where the client agrees to meet the cost of job elements that cannot be accurately calculated at the start of a project, for example if the design choice of certain items such as windows has not been finalised.

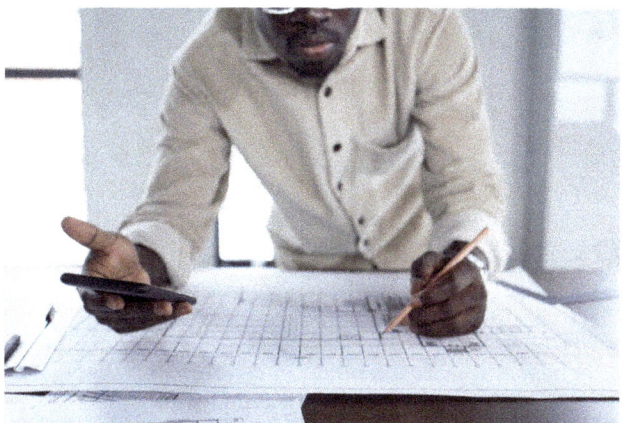

▲ Figure 6.8 Various techniques can be used to cost a project

Batch costing

This can be viewed as an extension of job costing, since a batch (like a job) is broken down into component materials to be costed individually. For example, concrete is manufactured from sand, aggregate and cement. The quantities of the three component materials can be costed separately and those costs totalled as a batch, usually expressed in cubic metres (m^3) in the case of concrete.

Batch costing is often used where the manufacturing process takes the form of a continuous flow of mass-produced construction materials, such as plasterboards or bricks.

Activity costing

This technique assigns costs to specific activities rather than whole jobs or processes. Identifying the many activities required to produce a building provides costing data that can feed into the overall costing process.

For example, the activity of building a brick wall can be costed by using specific information:
- ▶ The dimensions of the wall allow the number of bricks required to be accurately calculated.
- ▶ The quantity of sand and cement for mortar can be calculated.
- ▶ The hourly rate of pay for a bricklayer can be precisely set.
- ▶ The number of bricks laid per hour can be established from past work experience.

By identifying quantifiable activities, the associated costs for materials, labour, equipment and overheads can inform the costing process – both for the current project and also for future projects with similar characteristics.

Life cycle cost analysis

This technique is used to estimate the overall costs of a building throughout its entire life cycle. It allows for comparison of alternative building concepts at the design stage, so that an evaluation can be made of which design is most economically and environmentally beneficial over time. The analysis includes the cost of everything throughout the life cycle of the building, including:
- ▶ land acquisition
- ▶ construction
- ▶ energy
- ▶ water
- ▶ operation
- ▶ maintenance and repair
- ▶ replacement.

Some professionals consider life cycle analysis to be little more than a 'best guess' of costings for a building that could potentially be in use for 100 years or more. Over such an extended time period, circumstances may change unpredictably.

However, with increasingly demanding environmental regulations impacting on building design and operation, refinement of this technique could provide an important tool for costing new buildings that are designed to be more environmentally friendly.

> **Research**
>
> Search online for the 'Designing Buildings Wiki' website. Enter 'costing' into the search bar and research the differences between life cycle costing and whole-life costing.

> **Test yourself**
>
> Look at the costing techniques discussed in this section. Which technique would you use to cost the replacement of slates on a roof?

▲ Figure 6.9 A slate roof covering

> **Improve your English**
>
> There are other types of costing technique that mainly apply to manufacturing processes but can have an application in construction. Find out about target costing and write down your opinion on whether this technique could be used in a construction setting.

2 Standard units of measurement and measurement techniques

The UK construction industry uses **metric units** as standard. These are easy to use because they are based on a **decimal system** which provides a consistent relationship between different measurement units. The value of a measured number can be changed simply by moving the decimal point (see Table 6.2 later in this chapter).

▲ Figure 6.10 Metric units of measurement are easy to work with

> **Industry tip**
>
> You may hear **imperial units of measurement** being discussed on site. Material sizes are often referred to in imperial units, even though they are now sold in metric units. An example of this is 8 ft × 4 ft sheets of plywood or plasterboard, where the correct size is 2400 mm × 1200 mm.

> **Key terms**
>
> **Metric units:** decimal units of measurement based on the metre and the kilogram
>
> **Decimal system:** a number system based on the number ten, tenth parts and powers of ten
>
> **Imperial units of measurement:** units of measurement of the British Imperial System, used until 1965 when the metric system was adopted

2.1 Units of measurement

Some standard metric units of measurement and their abbreviations are as follows:
▶ Length
  – millimetres (mm)
  – centimetres (cm)
  – metres (m)
  – kilometres (km)

▶ Weight or mass
- gram (g)
- kilogram (kg)
- tonne (t)
▶ Liquid
- millilitre (ml)
- litre (L or l).

Table 6.2 shows the relationships between metric units of measurement and gives examples of how they are applied.

▼ Table 6.2 Standard metric units of measurement

Measurement	Relationship between units	Example
Length	10 mm = 1 cm	1 mm × 10 = 1 cm
	100 cm = 1 m	1 cm × 100 = 1 m
	1000 m = 1 km	1 m × 1000 = 1 km
Moving the decimal point changes the value: • 6250 mm can be shown as 6.250 m • 6250 m can be shown as 6.250 km.		
Weight or mass	1000 g = 1 kg	1 g × 1000 = 1 kg
	1000 kg = 1 t	1 kg × 1000 = 1 t
Liquid	1000 ml = 1 l	1 ml × 1000 = 1 l

Research

Research the units of measurement for length and weight used in the Imperial system. Create a table showing how they relate to each other, for example how many ounces in a pound. Can you see how much easier it is to use the metric system?

Improve your maths

Convert 1.26 m into millimetres. Work with someone else to think up some other examples to convert.

Units of area and volume are derived from units of length:
▶ area is measured in square units, for example square metres (m^2)
▶ volume is measured in cubic units, for example cubic metres (m^3).

Every building must be designed to withstand the loads and pressures created by:
▶ the weight of the materials it is constructed from
▶ the weight of equipment and occupants within it
▶ external forces, such as wind or the weight of snow.

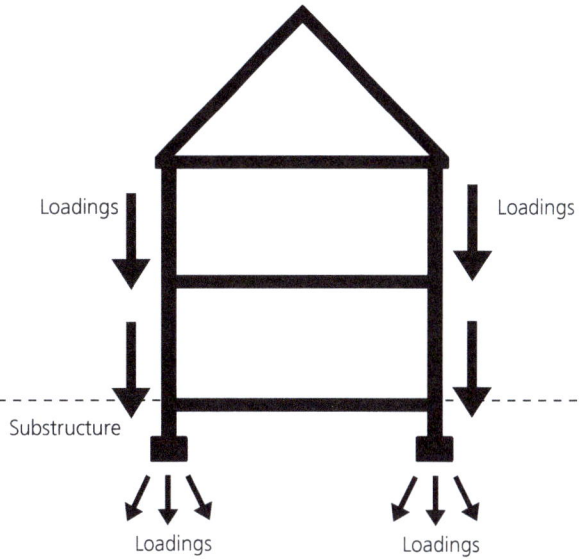

▲ Figure 6.11 The measurement unit for loadings is the newton (N)

These forces are measured in newtons (N). For example, the force exerted by pressure on concrete can be measured as newtons per square millimetre (N/mm^2). This measurement is used to classify the amount of crushing force that concrete can withstand.

Industry tip

Structural engineers must have a clear understanding of the definition of the newton: 'The force which gives a mass of 1 kilogram an acceleration of 1 metre per second, per second (1 kg m/s^2)'. This enables them to produce accurate and reliable structural calculations for a building.

Many of these units of measurement we take for granted, using them every day without thinking too much about how they fit together or have an influence in our lives.

In construction, these standard units of measurement are used in specific ways using appropriate equipment. Let us look now at some measurement techniques.

2.2 Measurement techniques

Measuring equipment used in construction work can be relatively simple or technically complex.

Simple equipment such as a tape measure can give accurate results when used carefully to measure:
▶ height
▶ length
▶ distance.

Onsite Construction T Level: Core

In any structure, it is important to measure heights accurately, for example of floors, ceilings, door and window openings, and roofs. Any inaccuracies could have serious consequences:

▶ If the height of a ceiling is not measured accurately during the building process, it may clash with the height measurement of the tops of doors or windows in the room.

▶ If the height of floors is not measured accurately during installation, a staircase will be placed at the wrong angle of slope, and it may not sit properly where the top of the stairs meets the upper floor.

▲ Figure 6.12 Getting a height measurement wrong could have serious consequences

Accurate measurement of length is also important if the design is to function as intended. Accuracy is vital when measuring the outline of the building, the position of internal walls and the position of doors and windows along an *elevation* of a building.

> **Industry tip**
>
> The length of a building is often expressed as an 'overall measurement' supported by intermediate dimensions referred to as 'running measurements', which show the position of internal walls, etc. The running measurements added together should equal the overall measurement.

> **Test yourself**
>
> Give two reasons why accurate measurement of height is important.

> **Key term**
>
> *Elevation:* a view of the front, back or sides of a building

Technically complex equipment can be used to measure height, length or distance to high levels of accuracy. Laser instruments can be used to measure differences in height and 3D laser scanning can be used to give accurate data about the height and length of existing structures during a survey.

▲ Figure 6.13 A 3D laser scanner

Traditionally, an instrument called a theodolite has been used to measure distance and height, using mathematical methods of trigonometry and triangulation to establish accurate angles and measurement details for setting out new buildings or surveying existing structures.

> **Improve your English**
>
> Search online for 'how to use a theodolite'. Write a short report on the steps for setting up the instrument.

The optical theodolite has evolved into an instrument known as a total station. This modern surveying instrument combines an electronic version of the theodolite (which can record location data using a global positioning system GPS) with an electronic distance meter to speed up the measurement process. Some versions can even be operated remotely by a single operator and the gathered data can be streamed to an office location for reliable storage.

Chapter 6 Measurement principles

▲ Figure 6.14 A total station surveying instrument

> **Test yourself**
>
> What is the name of the instrument that has been developed from the theodolite?

To ensure efficiency and productivity throughout the construction process, accurate measurements are important for calculating the quantities of materials required. Calculations of area and volume are used to establish the amount of plasterboard needed for walls and ceilings, bricks and blocks needed for walls, and concrete needed for foundations and solid floors. **Ratio** is often used to establish correct quantities of different components mixed together to produce materials such as concrete or mortar.

> **Industry tip**
>
> Although accurate measurement is important when calculating quantities of materials, sometimes to speed up a specific job it might be acceptable to use approximation. For example, you may see someone pacing out the length of an excavated trench to establish an approximate measurement in metres, to quickly calculate the volume of concrete required for a foundation.
>
> Note: Although 'approximation' can be viewed as an estimate of quantity, this is not the same as the estimate produced during the costing process for a project. See Chapter 4 for more on estimates.

> **Key term**
>
> *Ratio:* the amount or proportion of one thing compared to another

Accurate measurement is also important when establishing the weight or mass of materials and the capability of lifting equipment. There is usually no means of weighing things on site, but the weight of items should be labelled on packaging, and equipment for lifting and transporting materials should have a safe working load (SWL) clearly and permanently indicated on it.

> **Industry tip**
>
> While scales and weighing equipment are not routinely installed on a site, if large volumes of material such as topsoil are to be removed it may be considered economical to install a weighbridge to monitor and record the weight of materials being transported off site.

Legal requirements to make new structures more environmentally friendly have led to the use of specialised measurement techniques for calculating heat transfer and energy use. Energy used in the construction and operation of a building produces carbon. This combines with oxygen in the atmosphere to produce carbon dioxide (CO_2) gas, which contributes to climate change.

To calculate CO_2 emissions, the thermal transmittance of the building's structure must be established. This is known as the U-value.

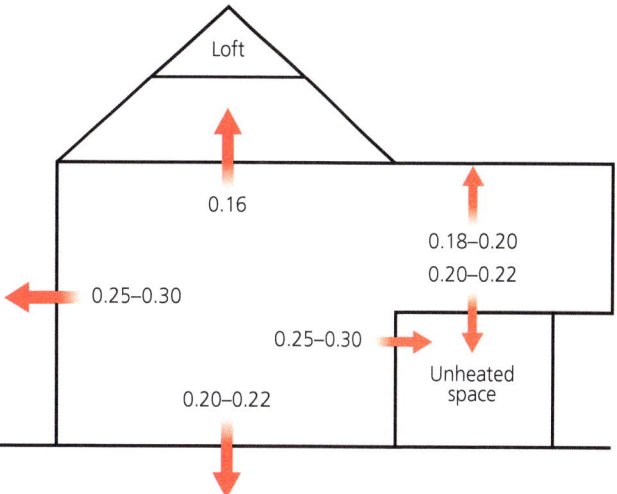

▲ Figure 6.15 Heat loss expressed as U-values in a typical house

To control heat transfer in a building, insulation must be used.

181

▲ Figure 6.16 Insulation must be installed to control heat transfer

▲ Figure 6.17 Construction drawings

Industry tip

It is pointless calculating the required level of insulation if operatives do not use appropriate techniques when installing insulation materials. Measures to limit heat transfer are only effective when insulation materials are properly installed.

Key terms

Conventions: agreed, consistent standards and rules

Scale: when accurate sizes of an object are reduced or enlarged by a stated amount

Working drawings, are commonly used to 'take off' information for calculating quantities of materials and establishing timescales for work programmes. Let us look at some points regarding standards and practice when using drawings.

Improve your English

Describe the difference between U-values and R-values.

3 Measurement standards, guidance and practice

Construction drawings are produced to a set of **conventions**, which include particular units of measurement, views and sheet sizes. Conventions in measurement standards include the consistent use and application of **scale** when producing drawings as information sources.

An architectural technician or a draughtsperson produces drawings of a building to scale. This means that large structures can be represented with accurate proportions on a document that is much smaller, so that it is easy to work with. It would obviously be difficult to work with a drawing of a house that was produced at full size.

Scale is shown using a ratio, such as 1 to 10. This would usually be written in the form '1:10'. Using this scale, a drawing of a feature that is 1 m (or 1000 mm) long in real life would be drawn 100 mm long on the paper. This is because 100 mm is a tenth of 1000 mm.

In some cases, more than one scale will be used on the same drawing, with different scales being used for different views, perhaps to allow for greater detail to be shown or a wider view to be given in context.

Test yourself

What name is given to a set of agreed, consistent standards and rules?

Chapter 6 Measurement principles

▲ Figure 6.18 A drawing showing a scale of 1:10

3.1 Common scales

Table 6.3 shows the scales commonly used for different types of drawing.

▼ Table 6.3 Types of drawing and common scales

Type of drawing	Description	Scales commonly used
Detail drawing	This shows accurate, large-scale details of the construction of a particular item, such as a timber-frame structural corner or the makeup of a suspended concrete floor.	1:1, 1:2 and 1:5
Floor plan	This shows the layout of internal walls, doors and stairs in a building. In a dwelling, it also shows the arrangement of special-use rooms, such as bathrooms and kitchens.	1:50 and 1:100
Elevation drawing	This shows the external appearance of each face of the building, with features such as slope of the land, doors, windows and the roof arrangement.	1:50 and 1:100
Sectional drawing	This is a slice or cut through of a structure to give a clear view of details that would otherwise be hidden. For example, on a working drawing for a house, a sectional drawing could allow us to clearly see the layout of the stairs within the building.	1:50 and 1:100
General arrangement drawing	Sometimes referred to as a location drawing, this can be used to show a single building element and what it should contain. It can also be used to show the main elements of a structure, such as the external walls, internal or partition walls, floor details and stairs.	1:50, 1:100 and 1:200, depending on the level of detail required
Site plan	This shows the proposed development in relation to the property boundary. It also shows the positions of drainage and other services and access roads and drives. It may show the position of trees and shrubs if they are required as part of the planning details.	1:200 or 1:500
Block plan	This shows the proposed development in relation to surrounding properties. It must be based on an up-to-date map and drawn at an identified standard metric scale. It usually shows individual plots and road layouts on the site as a simple outline with few dimensions.	1:1250 or 1:2500

Onsite Construction T Level: Core

> **Improve your maths**
>
> Show your understanding of how scale works by filling in the gaps in this table:
>
> | Scale size on drawing | Scale | Actual size (mm) |
> |---|---|---|
> | 10 mm | 1:10 | 100 mm |
> | 25 mm | 1:20 | |
> | | 1:50 | 300 mm |
> | 50 mm | 1:200 | |

In order to be used reliably during both the planning and construction stages of a project, drawings should comply with British Standards.

BS EN ISO 19650 sets out a comprehensive code of practice for managing the production, distribution and quality of construction information, including drawings. It allows for the co-ordination of a range of essential construction data and information.

When using a drawing to build a structure or install components, written dimensions and measurements on the drawing should be used to ensure the work is completed exactly as designed. Using a scale rule on a drawing to establish dimensions for a work task can lead to errors. However, a scale rule can be used to 'take off' measurements from a drawing when calculating quantities, if written dimensions are not shown.

▲ Figure 6.19 A scale rule

Costing a project from drawings must be done in accordance with 'rules of measurement'. The rules most commonly used are the New Rules of Measurement (NRM), produced by the Royal Institution of Chartered Surveyors (RICS). This is a standard set of measurement rules and essential guidance for the cost management of construction projects and maintenance works. Their use helps to prevent disputes and maintain efficiency and productivity.

3.2 Tolerances

As stated a number of times in this chapter, accuracy when using and interpreting information sources is essential if a building is to be completed exactly as designed. However, a building will often be constructed in conditions that are not perfect. In addition, many materials have characteristics that make it difficult to produce a flawless result.

For example, a concrete foundation or solid floor slab may easily be shown on a drawing with straight lines to indicate its depth and position, but the properties of the material do not allow it to be installed with a perfectly straight profile or edges.

▲ Figure 6.20 It is difficult to install concrete with a perfectly straight profile or edges

Allowances can be made for variations in the installation of certain materials and components in the form of **tolerances**, expressed as plus or minus (±) a stated amount. These tolerances will vary, depending on the nature of the material. For example, a double-glazed unit has a very limited range of tolerance – if it is larger than the window frame to be glazed by only 1 mm, it simply will not fit.

> **Key term**
>
> *Tolerances:* allowable variations between specified measurements and actual measurements

By contrast, when setting out a brick wall, it may be necessary to adjust the size of the mortar joints to fit whole brick sizes into the overall length of the wall, to avoid cutting bricks. In this case, a tolerance of ± 3 mm is allowable. This means that a 10 mm mortar joint can be enlarged by 3 mm to 13 mm or reduced by 3 mm to 7 mm to accommodate whatever is needed.

Chapter 6 Measurement principles

Tolerances may also be used to allow for acceptable variations in the strength of materials, temperature ranges in which materials can be used and many other situations.

> **Test yourself**
>
> Give one reason why tolerances are necessary when using certain materials.

> **Industry tip**
>
> Manufacturers often provide guidance on tolerances for the installation of components and materials they make. The Building Regulations and British Standards are another source of information on tolerances that are acceptable for a range of materials and associated construction activities.

Exceeding allowable tolerances for a given construction task will mean that the work does not meet the specification. The function of the structure could be compromised and other elements of the building may not be able to be installed properly.

Assessment practice

Short answer

1. What units are used when calculating the quantity of topsoil to be removed from a site?
2. What two expectations for a project must a contractor fulfil to satisfy the client?
3. Write a definition of the decimal measuring system.
4. Name two work activities when constructing a building that use linear measurements.
5. What does NRM stand for in relation to costing a project?

Long answer

6. Explain how accurately calculating quantities of materials and components assists in the creation of a realistic work programme.
7. Explain the reason for the use of tolerances and the implications of not meeting them.
8. Explain the possible consequences to the contractor and the client if a building is constructed out of square due to carelessness in measurement.
9. Describe how the installation of a staircase could be affected if floor levels in a building are not accurately set at the designed height measurement.
10. Explain the consequences if operatives fail to correctly install insulation.

Project practice

A new college is to be constructed on a site located on a hillside. The site has an old farmhouse and a number of farm buildings on it, which have to be demolished and removed.

Adequate parking must be provided for staff and students, and a sports and recreation area is included in the plans.

Working in a small group, list the sequence of work from start to finish and select a suitable costing technique for each stage of the work.

You will need to consider the work needed to:
- carry out demolition
- manage the flow of rainwater down sloping ground
- move soil and shape the landscape
- construct the parking area, college building and recreation facilities.

Chapter 7 Building technology principles

Introduction

The construction industry is evolving rapidly, with innovations in building technology to meet the demand for more affordable housing while reducing the impact of construction work and manufacturing on the environment.

In this chapter, we will look at the history of domestic buildings, forms of construction and different building materials. We will also compare traditional onsite construction methods with modern off-site manufacturing, and look at sources of building information and guidance.

Learning outcomes

By the end of this chapter, you will understand:
1. construction methods
2. forms of construction
3. key content and required notifications of UK building regulations and approved documents
4. building standards
5. regulatory bodies and guidance on technical safety and legislative aspects
6. manufacturers' instructions
7. building structure and fabric.

Chapter 7 Building technology principles

1 Construction methods

1.1 Onsite construction

In the early years of construction, all buildings in the UK were made on site, with materials often sourced locally from the surrounding land. This led to the development of low-rise dwellings with walls made from heavy timbers, **cob** or natural stone bonded together with lime mortar, or a combination of these materials. The walls were solid and often built directly on the ground with little or no foundations to support them. Roofs were often pitched with timber rafters and covered with slate or thatch to protect the occupants from the elements.

Over the years, construction methods have evolved, due to advances in construction design and technology, alongside the ability to use different building materials which are manufactured and transported to site from further away.

Table 7.1 illustrates how traditional and modern construction methods have transformed the way we build today.

> **Key terms**
>
> **Cob:** a blend of subsoil (clay or earth), sand and straw mixed with water to make an organic material historically used to construct walls for homes and agricultural buildings in Devon, Cornwall and Wales
>
> **Wattle and daub:** a building method whereby a woven matrix of wooden strips (wattle) is covered with a mud-based daub; this technique was used for the construction of non-load-bearing walls and infill panels in buildings for thousands of years, but since the development of plasterboard it has become obsolete and is only ever used in restoration of historical or eco buildings
>
> **Mullioned windows:** windows divided by vertical members, usually made of stone or wood
>
> **Pebbledash:** a coarse surface finish for external walls consisting of small shells, gravel or pebbles applied to cement render while it is still wet

▼ Table 7.1 Traditional and modern construction methods

Period	Build method
Tudor (1485–1603)	• Thatched or slate roofs • Exposed timber frame with **wattle and daub** walls
Stuart (1603–1714)	• Single-skin stone and brick walls, sometimes rendered • Granite-framed and **mullioned windows** with coloured glass • Slate floors
Queen Anne (1702–14)	• Terracotta tiles and panels • Single-skin red-brick walls • Windows with glazing bars
Georgian (1714–1837)	• Single-skin stone or handmade brick walls • Buildings constructed either directly off firm ground, or on stone, brick or concrete foundations • Vertical sliding wooden sash windows • Architecture, symmetrical facades
Victorian (1837–1901)	• Single-skin brick solid walls, built on concrete, hydraulic lime or stepped brick foundations • In some areas of the country, cavity walls were used for the first time in late Victorian dwellings • First use of a damp-proof course (DPC), made of materials such as lead or copper • Pointed arches constructed over doors and wooden or metal-framed windows (Note: most of the population still lived in small houses or cottages)
Edwardian (1901–10)	• External cavity walls used in some buildings • Introduction of electric lighting • Some timber framing • Hanging wall tiles • **Pebbledash** walls • Timber porches and balconies

▼ Table 7.1 Traditional and modern construction methods (continued)

Period	Build method
Addison Act (1919)	• Introduction of affordable council housing after the First World War • Commonly three to four bedrooms, with indoor toilets, baths and hot running water • Often constructed with brick, block or concrete walls • Window openings designed for lots of natural light
Semi-detached (1930s)	• Hipped roofs and pebbledash walls • Many homes still constructed with single-skin brick walls bonded with lime mortar • Recessed porches • Mock timber framing • Timber bay windows
Art Deco (1920s–40s)	• Flat roofs • Plain white walls • Metal-framed windows (known as 'Crittall' windows) • Open interiors with Egyptian influences used in the design • Other forms of foundations introduced, including piles and raft
Prefabs (1940s) after the Second World War	• Mass produced in factories and assembled on site • Precast concrete columns and metal tubing • Small windows
Terraced (1960s–70s)	• Integral garages • Clad with hanging wall tiles or weatherboarding • Single-glazed aluminium windows and doors • Polythene damp-proof membranes (DPM) used as a barrier at ground level for the first time to prevent moisture entering the building
New build (1990s)	• uPVC double glazing • Insulated roofs, floors and cavity walls with a rendered finish • Period features
New build (current day)	• Energy efficient, eco-friendly building materials • Open plan • Good use of glass for solar gain

▲ Figure 7.1 Victorian house

▲ Figure 7.2 Art Deco house

Chapter 7 Building technology principles

▲ Figure 7.3 A dwelling constructed from cob

▲ Figure 7.4 Wattle and daub

There are many different forms of construction used to build new dwellings.

Onsite construction using traditional methods such as brick and block cavity walls or timber frame can be slow, and progress can often be delayed by adverse weather.

In the pursuit of building sustainable homes more quickly, with less of an impact on the environment, alternative methods of construction are sometimes used. These might include the use of straw-bale walls or, less conventionally, shipping containers.

Although the idea of building with straw bales ticks a lot of sustainable-construction boxes, this method is unlikely to be used for large housing developments because of its limitations, for example it can only be used for low-rise buildings, as well as having a limited lifespan and often rustic appearance. On the other hand, the idea of using shipping containers to build is not as extreme as it may first appear, especially if they are converted into habitable spaces in factories and delivered to site in modular form.

▲ Figure 7.5 Modular container house

Besides the construction of most new dwellings on site, the services and fabric of existing buildings still need to be maintained, and at some point upgraded, while in situ. Cosmetic and structural renovation and refurbishment of buildings are also jobs that are often only able to be completed on site because of the nature of the work.

The use of 3D printing and robotics as methods of onsite construction are covered later in this chapter.

1.2 Modular construction

Modular construction combines pre-engineered units (or modules) to form major elements of a building. These modules are manufactured in factories and then transported to site where they are assembled.

After the Second World War, there was a housing shortage in the UK. In order to provide homes quickly, many temporary buildings known as **prefabs** were made using precast concrete sections. However, modern prefabricated buildings are far superior to historical examples, and modular construction is now used to build permanent structures that are comfortable for the occupants, attractive to look at and sustainable.

> **Key term**
>
> *Prefabs:* buildings manufactured using factory-made components or units that are transported and assembled on site

Approximately 13 per cent of all new houses built in the UK use modular construction, as it offers many benefits:
- Automated systems improve accuracy and quality control.
- Large numbers of components can be manufactured efficiently with reduced waste.

- By carrying out a large proportion of the construction work in a controlled factory environment, the delays associated with adverse weather conditions on site are avoided.
- Up to 95 per cent of a building can be constructed in a factory, therefore reducing the number of trades needed on site during the assembly and final finishing of the project.

Due to the cost savings from mass production in factories, standard houses using modular construction can be produced more cheaply and efficiently, with better quality control, than using traditional bricks and mortar. It is also possible to design and construct unique houses in modular form, although these are usually more expensive, depending on the specification.

Most modular houses are constructed using structural insulated panels (SIPs) or timber frame, although heavier and less-sustainable precast concrete can also be used. SIPs are made using two layers of oriented strand board (OSB) bonded to each side of a polyurethane insulation core to provide a strong, rigid and highly insulated panel.

There are several types of modular building, each providing a different level of finish:
- turnkey (pre-assembled) – fully finished with plastered walls, fixtures and fittings, such as bathrooms and kitchens
- shell only (pre-assembled) – walls, floors and roof assembled, with no fixtures and fittings or internal finishes
- panel system (flat pack) – the basic elements of the structure provided, ready to be assembled on site.

Further information on modular construction can be found in Chapter 3.

▲ Figure 7.6 Prefab dwelling

Test yourself

Explain the advantages of using modular construction.

1.3 First fix

The construction of a new building is broadly divided into three areas of work, known as structural, first fix and second fix. As the name suggests, structural work includes the main load-bearing elements of a building, such as the foundations, basement (where there is one), walls, precast concrete staircases, floors and roof.

First-fix building work has a different meaning for each trade, although it is generally considered to be the phase of work completed after the structure has been erected and before plastering commences.

Table 7.2 outlines some of the first-fix tasks completed by various trades.

▼ Table 7.2 Work undertaken at first fix

Construction operative/trade	Range of work undertaken
Plumber	Installation of pipes through the building for hot- and cold-water systems
Electrician	Installation of cables and back boxes for switches, electrical sockets, alarms, smoke detectors, CCTV, data-networking cables and television
Gas fitter/heating and ventilation fitter	Installation of pipes for gas distribution for boilers, underfloor heating, and ducting for heating, ventilation and air-conditioning systems
Carpenter	Installation of flooring, door frames, linings, staircases, stud partition walls and windows **Encasing services**
Dryliner/plasterer	**Drylining** ceilings and walls in preparation for plastering, or taping and filling the joints between the plasterboard sheets

Chapter 7 Building technology principles

> **Key terms**
>
> **Encasing services:** creating a framework of materials to cover pipes, cables and other services in a building; also referred to as boxing-in
>
> **Drylining:** fixing plasterboard to ceiling and wall surfaces, such as metal suspended ceilings, timber and metal stud partition walls, and masonry walls

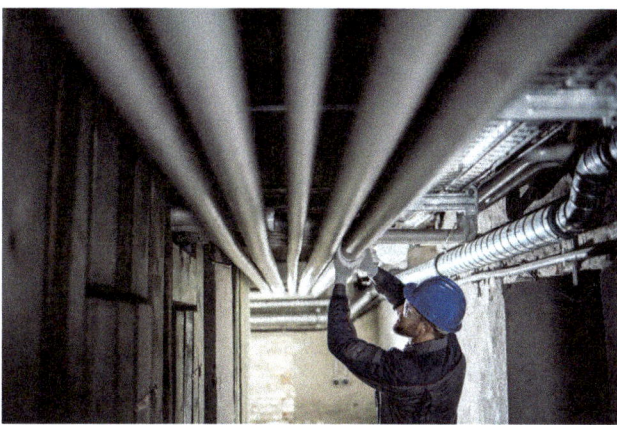

▲ Figure 7.7 First-fix plumbing

The first-fix stage of construction needs to be well planned and executed, especially when installing pipes, cables and ducting. If mistakes made at this stage are identified later in the building process, surface finishes on the ceilings or walls may need to be damaged to allow access to the hidden services.

1.4 Second fix

Once the various trades have completed their first-fix work and the plastering/drylining is finished, second-fix work can commence. Second fixing often involves the installation of items and equipment that could have easily been damaged or affected by earlier stages of construction work.

Table 7.3 outlines some of the second-fix tasks completed by various trades.

Items installed during second fix are still vulnerable to damage from other construction workers and equipment on site, for example tilers, painters and decorators; therefore where necessary adequate measures should be taken to protect them until completion of the project.

Finishing trades such as tilers and decorators are usually scheduled to work on site at the end of the second-fix stage. It is important that they co-ordinate the completion of their work with the other trades, to avoid getting in each other's way and to prevent unnecessary damage to their work.

Further information on first-fix and second-fix tasks undertaken by different trades can be found in Chapter 4.

▼ Table 7.3 Work undertaken at second fix

Construction operative/trade	Range of work undertaken
Plumber	Installation of fixtures and fittings, for example sinks, toilets, showers, baths, towel rails and radiators
Electrician	Installation of electrical fixtures and fittings, for example switches, sockets, fuse boards (power distribution units and circuit protection) and light fittings Testing and commissioning of new electrical systems
Gas fitter/heating and ventilation fitter	Installation, testing and commissioning of gas boilers and other heating, air-conditioning and ventilation systems
Carpenter and joiner	Fixing skirting, architraves, doors, kitchens and fitted wardrobes/cupboards

1.5 Self-driving vehicles

Fully **autonomous vehicles** can be found on construction sites today. Due to the unpredictable nature of the environment, for example the movement of workers and plant, current deployments of autonomous vehicles typically have the machines operating within a separate, fenced-off area away from people. Semi-autonomous vehicles, which depend on a small amount of human intervention, are already being used in some areas of construction in the UK, for example earth-moving plant used to prepare sites ahead of building work starting.

> **Key term**
>
> **Autonomous vehicles:** vehicles able to operate without human intervention

In other countries, semi-autonomous and fully autonomous excavators, trucks, diggers and cranes are also being used successfully for simple but time-consuming and otherwise dangerous construction tasks, such as digging, trenching, drilling and excavating the ground on site.

Using autonomous vehicles for construction projects can reduce costs by improving productivity, machine utilisation and site progress, because they are able to operate 24 hours a day without a driver. They also produce fewer carbon emissions than vehicles driven by humans, due to less erratic movements and more efficient use.

▲ Figure 7.8 Self-driving (autonomous) vehicle

1.6 Computer-controlled manufacturing robots

While manual labour will likely always be a huge factor in building projects, the development of robots to automate construction tasks has led to improvements in productivity. They have been designed to undertake repetitive and labour-intensive tasks, such as demolition, laying bricks and blocks and plastering.

Advantages of using robots include:
- protecting workers from hazards and reducing workplace injuries
- reducing operating costs and waste
- increasing production rates and efficiency
- ensuring accuracy.

▲ Figure 7.9 Computer-controlled manufacturing robot

1.7 Large-scale 3D printers

Large-scale 3D printers can produce perfectly formed walls for large buildings on demand. They function by controlling a mobile robotic arm on site to 'print' cement-based mortar walls, layer by layer, until the full height of the walls is reached.

The advantages of using 3D printers on site are similar to those for using robots; however this method also produces seamless walls.

▲ Figure 7.10 Large-scale 3D printer

1.8 Drones

Recent innovations in drone technology have provided the construction industry with a resource that can be used to plan, manage, report and communicate efficiently through a number of digital platforms, for example Building Information Modelling (BIM), Auto CAD and mobile apps on site.

Chapter 7 Building technology principles

With the right software, drones can reduce costs and labour needs, and improve the efficiency of construction projects. Typical uses of drones include:
- surveying land using laser scanning (topography, volume and measuring distances)
- producing 2D and 3D maps
- identifying improvements that can be made to infrastructure
- pre-construction site planning, for example traffic-management systems and welfare facilities
- performing maintenance inspections that may otherwise be difficult to complete without the risk of harm, working at height or the use of costly access equipment, for example inspecting a roof
- producing thermal imaging to identify hot spots in a building which could be potential hazards, or cold spots resulting in heat loss and poor energy efficiency
- streaming live footage to the project team with the use of virtual reality (VR) glasses
- inspecting work for quality control
- monitoring progress on site in real time and sharing updates with the project team
- identifying hazards and improving site security.

The law states that drones used for business purposes must be flown responsibly, following Civil Aviation Authority (CAA) regulations. From 2021, drone operators must also have several licences to fly them safely and confidently. To obtain these, operators must successfully complete both of the following qualifications:
- A2 Certificate of Competency (A2 CofC)
- General Visual Line of Sight Certificate (GVC).

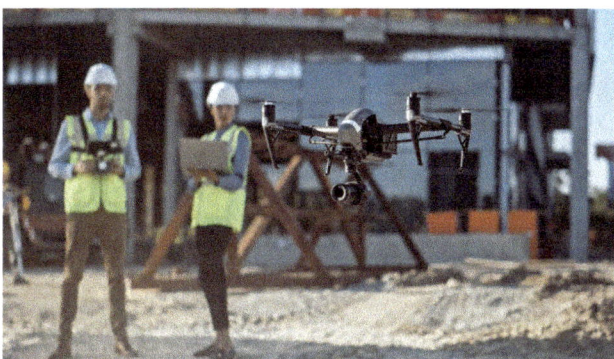

▲ Figure 7.11 Drone in use on a construction site

> **Health and safety**
>
> Drones can be used for mitigating health and safety risks, by reducing the exposure of workers to hazards. They can also be used during site inductions and training sessions, to navigate the construction site without risk of harm.

> **Test yourself**
>
> Describe a situation in which a drone could be used by a building services engineer.

2 Forms of construction

In this section we will be looking at the current forms of construction and their use and suitability for both built environment and civil engineering structures.

In the construction industry, the structure of a building is looked at in two parts:
- The substructure is the section of a building extending below the horizontal damp-proof course (DPC), including **basements** and **retaining walls**.
- The superstructure is the section of a building from the level of the DPC upwards.

> **Key terms**
>
> **Basements:** habitable rooms or spaces constructed below ground level
>
> **Retainer walls:** structural walls designed to bear the weight of lateral loads imposed from one side; they are built below or above ground to hold back earth, and are often **battered** in form
>
> **Battered:** a type of retaining wall designed with a sloped/leaning surface to provide increased structural support

▲ Figure 7.12 Substructure and superstructure of a building

2.1 Substructure

The substructure of a building is designed to bear (support) the weight of the superstructure by transferring the imposed loads down to the ground below. It includes a building's foundations. Designers, such as structural engineers and architects, usually determine the type and size of foundations needed by considering the following factors:
- type of building (industrial, commercial or domestic)
- size and weight of the superstructure
- build method, for example timber frame or brick and block
- number of levels
- location, for example exposed position where there are high winds or an area that is prone to flooding
- topography of the land
- load-bearing capacity of the ground on site
- impact on the environment and sustainability
- construction costs.

The weight of materials used to construct a building is referred to as the dead load. This remains constant and can easily be calculated for the design of building foundations. As well as this weight, designers also have to consider variable imposed loads, known as live loads, such as people, furniture, snow and wind.

If the foundation of a building is inadequate or has been poorly constructed, it will often fail and result in subsidence. If this occurs, the building will partially sink into the ground, either evenly or unevenly:
- A small amount of even subsidence is known as settlement, and this sometimes happens over a short period of time after a building has been newly constructed. Providing this type of subsidence is only short lived, it should not cause any structural damage to the building.
- Where uneven subsidence occurs, it often causes structural damage, which may result in cracking of hard building materials, such as brick and block walls. Cracks in masonry walls usually occur between the openings for doors and windows, where they follow the line of the mortar joints from the ground to the top of the building.

Health and safety

Any subsidence identified in a building should be inspected and carefully monitored by a building surveyor to ensure it does not get worse over time. Subsidence can lead to the collapse of part or all of a structure.

▲ Figure 7.13 Subsidence in a building

It is sometimes possible to undertake retrospective structural work, known as 'underpinning', on existing foundations when they are unable to support current loads or proposed additional construction work, for example an extension or loft conversion. This involves digging underneath the foundations, one section at a time, to enable deeper concrete foundations to be created. These are often reinforced with steel.

▲ Figure 7.14 Retaining wall

Improve your English

Look up the meaning of the word 'lateral' and include it in a sentence about part of a building other than a retaining wall.

The term 'footings' was traditionally used to describe a shallow foundation of a wall constructed of stepped brickwork. It can now be used to mean any building foundation below DPC level.

Foundations are broadly divided into two categories, determined by their depth:
- shallow foundations (less than 2 metres)
- deep foundations (2 metres or more).

Chapter 7 Building technology principles

Shallow foundations

There are several types of shallow foundation, used to construct a variety of buildings:
▶ narrow strip and trench fill
▶ wide strip
▶ raft
▶ pad.

Narrow strip and trench fill foundations

Narrow strip foundations (also referred to as 'strip foundations') are the most common type of foundation for **low-rise buildings**. They are constructed by digging a narrow trench approximately 600 mm wide (the width of a digger bucket, although the width will depend on the thickness of the wall) and the length of the building's load-bearing walls. Approved Document A provides the minimum widths of strip foundations determined by the type of ground. The foundations have to be deep enough so that frost does not affect them, usually at least 1 metre deep or until load-bearing soil is reached, and approved by a building control officer.

Traditional strip foundations are built with a layer of concrete filling the bottom of the trench, to provide a strong base for the blocks laid below ground level. Trench fill foundations are very similar to traditional strip foundations, but they have a deeper strip of concrete poured in the trench, just a couple of brick courses below ground level. Trench fill foundations are often cheaper and safer to build than traditional strip foundations, because they are quicker to construct and avoid the risk of the sides of the trench collapsing while bricklayers are working below ground level.

When the concrete has cured, brickwork is laid directly on top of the foundations up to the damp-proof course (DPC) 150 mm above finished ground level.

Where buildings are constructed on sloping sites, the narrow foundations are usually stepped up the gradient. This avoids unnecessary deeper foundations on one side of the building and reduces the volume of concrete needed.

▲ Figure 7.15 Narrow strip foundation

Key term

Low-rise buildings: buildings with up to four storeys

Health and safety

Nobody should enter or work in a deep excavation for a foundation unless it has been adequately supported using timbering or trench supports and risk assessed, with satisfactory control measures in place to protect people from harm. Further information on excavations and confined spaces is covered in Chapter 1.

Onsite Construction T Level: Core

> **Industry tip**
>
> Wet concrete usually contains small pockets of trapped air that can weaken the foundation if they are not removed before it cures. There are several methods to remove air from wet concrete, including:
> - tamping – tapping the surface of the wet concrete with a timber batten to release the trapped air to the surface; this method is quite basic and often does not remove all of the air trapped at the bottom of deeper concrete foundations; therefore it is only suitable for floor slabs
> - using a vibrating concrete poker – a mechanical tool with a vibrating head; by partially inserting the poker into the wet concrete, the vibration from the tool head causes trapped air to be released deep within the concrete.
>
>
>
> ▲ Figure 7.17 Vibrating concrete poker
>
>
>
> ▲ Figure 7.16 Tamping

> **Health and safety**
>
> Extended use of a vibrating concrete poker could be harmful to health, for example causing vibration white finger. An appropriate risk assessment must be carried out and the operator must be trained.

Wide strip foundations

Wide strip foundations are similar in construction to narrow strip foundations. However, they are normally at least 1.5 metres deep and much wider. This provides enough space for bricklayers to work safely in the trenches to build their brick and block walls to the DPC level. They are used to support the superstructure of buildings with heavier loads or where the soil has a lower bearing capacity, for example soft sandy clays, by distributing the weight through the reinforced concrete foundation over a wider area.

▲ Figure 7.18 Wide strip foundations

Raft foundations

Sometimes it is impractical or uneconomical to keep digging into the ground until firm soil with load-bearing capacity is found for strip foundations; therefore other types of foundation may need to be considered.

Raft foundations are often used for small low-rise domestic buildings, because they are relatively quick to construct and use less concrete compared with other methods.

> **Improve your maths**
>
> Calculate the volume of concrete needed for a narrow strip foundation measuring 7.5 metres × 4.25 metres, with a depth of 1 metre and a width of 0.6 metres.

Chapter 7 Building technology principles

A raft foundation consists of a slab of concrete reinforced with steel frame, constructed under the entire footprint of a building. The concrete slab acts like a raft would on water, by distributing any imposed loads over a large area to the ground below. Some areas of the raft will have additional loads imposed directly on them from the load-bearing walls above, for example the edge (also known as the toe). At these points, the depth of the foundation is increased to create a ring beam in the slab, which improves the load-bearing capacity of the raft and prevents cracking.

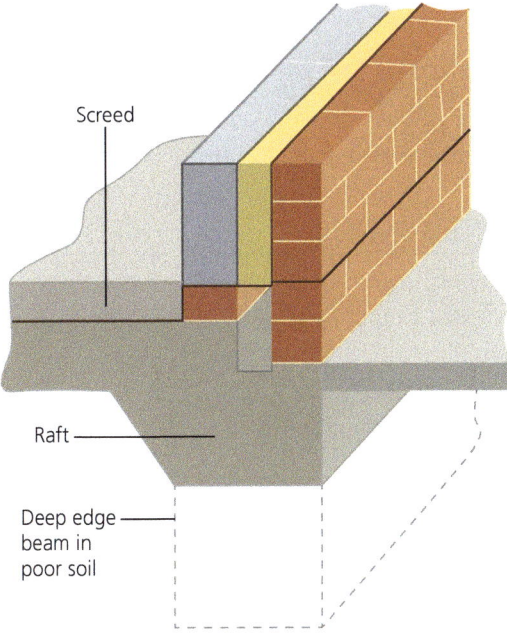

▲ Figure 7.19 Raft foundations

Pad foundations

The methods used to construct industrial units and commercial buildings are very different to those used for domestic houses. This is because they are designed to be erected quickly and efficiently, by maximising the use of large open spaces for industrial and commercial purposes rather than as places to live. These types of building are usually constructed around a steel **portal frame**, with suspended external cladding and low-height walls. Most of the weight of these buildings is transferred from the superstructure down through the steel frame to single point loads at ground level. At these positions, square or circular concrete pad foundations are built to support the structure by distributing the imposed loads down to the ground below.

Pad foundations can also be used to support ground or ring beams. These connect the pad foundations together, so that the weight of the structure can be distributed along the beams and down to the pads in the ground below.

> **Key term**
>
> **Portal frame:** a large structural frame made from load-bearing timber and steel beams and columns

▲ Figure 7.20 Portal frame

Deep foundations

In ground conditions where the surface layer of soil is unable to support the superstructure of a low-rise building with shallow foundations or the weight of a high-rise building, deep pile foundations have to be constructed. This involves driving long, **pre-cast** cylindrical steel and concrete piles deep into the ground until they reach rock or strong load-bearing soil; these are referred to as driven piles. This method displaces the soil as the piles are driven into the ground to provide a strong foundation resistant to downwards and lateral forces.

Piles that rely on the firm soil around them to support any imposed loads are referred to as friction piles, whereas piles that are supported from below by bedrock or another solid surface are known as end bearing piles.

Piles are often designed to be evenly spaced at strategic positions around the footprint of the structure, where they are cut off to the same level and connected at the surface with a concrete and steel pile cap, also known as a ring beam.

> **Key term**
>
> **Pre-cast:** formed into a shape in a factory before being delivered for use on site

197

Driving deep piles into the ground with heavy plant is an extremely noisy process, which could be a nuisance to the public. Furthermore, the amount of vibration caused in the ground could cause damage to the structure of nearby buildings. To overcome these issues, piles can be cast **in situ** by boring holes into the ground with long augers (drill bits) and then filling these holes with wet concrete, reinforced with steel; this method of construction is referred to as replacement.

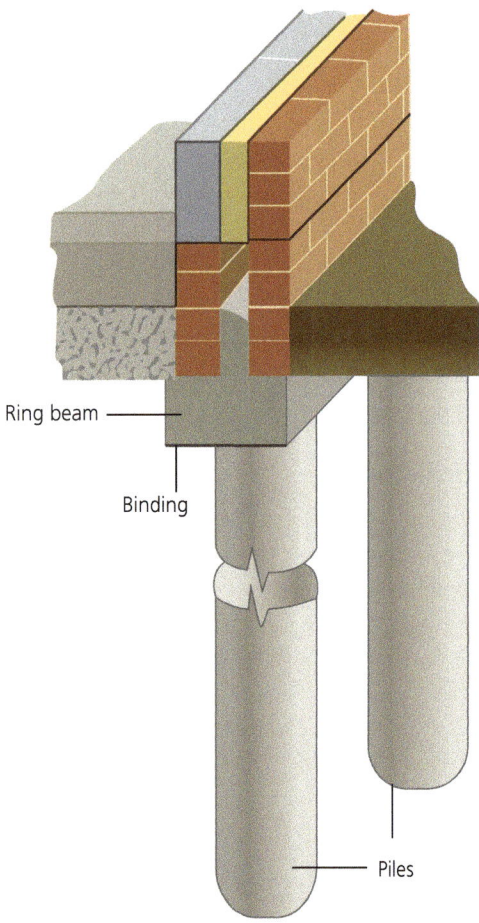

▲ Figure 7.21 Pile foundation

Key term

In situ: in position

Health and safety

Steel reinforcement bars are often left exposed at the top of pile foundations until the pile cap has been cast. These can be extremely hazardous to construction workers, so bar protection caps must always be fitted over them to reduce the risk of injury.

Test yourself

List the four main types of building foundations.

Research

Research screw pile foundations and explain when and why they may be used.

Research

Research alternative types of building foundations and explain why they might be preferred over conventional types.

2.2 Superstructure

As previously mentioned in this chapter, the superstructure of a building is everything above the DPC level on the foundations, including:
- floors
- walls
- roofs
- windows
- frames and doors.

Roofs

Roofs protect occupants of a building (and their possessions) from the elements.

Pitched roof

A basic pitched roof on a domestic building is traditionally constructed with pairs of **common rafters**, connected to a spine (known as a ridge board) at the **apex**. Towards the bottom (foot) of each rafter, a bird's mouth joint is cut to fit over the wall plates, which are secured to the top of the inner skin of the external walls. These joints provide secure fixing points for the rafters, so that the imposed loads from the roof can be transferred down through the load-bearing walls to the foundations below.

Key terms

Common rafters: rafters that run from a ridge board to the wall plate at 90 degrees on plan

Apex: the top of a pitched roof

Chapter 7 Building technology principles

The sectional sizes of the rafters are determined by the distance they have to span and the weight of the roof covering; guidance can be found in Building Regulations Approved Document A: Structure or the Timber Research and Development Association (TRADA) tables. If the sections of the rafters become too large, it may be more economical to reduce them in size and support them from below, mid span, with a heavy structural beam known as a purlin. These days, metal purlins (and sometimes ridges) are preferred rather than traditional timber ones, because they are stronger, lighter and more durable.

The process for constructing a basic pitched roof is referred to as a 'cut roof' or 'cut and pitch'. This is because all the timber components are marked out and cut (sawn) on site by highly skilled carpenters. This method has now been largely replaced with prefabricated **roof trusses** manufactured in factories and delivered to site where they can be erected much more quickly than cut roofs.

> **Key term**
>
> **Roof trusses:** prefabricated roof sections, held together with gang nail plates at the intersections (nodes)

> **Improve your maths**
>
> Use a formula or scaled drawing to calculate the true length of a rafter with a rise of 1.8 metres and a run (half the span of the roof) of 2.4 metres.

▲ Figure 7.23 Typical section through a pitched roof

Roof trusses

Each member of a trussed roof is specifically designed and calculated to suit its intended use, for example roof shape, pitch, span and weight of the roof covering. There are many advantages of using trussed roofs, including:

- less waste
- greater energy efficiency
- greater accuracy
- reduced material costs, building times and onsite labour costs
- less-skilled labour required
- lighter and more sustainable materials.

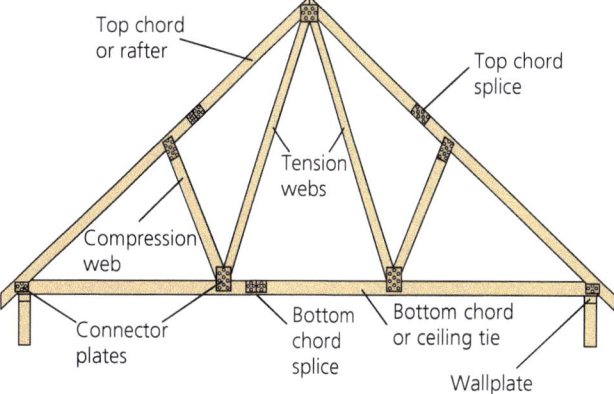

▲ Figure 7.24 Roof truss

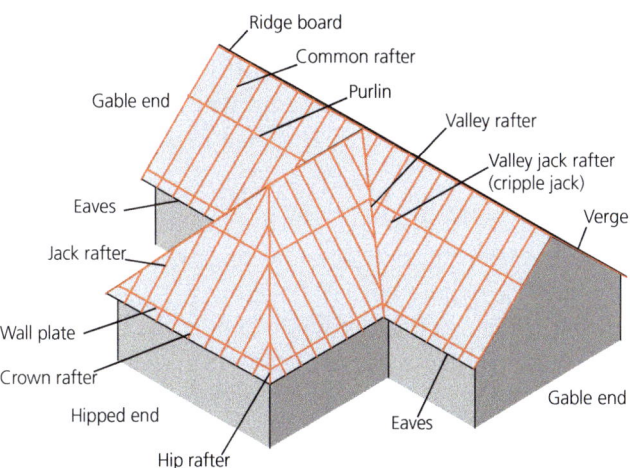

▲ Figure 7.22 Traditional cut roof

▲ Figure 7.25 Nail plate

Loft space

The loft space created in a domestic pitched roof can be utilised as living areas, such as bedrooms, bathrooms and home offices. These are either designed into the building when it is initially constructed or completed retrospectively at a later stage. Constructing a loft conversion in a cut roof is generally a simple process, because the arrangement of the rafters and supporting roof members leaves open spaces for the additional rooms, which will require additional steelwork to support a subfloor. However, converting a trussed roof is much more difficult, because the supporting braces have to be removed and the smaller sections of timber trusses result in a loss of structural integrity.

Where a loft has been designed for a living space, it may be restricted in headroom in some areas because of the triangular shape of the roof. To overcome this issue and to create a source of natural light and ventilation in the loft, the roof may be extended to the back to create dormers. When it is not possible to use dormer windows because of planning restrictions or a tight budget, natural light and extra headroom can be created with standard roof windows, known as skylights, or balcony windows.

▲ Figure 7.26 Dormer

▲ Figure 7.27 Roof window

▲ Figure 7.28 Balcony window

> **Research**
>
> Research the planning permission and building regulations needed for a loft conversion in a domestic property.

Flat roofs

When a cheaper form of roof construction is needed or has been designed for all types of structures, the pitch of a roof can be lowered to less than 10 degrees to form a flat roof.

Flat roofs are constructed with a slight fall (pitch or angle) to allow water to run towards the guttering on one side of the roof and the drainage system at ground level. If a flat roof does not have the correct fall, it leads to rainwater sitting on the roof (referred to as 'ponding') for long periods of time which could promote weed and moss growth.

Due to the low angle of a flat roof, some roof coverings, for example slates and tiles, are unsuitable because driving wind will cause them to lift and rainwater could be blown underneath.

Chapter 7 Building technology principles

▲ Figure 7.29 Typical section through a flat roof

Figure 7.30 shows a range of basic roof shapes used in the construction of domestic houses.

▲ Figure 7.30 Roof shapes

Protecting roofs

The timber framework for a pitched domestic roof is usually protected with a layer of breathable membrane to prevent water entering the building; it also allows any moist air to be ventilated from the loft space. If moist air does become trapped, it will increase the moisture content in the roof timbers above the 20 per cent threshold, leaving it susceptible to an attack of dry rot and possible collapse of the structure.

A range of products and materials can be used to provide a waterproof covering to protect different-shaped roofs from the elements. For pitched roofs, these include:
- clay and concrete tiles
- natural and synthetic slates
- metal, for example lead, zinc and copper
- fibreglass
- glass
- asphalt shingles
- timber shingles
- thatch
- solar tiles.

For flat roofs, the following might be considered:
- metal, for example lead, zinc and copper
- fibreglass
- glass
- rubber
- green roofs.

Warm and cold roofs

To maintain the energy efficiency of a building and prevent heat loss, all types of roof must be highly insulated to conform to building regulations. The location of insulation in a roof is important, because it determines the dew point. This is the place where warm moist air rising in a building meets cold air from outside to create condensation.

Pitched and flat roofs insulated between ceiling joists/rafters are referred to as cold roofs. They must be adequately ventilated along the **eaves** on both sides of the roof or the ridge where the insulation is between the rafters, to allow moisture to escape. If the moisture does not escape, the dew point may be within the structure of the roof. This is known as interstitial condensation of the roof and can cause rot in the timber rafters and joists.

The thickness of insulation required in a cold roof is usually determined by the type of insulation; however, where it is positioned between the joists/rafters, a gap of at least 50 mm has to be maintained to provide adequate ventilation.

> **Key term**
>
> **Eaves:** the part of a roof that overhangs the internal skin of the external walls

Alternatively, a warm roof can be used. This is where reflective foil rigid insulation is placed over the top of the rafters or flat-roof joists with an external grade **decking** fixed either below or on top, depending on the specification. Warm roofs do not need to be ventilated because the dew point occurs outside the building, and although they are initially more expensive to construct, they are more energy efficient compared to cold roofs.

> **Key term**
>
> **Decking:** timber boards or sheet materials such as chipboard or plywood used to cover structural joists

> **Test yourself**
>
> Explain the difference between a traditional cut roof and a trussed roof.

External walls (load-bearing)

Solid walls

Internal and external walls are classified as either load-bearing or non-load-bearing, depending on the method of construction and materials used. Traditional load-bearing external walls were often made of natural resources, such as stone and cob, because these materials were readily available on the land where they were being used. Although the thermal mass of solid walls kept buildings relatively warm in the winter and cool in the summer, they did not achieve the same levels of energy efficiency as the methods we use now.

▲ Figure 7.31 Solid brick wall

Cavity masonry walls

A cavity masonry wall actually consists of two walls (referred to as leaves or skins) constructed with a space between them, known as a cavity. The external skin of a cavity wall is usually built with attractive clay facing bricks **bonded** with 10 mm mortar joints. The internal walls are usually constructed with PFA (pulverised fuel ash) blocks, although there are still occasions when concrete blocks are used for internal leaf or skin masonry with suitable insulation. The internal walls are often covered with plaster or plasterboard on one face, to provide a smooth surface on the inside of the building.

> **Key term**
>
> **Bonded:** the arrangement of staggered joints

The internal and external skins of a cavity wall are connected with metal wall ties, built across the cavity into the mortar joints as the walls are constructed. Wall ties improve the load-bearing capacity of the walls by preventing the skins from moving apart and possibly collapsing under the weight from the load above.

During the 1990s, building regulations changed, making it compulsory to include insulation in all cavity walls to reduce heat loss and improve the energy efficiency of buildings. There are several types of insulation that can be used in a cavity wall; some partially fill the cavity and leave a compulsory minimum 50 mm void to comply with building regulations, while others completely fill it. Where an existing cavity wall does not contain insulation, loose-fill can be retrofitted by blowing it into the cavity through holes drilled into the external wall. Blown-in insulation uses small particles of recycled materials such as fibreglass, mineral wool and cellulose (made from old newspapers), which has less of an impact on the environment than alternative types of insulation.

If insulation can be built into the walls as they are being constructed, full-fill **fibreglass batts** could be used in the cavity. These provide low-cost insulation with excellent sound resistance. However, over time the fibreglass batts can settle in the cavity resulting in cold air pockets in the wall, which could result in heat loss. To maintain the air gap in a cavity wall and avoid moisture **bridging** from the external skin to the inner skin, rigid reflective insulation is often preferred. As a cavity wall is built, the reflective foil sheets are positioned in the cavity and held tight against the internal wall with plastic retaining

Chapter 7 Building technology principles

clips fixed over the wall ties. Although this type of insulation is slightly more expensive, the reflective foil faces are highly effective at preventing heat loss from a building.

▲ Figure 7.32 Cavity walls with loose fill, fibreglass batt and reflective foil insulation

Key terms

Fibreglass batts: pieces of fibreglass insulation made from sand and recycled glass; besides providing good heat insulation, they are also fire resistant, sound absorbent and water repellent

Bridging: the action of water travelling across materials

Industry tip

Bricklayers should avoid building cavity masonry walls any higher than 1.5 metres in one day to avoid overstressing the mortar joints and making the wall unstable.

Industry tip

Frog bricks should be laid with the frog facing up, so that it can be filled with mortar to improve the strength of the wall and make it more resistant to sound transmission.

Industry tip

The internal and external skins in a cavity wall should be constructed at the same time to keep the mortar joints (bed) correctly aligned and prevent weakening a single skin if left unsupported.

Lintels

Wherever an opening is created in a masonry wall for a window or door, the brickwork over it has to be supported either with a lintel or by forming a self-supporting brick arch. Lintels are structural beams manufactured from steel, concrete or reinforced brickwork. Alternatively, when a feature is required, natural stone is often used, although this is generally more expensive. Traditionally, heavy timber beams were also used for lintels but these are not as durable as metal, stone and concrete and are therefore rarely specified for external masonry walls on new builds unless they are protected from the weather.

The sectional size of a lintel and the amount of bearing that it has on the masonry wall either side of an opening will determine its load-bearing capacity. Lintel sizes are usually calculated and specified by the project designer. However, where they span distances up to 1.2 metres, they must have at least 100 mm bearing on the wall each side of the opening and 150 mm where the opening is greater than 1.2 metres.

▲ Figure 7.33 Parts of a brick

Onsite Construction T Level: Core

> **Health and safety**
>
> Lintels can be extremely heavy; therefore mechanical lifting equipment should always be used whenever possible to prevent manual handling injuries.

▲ Figure 7.34 Metal lintel

▲ Figure 7.35 Natural stone lintel

Weep holes

Moisture can sometimes penetrate porous brickwork or build up in a cavity wall. To prevent this happening, weep holes are built into the external skin above window and door openings and towards the bottom of the walls, just above ground level. Weep holes are plastic vents, positioned in the vertical mortar joints (known as perps, short for perpendicular) in the external skin of a wall. They allow any build-up of water to escape the cavity to prevent water ingress.

Closing-off

Building regulations state that all cavity edges in external masonry cavity walls must be closed off where they are exposed, for example around window and door openings, around vents and at the top of the wall. Closing the cavity at these points improves the energy efficiency of the building by reducing heat loss. It also prevents smoke, dangerous gases and flames travelling up the cavity like a chimney in the event of a fire.

There are several methods that can be used to close off a cavity:
- ▶ Cavity closers can be built into the brickwork around window and door openings. They are available in a range of sizes to suit the width of standard cavity openings.
- ▶ The blockwork on the internal skin of the cavity wall can be returned, to reduce the width of the cavity. A gap of at least 50 mm must still be maintained between the internal and external skins to prevent 'cold bridging'. This joint is then filled with an insulated vertical damp-proof course (DPC), to stop water ingress around the opening in the wall and improve its thermal efficiency.
- ▶ Cavity barriers, also known as cavity stop socks, are often used to seal the cavity along the top of the wall to provide fire resistance.

Note: building regulations will determine the exact requirements for cavity closers and cavity barriers in a cavity wall.

▲ Figure 7.36 Vertical DPC in a cavity wall

Chapter 7 Building technology principles

Internal walls

Concrete block

An internal wall used to divide a space is known as a partition. These can either be load-bearing or non-load-bearing, depending on their method of construction.

Solid concrete blocks are often used to build load-bearing internal walls, because they are cheaper and quicker to build than bricks and also provide excellent sound insulation and fire resistance. The exposed faces of a concrete block wall are traditionally covered with cement render as a base coat and finished with gypsum plaster to provide a flat, smooth surface ready for decoration.

One disadvantage of this method is the time it takes for the cement render to set; therefore faster-drying backing plasters are now preferred to reduce drying times. Alternatively, the masonry of solid internal walls can be covered with sheets of plasterboard bonded to the wall with **drywall** adhesive, in a process known as dot and dab. The long edges on the plasterboard sheets used for this process are slightly tapered. This allows for the joints between the boards to be covered with jointing tape and made flush with drywall joint filler.

> **Key term**
>
> **Drywall:** another name for plasterboard

> **Test yourself**
>
> Explain the purpose of cavity walls compared to solid walls.

Timber partition walls

Non-load-bearing timber partition walls are more commonly known as stud partitions, because they consist of vertical members known as studs. Studs are usually spaced equally with 400 to 600 mm between the centres of the uprights to suit the size of plasterboard wall covering and to provide support for fixing abutting edges, with extra studs being used to form corners and junctions.

Each stud in a partition is fixed into a head plate at the top of the wall and along the floor into a sole plate, with two wire nails in each butt joint. The studs are strengthened further with short noggins fixed between them, in positions determined by the height of the wall or where the wall is most vulnerable, for example waist height in a school or hospital. Further noggins may also be built into a partition before it is covered with plasterboard, to provide additional fixing points for heavy items, such as kitchen units, radiators or electrical.

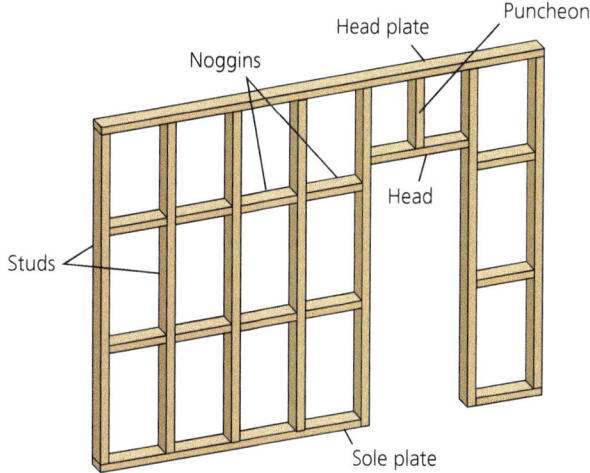

▲ Figure 7.37 Non-load-bearing stud partition wall

The arrangement of timbers in load-bearing partition walls is similar to non-load-bearing walls. However, the timbers used must be **stress-graded**. Two head plates must be fixed in these types of walls to spread the weight of any imposed loads and there must be additional structural timbers (lintels) around any openings for doors. Further timber studs should also be used where there are specific heavy loads, for example directly above load-bearing walls.

> **Key term**
>
> **Stress-graded:** timber that has been visually or machine-assessed for its strength and stiffness, and certified according to its structural classification to British Standard rules

> **Research**
>
> Research the methods for stress grading timber and the grading system used for each class.
>
> Identify which stress-graded timber classifications are most commonly used and explain why you think this is the case.

> **Industry tip**
>
> Masonry walls should not be supported by timber floors or partition walls, because any movement in the timber will crack the masonry.

Metal stud partition walls

Metal stud partitions are an alternative solution for non-load-bearing walls. They are often used to divide spaces in domestic and commercial buildings.

The low-cost, lightweight metal sections of this system are easy to cut and quick to install. In positions where additional support is needed, such as doorways or hatches, the hollow sections of the metal studs are filled with timber to strengthen the wall. The timber inserts also provide strong fixing points for any door or hatch linings that may be fitted after the wall has been constructed.

Each metal stud section is designed and manufactured with elongated holes in it. These are used to route services through the walls and prevent the need for tradespeople to create holes and notches in the metal studs themselves, as they would have to in timber.

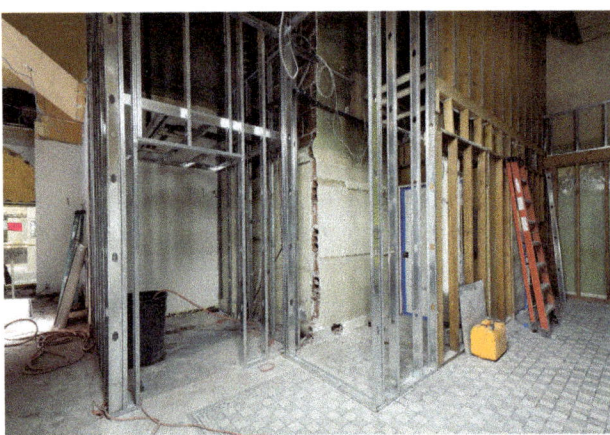

▲ Figure 7.38 Metal stud partition wall

Health and safety

You should always wear gloves when handling metal stud-wall components to protect your hands from the razor-sharp edges.

Improve your English

Non-load-bearing walls are not limited to timber and metal stud partitions. Research proprietary partition walls and write a definition in your own words.

Floors

Floors are designed to provide a level surface for the occupants to live and work, while preventing moisture rising from the ground and weed growth. Modern floors are also designed and constructed to:
- prevent heat loss
- prevent the transfer of sound from one level to another
- slow the spread of flames in the event of a fire.

In domestic buildings, the methods used to construct ground floors are often different to those used for upper floors, mainly due to the increased risk of moisture affecting materials in floors closest to ground level.

There are two types of floor construction:
- solid floors
- suspended floors.

Solid floors

Solid floors are used at ground level. They are constructed using different layers to make a solid concrete base:
- The first layer is compacted hardcore; this consists of materials that are **chemically inert** and not affected by water. Clean, broken bricks and aggregates or similar **unbound** materials can be used as hardcore, because they easily drain water and are not affected by the chemicals in other materials in the floor.

Key terms

Chemically inert: will not react with chemicals

Unbound: not bonded together with cement

- A thin layer of sand blinding is used for the next layer, to protect the polythene damp-proof membrane (DPM). The DPM is a protective barrier to prevent moisture penetrating the floor, which could lead to rising damp in other areas of the building such as the walls.
- Concrete is poured over the DPM to form the floor slab. Also referred to as oversite concrete, this layer provides most of the structural integrity in a solid floor and should not be confused with the finished floor level.
- Sheets of rigid reflective foil insulation are laid over the concrete slab to prevent heat loss through the floor. If an underfloor heating system is being used in the building, it will be fixed either in or on top of the insulation at this level.
- The floor insulation is protected with a thin-polythene sheet, before the final layer of cement screed is laid.

Chapter 7 Building technology principles

> **Test yourself**
>
> What type of timber should be used in load-bearing partition walls?

> **Research**
>
> Research dry and liquid screeds used in floor construction and explain the benefits of each system.

▲ Figure 7.39 Solid floor

Suspended ground floors

Traditionally, ground floors were built using timber joists suspended between load-bearing walls, with low-height sleeper walls supporting them mid span. Solid timber floor boards were laid across the joists at right angles and fixed in position with nails. The void underneath the floor was ventilated with air bricks built into the external walls; this created airflow around the timber floor and reduced the risk of rot.

Suspended timber floors at ground-floor level are not commonly used in modern construction, because the ends of the joists built into the external walls would often suffer with decay from moisture penetration over a period of time, unless they were supported with joist hangers. These days, a durable system known as block and beam is preferred. This involves using reinforced concrete beams infilled with blocks. The floor can be used as soon as the blocks and beams have been positioned on the DPC on the inner skin of a cavity wall; therefore construction work can continue without delay. The exposed block and beam floor is completed with insulation and screed in the same way that a solid floor would be finished.

Suspended upper floors

The block and beam system used for ground floors can also be used for upper floors in domestic buildings. However, it is more expensive than suspended timber floors and the additional weight would have to be supported with deeper foundations, which would increase construction costs. For these reasons, suspended timber floors are often chosen by designers for use in low-rise domestic buildings.

Besides traditional solid timber beams, engineered joists such as I-beams and Eco-joists have now been developed to span longer distances unsupported and without the natural defects sometimes found in solid timber, for example twisting, springing and large knots. These joists are wider and lighter than solid joists, making them easier to handle, and stronger joints are formed with the chipboard floor covering. Eco-joists are a type of engineered joists designed with a metal web on each side of the beam so that services such as pipes and cables can easily be laid through them, without weakening them by drilling holes and notches.

▲ Figure 7.40 Suspended timber upper floor with solid joists

▲ Figure 7.41 Suspended timber upper floor with Eco-joists

207

▲ Figure 7.42 Suspended timber upper floor with I-beam joists

▲ Figure 7.43 Pre-stressed concrete slabs

▲ Figure 7.44 Lightweight concrete floor system

Where timber suspended floors are unsuitable, for example in multi-storey buildings, **pre-stressed concrete** slabs manufactured off site can be used. This type of floor is quick to install with the use of a crane. It also has fewer joints than the block and beam system, meaning it has better fire resistance.

Alternatively, a floor constructed from a concrete slab reinforced with steel can be cast in-situ using formwork (a temporary mould), which avoids the need for expensive cranes. Suspended floors can also be cast in-situ using a lightweight concrete floor system. Rather than formwork, this method uses a steel deck which forms a permanent part of the floor.

> **Key term**
>
> **Pre-stressed concrete:** a type of concrete that has been compressed during production to improve its strength; this is normally achieved by 'tensioning' (stretching) high-tensile steel wires in the concrete

Windows, doors and frames

Windows, doors and frames can allow heat to escape a building if they are not correctly designed and installed; hence they are subject to planning permission and inspection by the building control officer after they have been fitted.

High-performance, energy-efficient doors, windows and frames can be made from hardwood, softwood, unplasticised polyvinyl chloride (uPVC) and aluminium. Although timber frames are not used as often as the other examples, because they have to be maintained more regularly, they are sustainable and therefore have less of an impact on the environment compared with other materials.

The type of glass used is just as important as the frames. Double and triple glazing is mainly specified for new construction projects, because the high-performance, **low-emissivity (low-E) glass** and **argon gas** filled units act as insulators to reflect heat back into a building to reduce energy bills.

Chapter 7 Building technology principles

> **Key terms**
>
> **Low-emissivity (low-E) glass:** a type of thermally efficient glass that is covered with a microscopic coating on its surface to minimise the amount of infrared and ultraviolet light that can pass through it to keep a building warm in the winter and cool in the summer
>
> **Argon gas:** a colourless and odourless gas with excellent thermal properties; often used to fill the voids between the panes of double and triple glazing to improve the energy efficiency of a building

> **Test yourself**
>
> Describe where a DPC is used in the superstructure of a building.

2.3 Infrastructure

Chapter 3 looks at how well-designed infrastructure, including roads, sewage systems, railways and bridges, can contribute to the community and the success of construction projects.

2.4 External work

External construction work could involve the following:
- paving
- boundaries
- drainage
- parking.

The infrastructure listed in section 2.3 is subject to planning restrictions and building regulations. Many of the main sewer systems in the UK were initially built in the late 1800s, since when the population has grown substantially, meaning that some of them no longer work efficiently.

Besides raw sewage, surface water from paving and driveways was directed into the main sewer system until 2008, when new regulations were introduced to prevent flooding and other environmental problems in populated urban areas. The law states that planning permission is not needed if a new or replacement driveway at the front of a property is made from porous materials (sub-base materials and surface finishes) such as permeable concrete, block paving or resin-bound gravel. These regulations do not apply to other parts of a property. Work in or around listed buildings may also need consent.

Where a kerb lining a path is raised across a new driveway, permission will have to be granted before it can be lowered. The local authority is responsible for public roads and paths and should therefore undertake this task, to ensure that no damage occurs to services under the paving surface and that the lowered kerb is adequately supported.

Walls that form a boundary between two or more properties are known as party walls. The Party Wall etc. Act 1996 is a framework of legislation that aims to prevent and resolve disputes over boundaries. The law states that you must inform your neighbour if you are intending to build on the boundary, work on an existing party wall or excavate near and below the level of their foundations. Further information on party wall agreements can be found on the following website: www.gov.uk/party-walls-building-works

3 Key content and required notifications of UK building regulations and approved documents

3.1 Approved documents

Virtually all new construction work and alterations to existing structures must comply with the Building Regulations 2010. These regulations are developed by the government and approved by parliament. Their purpose is to define the minimum standards of design, building materials and work in the UK.

To help people understand and comply with the building regulations, the Ministry of Housing, Communities and Local Government publishes general guidance on each part of the law in a range of approved documents. These documents provide practical solutions and examples of ways to comply with the building regulations in some common situations, for example water efficiency in a dwelling.

Table 7.4 contains a list of the approved documents and an outline of their key content.

> **Test yourself**
>
> Explain the purpose of approved documents.

> **Research**
>
> Find out about the Building Regulations 2010 by visiting www.legislation.gov.uk/uksi/2010/2214/contents/made
>
> Explain the type of construction work that is exempt from legislation.

▼ Table 7.4 Approved documents

Approved document	Heading	Key content
A	Structure	This document covers the **loadings** on a building and the construction of structural elements such as foundations, walls, floors, roofs and chimneys.
B	Fire safety	This document covers fire safety precautions in new and existing residential buildings, flats, offices, schools and colleges.
C	Site preparation and resistance to contaminants and moisture	This document covers the removal or treatment of site waste materials, subsoil drainage and resistance to contaminants. It also provides details about the resistance to moisture of roofs, floors and walls, with damp-proofing and ventilation solutions.
D	Toxic substances	This document provides health and safety guidance on the use of toxic substances during building works, for example the prevention of harmful fumes produced by urea-formaldehyde in cavity-wall insulation from entering a building.
E	Resistance to the passage of sound	This document covers resistance to the passage of sound in new domestic buildings, flats and schools, and buildings that have been converted. It also provides information on particular areas of concern in structures, for example between connecting buildings.
F	Ventilation	This document provides guidance on the design, installation, inspection, testing and commissioning of ventilation systems in new dwellings and other types of building. It also covers air quality and the provision of adequate ventilation to prevent condensation.
G	Sanitation, hot water safety and water efficiency	This document sets out standards for the supply of softened wholesome water in any area of a building for the purpose of washing, for example bathrooms, kitchens, showers, sanitary conveniences and any sinks provided in food-preparation areas. Its purpose is to ensure water is sanitary, safe and used efficiently. It also explains where alternative sources of water can be supplied to sanitary conveniences for flushing, for example grey water and rain water.
H	Drainage and waste disposal	This document provides guidance on surface- and foul-water drainage above and below ground and includes information on: • pipe sizes • protection of pipes • treatment and disposal of waste • sewage infrastructure and maintenance • refuse storage • hygienic pipework, discharge and cesspools • pollution prevention.
J	Combustion appliances and fuel storage systems	This document covers the supply of air, discharge of combustion products and protection of buildings for oil, gas and solid-fuel appliances. It also provides guidance on the safe installation of hearths, fireplaces, chimneys and flues.
K	Protection from falling, collision and impact	This document includes guidance on designing and fitting staircases, ladders, ramps, guarding and vehicle barriers in and around all types of building, in order to avoid falls, collisions and impacts. Note: Approved Document N (Glazing) has now been combined with Approved Document K.
L	Conservation of fuel and power	This document provides guidance on the energy efficiency of different structures. It is updated regularly to reflect developments in building materials and new technologies; therefore people must ensure they are using the most up-to-date version. Approved Document L is divided into four parts, with each providing specific details on a particular building type: L1A new dwellings, L1B existing dwellings, L2A new buildings other than dwellings, L2B existing buildings other than dwellings.

Chapter 7 Building technology principles

Approved document	Heading	Key content
		The documents cover the energy efficiency of heating, ventilation and air-conditioning systems, insulation, boiler productivity, lighting and hot-water storage. Further information on Part L can be found in Chapter 3.
M	Access to and use of buildings	This document looks at dwellings in three different categories: • Category 1 – Visitable dwellings • Category 2 – Accessible and adaptable dwellings • Category 3 – Wheelchair user dwellings. It provides information about ease of access to, and use of, buildings, including facilities for disabled visitors or occupants, and the ability to move through a building easily. This includes accessible stairs, corridors and lifts.
O	Overheating	This document covers the overheating mitigation requirements. It includes guidance on providing means of removing excess heat from residential buildings (opening windows, ventilation systems, cooling systems). This document applies to new residential buildings only.
P	Electrical safety	This document states that reasonable provision must be made in the design and installation of electrical equipment to protect people operating, maintaining or altering the installations from fire or injury. Only competent people should complete electrical work in dwellings, or on land where the supply comes from the same source. This document also provides information on the range of work that is considered notifiable under the building regulations.
Q	Security – dwellings	This document details the standards required for the design, installation and fixing of doors/**door sets** and windows in new dwellings and flats, to resist a physical attack by a 'casual or opportunist burglar'. To achieve this aim, doors and windows should be sufficiently robust and fitted with appropriate **ironmongery**.
R	Physical infrastructure for high-speed electronic communications networks	This document provides guidance on the design and installation (physical infrastructure) of high-speed electronic communications networks within a new building or one that is undergoing major renovation works. Copper, fibreoptic or wireless devices must be capable of delivering broadband speeds of no less than 30 mbps. Note: some buildings are exempt, for example in isolated areas where it is too remote to connect to a high-speed network.
S	Infrastructure for the charging of electric vehicles	This document provides guidance on the installation of electric vehicle charge points within buildings with associated parking (new build or undergoing major renovation works). Electric vehicle charge points should have a minimum nominal rated output of 7 kW and be fitted with a universal socket as well as a visual display or indicator to show the equipment's charging status.

Key terms

Loadings: the application of a mechanical load or force on a structure

Door set: a combination of a door, frame and any associated ironmongery

Ironmongery: metalwork such as locks, latches and handles

Research

In the event that electrical work is notifiable, who needs to be informed?

4 Building standards

The British Standards Institute (BSI) produces agreed standards across a wide variety of industry sectors. In construction, these relate to structures, materials and sustainability.

British Standards (BSs) are used as common minimum standards for **public-sector projects**, covering areas such as BIM, fire safety and waste management. Their purpose is to provide general and specific guidance for the construction industry to improve working practices.

The BSI also develops Publicly Available Specifications (PASs). These are fast-tracked standardisation documents produced to meet an urgent market need.

The International Organization for Standardization (ISO) is an independent, non-governmental organisation that develops and publishes international standards.

Table 7.5 outlines some of the standards used in construction and renovation.

> **Key term**
>
> **Public-sector projects:** projects funded by the government

▼ Table 7.5 Standards used in construction and renovation

Building standards	
BS 1192-4:2014	Collaborative production of information – Fulfilling employer's information exchange requirements using COBie. Code of practice
BS 7000-4:2013	Design management systems – Guide to managing design in construction
BS 7913:2013	Guide to the conservation of historic buildings
BS 8536-1:2015	Briefing for design and construction – Code of practice for facilities management (Buildings infrastructure)
BS 8541	Library objects for architecture, engineering and construction
BS 9999:2017	Fire safety in the design, management and use of buildings. Code of practice
BS ISO 55000:2014	Asset management – Overview, principles and terminology
ISO 16739-1:2020	Industry Foundation Classes (IFC) for data sharing in the construction and facility management industries – Data schema
ISO 14001:2015	Environmental management systems – Requirements with guidance for use
ISO 19650	Organization and digitization of information about buildings and civil engineering works, including Building Information Modelling (BIM) — Information management using Building Information Modelling
ISO 9001:2015	Quality management systems – Requirements
ISO 50001	Energy management systems
PAS 91:2013+A1:2017	Construction prequalification questionnaires
PAS 180:2014	Smart cities. Vocabulary
PAS 2035/2030:2019	Retrofitting dwellings for improved energy efficiency. Specification and guidance.
PAS 2038:2021	Retrofitting non-domestic buildings for improved energy efficiency. Specification
PAS 2080:2016	Carbon management in infrastructure
PAS 8811:2017	Temporary works. Major infrastructure client procedures. Code of practice
PD 7503:2003	Introduction to knowledge management in construction

5 Manufacturers' instructions

Any product that has been designed and manufactured in the UK, or imported into the UK, must conform to section 6 of the Health and Safety at Work etc. Act 1974 (HASAWA) – 'General duties of manufacturers etc. as regards articles and substances for use at work'. The legislation states that:

- products must be designed and constructed to eliminate risks, so that they are safe while being used, cleaned and maintained
- where necessary, products must be calibrated, tested, inspected and certified to meet with the relevant product regulations
- adequate information must be provided for the safe maintenance, operation and disposal of products
- as far as is reasonably practicable, revisions of information must be supplied if a serious risk to health or safety concerning a product has been discovered.

Further information on HASAWA can be found in Chapter 1.

Manufacturers of new products, importers of products and suppliers of significantly refurbished products must provide adequate information concerning safety and the absence of risks to health, and other issues such as the protection of the environment.

Product information is usually supplied with goods in the form of printed manufacturer's instructions, with further copies available on the manufacturer's website. This information will reflect the type of product, article or substance. For example, if the product is a new energy-efficient boiler, manuals should be provided to explain how it must be installed, operated, serviced and maintained to meet the minimum standards laid out in building regulations.

Industry tip

QR codes linking to manufacturer's instructions can be found on the body of some tools and equipment. This avoids having to keep paper copies of the instructions and can allow people to find specific product information much more quickly.

Test yourself

Under what regulation do manufacturers have to provide information on their products?

Case study

Mia is a self-employed gas engineer and is registered Gas Safe. She has been asked to install a new energy-efficient boiler in a three-bedroom semi-detached dwelling for a domestic customer. A hole needs to be bored in the fabric and structure of the building in an external wall for a new boiler flue.
- Explain the potential issues that could be caused for the occupants of the dwelling as a result of Mia drilling a hole for a flue in an external wall.
- Explain the processes and procedures that should be followed to complete the task.
- List the approved documents and guidance that Mia should refer to before installing the new boiler.

Assessment practice

Short answer

1. Which approved document provides advice and guidance on fire safety?
2. Which regulatory body must gas engineers be registered members of?
3. What type of foundation is often used to support metal portal frame structures?
4. What is the minimum gap permitted by building regulations between the skins in a cavity wall?
5. Name one type of fixing that can be used in a masonry wall.

Long answer

6. Explain the purpose and benefits of off-site construction.
7. Explain the difference between first fix and second fix.
8. List the benefits of using autonomous vehicles on construction sites.
9. List the materials used to construct a solid floor and explain the purpose of each component.
10. Describe the implications of not following manufacturers' instructions.

Project practice

Your employer has asked you to design a free-standing modern garden room for one of his customers.

The building should be made from as many sustainable materials as possible, with a single pitched roof and composite cladding. It can be constructed under permitted development rights and therefore does not need planning permission.

- Design a garden room to meet the customer's requirements using computer-aided design (CAD) software such as SketchUp.
- Explain the type of foundations you would recommend and why.
- Calculate the total cost of materials needed for the project and present them in a suitable digital spreadsheet.

Chapter 8 Information and data principles

Introduction

This chapter explores the meaning of the term 'data' and how data is used in planning, costing, constructing and operating a building. We will discuss how data can be processed and organised into useful and valuable information using a range of methods.

In the construction industry, expert analysis and evaluation of data can allow informed decisions to be made when bringing a project to a successful completion. Projections can be made about efficient operation of a building throughout its life cycle.

Being able to interpret information gained from data sources is a valuable skill for those employed in the modern construction workplace.

Learning outcomes

By the end of this chapter, you will understand:
1. data
2. sources of information
3. data management and confidentiality

1 Data

Humans have been gathering and processing data for thousands of years to refine their understanding of the universe. In modern times, working with data has developed into a science that follows defined methods to perform specific tasks.

Data and information are closely linked. Data is a collection of facts, such as numbers, measurements, words or descriptions, that can be analysed and processed so that meaningful information is created.

> **Research**
>
> Watch an introductory explanation of what data science can be used for (for example, search for 'Intro to data science' on YouTube).

1.1 Key elements of data

Every day, vast quantities of data are generated, often referred to as 'big data'. However, simply collecting masses of data is not the key to creating useful information. Data must be correctly stored, 'cleaned' to remove inaccurate or imprecise records, and identified to assign it for use.

▲ Figure 8.1 Vast quantities of data are generated every day

Managing data in this way can involve many complex data-processing systems working together, requiring them to be operationally compatible. This is known as **data interoperability**.

Complex processing of data can involve one system describing or analysing data in another system, for example to inform an architect regarding multiple intricate design elements of a building.

When a set of data is used to describe or analyse other data, this is referred to as **metadata**.

> **Key terms**
>
> **Data interoperability:** the ability of data systems to exchange and use information
>
> **Metadata:** a set of data that describes or analyses other data

> **Test yourself**
>
> How must data be processed to make it usable?

The level of detail in a set of data (or dataset) must be regulated or even limited, so that it can clearly inform users according to the required purpose of the information.

Different data types that can be analysed include:
- numerical – consisting of numbers, such as dimensions and measurements
- categorical – consisting of options, such as 'type A, B or C' or 'up, down, left, right'
- ordinal – consisting of steps or sequencing (ordering) of information.

To allow the extraction of useful information from large datasets, a process known as **generalisation** can be used. This is essentially a process of 'pulling back' from the mass of raw data to give a broader, more general view and reduce the extent of analysis required. This makes information extraction more manageable.

For example, rather than stating the floor area of dozens of individual rooms in a large multi-storey building, the data could be arranged in bands to categorise room volumes, such as 7–10 m^3, 11–15 m^3 and so on. This reduces the volume of data while still providing useful information, to allow analysis of requirements such as heating and ventilation systems or optimal occupancy levels in the completed building.

> **Key term**
>
> **Generalisation:** (in data processing) creating layers of summarised information from mass data

1.2 Different sources of data

Data are generated from many different sources during the life cycle of a building.

Data generated during design and construction

When a building design is conceived, decisions must be made about the shape and size of the structure. Therefore, early in the design process, measurements and dimensions must be established to ensure the design will fulfil its purpose and the completed building will function well for its occupants.

Quantities of data multiply rapidly as floor areas and volumes of rooms in the structure are calculated, energy requirements are assessed, and thermal performance of the structure is analysed. Calculations will be made to establish structural strength and the forces generated by loadings, to make sure the building will be structurally sound.

> **Improve your English**
>
> Using data to calculate loadings is the job of a structural engineer. Research and write a short report on the role of a structural engineer working on a construction project.

This means that as soon as the design process begins, data are being generated which can be used by other members of the construction team, such as quantity surveyors preparing costings, contractors and subcontractors preparing tenders, and systems designers planning building services requirements.

> **Industry tip**
>
> While a large project is costed by professionals using large volumes of complex data, successfully costing even a simple project involving a small number of workers will rely on careful use of reliable data sources, such as up-to-date lists of material prices.

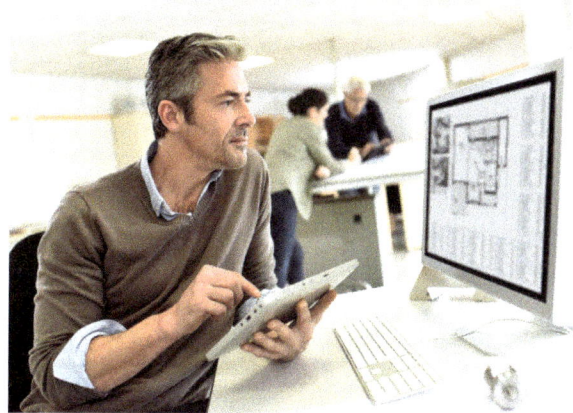

▲ Figure 8.2 Data are generated right from the beginning of the design process

Building services systems require extensive data calculations to establish **dynamic** elements, such as:
- air flow for ventilation systems
- gas supply rates for heating systems
- water flow rates and pressure levels for plumbing systems
- voltage and current ratings for electrical systems.

▲ Figure 8.3 Data are used to calculate dynamic elements in building services systems

> **Key term**
>
> **Dynamic:** characterised by frequent change or motion

> **Test yourself**
>
> What kinds of data can be generated during the design stage of a building project?

As the construction process proceeds on site, data regarding productivity and meeting deadlines are constantly scrutinised to manage efficient interactive working between site personnel.

Data generated and recorded during a project are valuable. Information about factors such as speed of operations and the effects of delays caused through bad weather or supply issues can be used in designing and building future projects more efficiently and economically.

A system that is increasingly helpful in managing large volumes of data through the complete life cycle of a building is Building Information Modelling (BIM), which integrates many data sources in a digital format. The system allows authorised users to access a range of important information during the design, construction, occupation and even demolition of a structure. There is more information on BIM later in this chapter, and in Chapter 4, section 7.

> ### Test yourself
> During which phases of a building's life cycle can BIM be used?

> ### Industry tip
> BIM is used mainly for large-scale complex projects. However, the BIM principle of having a central source of accessible and accurate data and information can be applied to good effect even during a project with just a few workers.

Structures within the built environment must have provisions for access by pedestrians and vehicles, along with reliable connections to services and telecommunications. These requirements mean that extensive transport systems and services delivery infrastructure must be designed, constructed and maintained.

Useful data can be generated by analysing vehicle and pedestrian movements, metering services provision and examining telecommunications activity. Information harvested from careful analysis can point to emerging trends and help in planning the future infrastructure needs of society.

▲ Figure 8.4 Analysis of infrastructure data can help in planning for the future

Data generated on completion and handover of a building

When a building is handed over from the contractor to the client for occupation and brought into use, a great deal can be learned from the analysis of data generated by a number of considerations:
▶ How successful was the delivery of the project by the contractor?
▶ Was the project delivered and handed over on time? If not, why not?
▶ Does the building meet its design brief fully now it is in use?
▶ What changes could be made to improve performance?

Gathering data to answer questions such as these is referred to as post-occupancy evaluation (POE). It may be undertaken by an independent consultant or by the client's own team and can provide valuable data to contractors who build repeat structures or specialise in specific project types.

Chapter 8 Information and data principles

POE is an increasingly important tool in analysing data to support the construction industry in meeting more stringent environmental targets, and more rigorous regulation connected with energy use. The design of systems delivering utilities such as electricity, gas and water, along with waste-disposal systems, can be refined and improved by analysis of performance data and metering of usage in completed buildings.

In most buildings, there are systems that automatically control various functions, such as heating thermostats or alarm motion sensors. In large structures, sophisticated computer-based building management systems can be installed to automatically control and monitor aspects such as general energy consumption and distribution, lighting levels, and water management and consumption rates.

> ### Key terms
>
> **Parameters:** limits which define the conditions of operation for a system or process
>
> **Software:** a sequence of digital instructions designed to operate a computer and perform specific tasks

> ### Research
>
> Find out the key difference between enterprise asset management (EAM) and a computerised maintenance management system (CMMS).

▲ Figure 8.5 Computer-based systems monitor and control a range of functions

Data generated by these systems can be used by estates and buildings managers to understand how buildings are operating, adjust and control systems to optimise their performance, and create reports or set alarms to give an alert when operational **parameters** are exceeded.

In the case of large estate facilities, such as a hospital complex or university campus, specialised computer **software** referred to as enterprise asset management (EAM) can be employed to manage the flow of data. This can inform accounting and purchasing decisions to sustain economic day-to-day operations and support planning for efficient use of individual buildings within the estate.

Data generated during maintenance

EAM can be used to collect data about efficiency, reliability and repair costs, allowing planning and work scheduling for maintenance. As inspections take place, data can be logged in real time, and work orders can be sent to personnel already on site to prioritise maintenance or repair tasks.

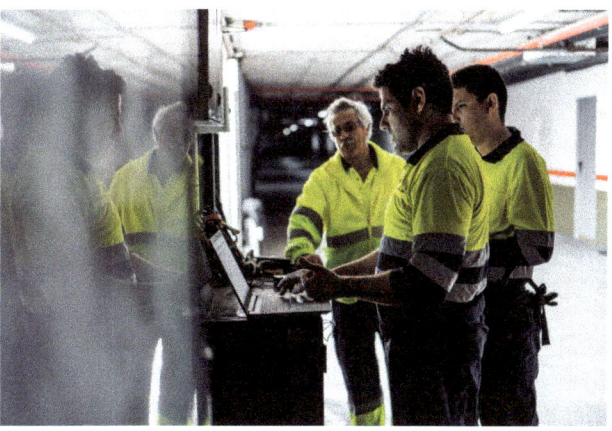

▲ Figure 8.6 EAM can create 'intelligent data' to manage the maintenance of a building

The real-time capture of data using systems like EAM can be linked with BIM systems to create so-called 'intelligent data' that can lead to improved efficiency throughout a building's life cycle. Benefits of this type of interoperability include:
▶ performance projections for long-term planning
▶ reliability of building services
▶ operational cost monitoring.

Data generated, processed and stored using information and communications technology (ICT) that is managed using equipment in multiple locations allows information-rich activities to be conducted in a collaborative way. Data are a key part of the construction process from design to demolition, used by both personnel in construction company offices and workers on construction sites.

> **Test yourself**
>
> What are the benefits of using 'intelligent data'?

1.3 How data can be used

Table 8.1 identifies how data can be used to complete construction projects successfully.

> **Key term**
>
> **Procurement:** the process of agreeing business terms and acquiring goods, products and services from suppliers

▼ Table 8.1 How data can be used in construction projects

Data use	Applications
Understanding behaviour	Data can be used to analyse and understand the behaviour of personnel working on a project. Efficient deployment of skilled workers and the creation of workforce motivation to achieve operational efficiency are valuable management skills. Data regarding personnel numbers on site can be referenced to output over time to establish optimum levels of staffing, enabling managers to support efficiency in office environments and on site. In terms of data use, the reference to 'behaviour' could be extended to factors that might impact on project success, such as the behaviour of certain materials and equipment in variable conditions. With the emergence of new materials and the progressive adoption of new construction methods, data about how these materials and methods function are of real value.
Performance assessment	Key performance indicators (KPIs) can be linked to appropriate data to measure performance in areas such as: • monitoring project costs • tracking project progress over time • identifying company strengths and weaknesses • confirming client satisfaction. Uses for performance data include: • tracking profitability by comparing cost with budget • highlighting trends in complaints from clients • evaluating quantities of waste with a view to improving recycling.
Improving market competitiveness	When companies bid for contracts, they need to be well informed about workforce availability, materials and components procurement, current regulatory requirements and fluctuating economic conditions. Accurate and current data are critical in providing reliable information, so that a company can improve and maintain its success in competition with other contractors. Fully understanding the area of construction activity that a company works in through efficient data analysis is key to making decisions on future company development. Decisions can be made regarding investment in training and equipment needs. Assessment of which types of construction project match the capabilities and experience of a company will help to maintain financial profitability and a reputation for high-quality work.
Allocation of resources	A company may work on a number of projects simultaneously. This means that assets and resources will need to be actively managed to make best use of them across the range of work being progressed. Data on factors such as personnel placement, work activities, plant and equipment usage, and temporary accommodation on site allow analysis of resource needs and efficient allocation across the range of project commitments.

Chapter 8 Information and data principles

Case study

After analysing footfall data in a town-centre pedestrianised area, the local authority decided that improvements were required to footpaths, landscaped areas and lighting. The works involved the installation of gravel and paved paths, granite kerbs, stone steps, play and sports equipment, and street furniture.

During a site survey as part of project planning, some challenges were identified, including:
- a lack of site access, restricting movement of materials to the work location from delivery points
- a lack of onsite storage.

Management discussions took place and research was undertaken into available options. Based on data made available by local employment agencies and advice from local plant hire companies, the decision was made to:
- temporarily increase the workforce
- hire materials conveyers to move materials from the restricted delivery bays.

Would you agree that these are effective strategies?

If you agree, state why.

If you disagree, state your alternative proposals.

Digital data networks that transfer information efficiently between users have become integral to our increasingly technology-based communities and now have an important role in the successful completion of construction projects.

▲ Figure 8.7 Efficient digital data networks have an important role in everyday life as well as construction

Improve your English

Make a list of the ways that data are transferred using digital networks. Write a short description of each way that you identify. An example you could start with is streaming films over the internet.

Test yourself

How can data be used to improve a company's market competitiveness?

2 Sources of information

So far in this chapter we have considered how data are the raw material that can be processed to provide information. The resulting information can then be presented in various ways and used by construction personnel in a range of activities.

In this section, we will consider different sources of current and accurate information and how they can be interpreted for use in the workplace.

2.1 Interpreting data sources

Specific data sources may provide information for workers with designated responsibilities.

For example, bricklayers working on the construction of a building's superstructure would not require information about which size and type of cables the electrical installation requires. However, they would need information on the type of bricks and blocks that are required and the type of mortar to be used.

We can identify specific sources of information and which personnel groups or individuals could use them.

Product data

Product data provides information or instructions about how to use or install a product correctly. For example, a manufacturer has to supply information about how to handle and store a product safely in accordance with health and safety regulations, such as the Control of Substances Hazardous to Health (COSHH) Regulations 2002. There is more on these regulations in Chapter 1.

▲ Figure 8.8 COSHH symbols that could appear on a product label

The correct and safe installation or use of materials or components could be made clear by using illustrations, diagrams, performance charts, or clear written instructions.

> ### Health and safety
> Many materials and substances used on site have the potential to cause harm or injury. Always check for product data labels and examine them carefully.

Product data are closely tied to manufacturers' specifications. These provide very specific information on performance data, such as the temperature range within which construction work must take place or the type of surrounding environment in which the product will operate successfully.

Manufacturers' specifications may include key data and information on how their product should be assembled, dismantled, maintained, repaired, examined or inspected.

Product data can be used by building designers working in an office environment or by onsite personnel installing a range of systems. Trade workers might use product data to confirm that the materials they are working with are in accordance with specifications for the work task.

> ### Test yourself
> What key information can be provided by a manufacturer's specification?

Client specification

In the Royal Institute of British Architects' (RIBA) plan of work 2020, the client specification has been defined in part as 'a statement or document that defines the project outcomes and sets out what the client is trying to achieve'.

The client's needs will obviously depend on the type and use of the required structure, for example:
- A client requiring construction of a dwelling may specify the floor areas needed to satisfy the living requirements of the intended number of occupants.
- A client who has commissioned the construction of an office block may specify the levels of light within the building to allow comfortable work in office spaces.
- The specification for a retail development may centre on the project being attractive and convenient for customers to use.

Information in the client specification must be interpreted carefully in order to produce a design brief that meets the objectives as fully as possible. Professional personnel such as architects, quantity surveyors and engineers will give careful attention to the client specification when performing their work.

Building Information Modelling (BIM)

As mentioned earlier in this chapter, BIM is a structured system that allows authorised users to access a range of important information at all stages of construction and beyond. It uses digitally processed information to analyse design elements of a building, including 3D modelling.

Using BIM, complex design ideas can be transformed into a medium that is easier for all personnel to work with. It allows collaboration between all the designers, engineers and contractors working on a project, providing comprehensive information about each role's workflow. This enhances the process of design

Chapter 8 Information and data principles

and construction and allows exploration of alternative design possibilities before work on site begins.

BIM allows digital data describing internal building engineering services to be presented and analysed visually, so that the way they interact with each other can be seen. Clashes in systems can then be identified at an early stage in the design process; this is known as 'clash detection'.

An example could be drainage system pipework within a building interfering with the route of heating ducts. In the past, this could require extensive redrawing of plans and cause expensive delays if the 'clash' was discovered after work had commenced.

▲ Figure 8.9 BIM allows possible clashes in the systems of a building to be detected before work begins

> **Research**
>
> Find out the history of BIM. When was it developed and what did the first BIM system contain?

The Common Data Environment (CDE) is a single central source of information used within the BIM system. Relevant documents and data are brought together in a shared digital environment that can be accessed by all authorised personnel collaborating on the project.

The CDE can contain different types of digital information, including schedules, contracts, registers, reports and 3D models, forming the foundation of shared information on which collaboration can take place. Large amounts of digital data and information can flow in a controlled way through the CDE during the development of a project, making it possible to reduce mistakes and avoid duplication.

> **Test yourself**
>
> What function does the CDE perform in a BIM system?

Work programme planning

To ensure work can be completed on schedule and within budget, careful prior planning of the construction process is required. Planning a programme of work is often undertaken using charts that provide data and information about the sequence of activities.

There are two main documentary methods of planning the sequence of work in construction – Gantt charts and critical path analysis (CPA) (see Chapter 3, section 4.2).

> **Improve your English**
>
> Search online for the origins of the Gantt chart and write a short report.

> **Research**
>
> Look online for examples of critical path analysis. Once you understand how they work, create your own example for a simple activity, such as making a cup of tea. Ask one of your peers to do the same and compare your results.

Commissioning and certification

Commissioning is the process of confirming that all building systems are installed, tested, operated and maintained according to predetermined operational requirements. Data and information linked to engineering techniques and procedures are used to check the correct installation and function of systems such as heating, ventilation and air conditioning.

Systems must be confirmed as fit for purpose before complete handover of the building to the client. However, as part of the contract arrangements, there are circumstances where performance testing of systems may continue after the client has accepted handover. For example, some systems are dependent on different weather conditions, and commissioning inspections and adjustments may take place over the first year of occupation.

Commissioning data and information form a checking and testing format to bring systems into operation and verify that they are in good working order. The testing format confirms specific details and responsibilities, such as:

- the party or parties responsible for each aspect of commissioning
- whether the commissioning process and outcomes must be witnessed (this is a legal requirement in some cases)
- the standards that should be adhered to
- the specific documentation that is required, which can include a record of operations, maintenance and future works.

A certificate is issued to confirm that installation and operational standards have been met, which must comply with technical benchmarks and criteria set by appropriate organisations. The standards could be set by Building Regulations, the British Board of Agrément, the manufacturer or another recognised certification body.

> ### Research
>
> Visit the British Board of Agrément (BBA) website at **www.bbacerts.co.uk**
>
> Search the site and identify the five key areas that the organisation works in. Find a case study and note how certification benefited the company working under BBA guidance.

Even after a building's systems have been commissioned and certified, ongoing inspection and testing are required throughout the life cycle of the building. A schedule for this will include test data that need to be recorded to ensure optimal operational efficiency is maintained.

▲ Figure 8.10 Testing an electrical system

Records of test data from inspections are a valuable source of information for new owners, if the building changes ownership. This is especially true in the case of a large complex structure that may be owned by **commercial** investors. A building is an expensive asset that must be properly maintained to retain its value.

In the case of a smaller **domestic** property, a condition report can be prepared for a new owner or homebuyer. This takes the form of a visual inspection to generate data and information about factors such as the condition of the building's structure, heating and lighting systems, exterior roof surfaces, attic roof spaces, floor spaces and underfloor spaces.

Efficient commissioning and reputable certification of a building's systems, coupled with planned and effective maintenance, will contribute to greater efficiency throughout the life cycle of a building. Operational efficiency can translate into reduced energy needs, which is an increasingly important factor in the continuing drive to reduce carbon emissions. Accurate and appropriate data and information are essential to the success of energy-saving strategies.

> ### Key terms
>
> **Commercial:** relating to buying and selling
>
> **Domestic:** relating to a dwelling or home

> **Improve your English**
>
> Using the definitions of 'commercial' and 'domestic' supplied, write a summary of the ways that data can be used for each category of building in terms of commissioning and certification.

2.2 Using data to calculate outcomes or cost

The term 'calculation' can be defined in different ways, according to the context. It can be used to describe the process of:
- using information to make a judgement or plan something carefully and intentionally, for example to calculate outcomes
- determining something by mathematical or logical methods, for example to calculate costs.

Data can be used to provide results that meet both of these definitions.

Calculating outcomes

When calculating outcomes, data can be used to analyse past or current activities to model what could happen in the future. A familiar example of this is forecasting the weather. Attempting this requires the analysis of past data related to a huge number of variables which, as we see daily, makes a reliable prediction difficult.

Something more straightforward might be calculating the outcome of using timber of specific dimensions to produce a load-bearing element of a building, such as a suspended floor. Consistent data have been recorded over many years which can be applied to produce reliable structural calculations, ensuring that the proposed design is safe and capable of bearing the required loads.

> **Test yourself**
>
> What are the two definitions of the term 'calculation'?

Building services engineering elements of a structure produce a wealth of data that has been harvested in differing types of building under varying conditions. This has led to the compilation of charts and tables that can be referred to when designing similar structures. Data from past experience can be used to calculate acceptable operational tolerances in new designs in order to avoid 'over-engineering', which could be wasteful and add to costs.

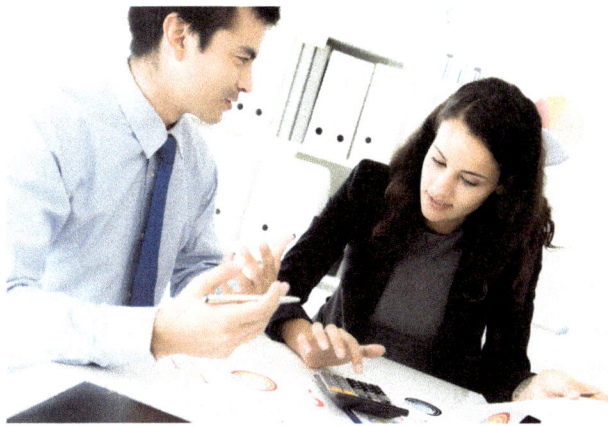

▲ Figure 8.11 Analysing data from past projects supports efficient design in new projects

Calculating costs

Many databases are used to help cost projects of all types and sizes. They draw on historical costing data and are often designed to track current prices of materials and components, almost in real time. Using up-to-date data for costing a project is vital to avoid cost overruns and accurately assess financial risks in a competitive construction environment.

Data from across the sector are collated, analysed, modelled and interpreted by the Building Cost Information Service (BCIS) of the Royal Institution of Chartered Surveyors (RICS). This valuable source of data is used by many contractors for projects of different sizes. There are also many other companies and institutions that specialise in providing data for costing a project, such as PropertyData and Bestdata.

A major project with complex design elements needs accurate data to produce costings that will result in profitability for the contractor and a fair price for the client. Of course, profitability is key to the success and financial health of any size of construction company, and small and medium-sized enterprises (SMEs) also rely on accurate data for success. A small contractor or even a sole trader benefits from keeping accurate project records to provide data for costing future work.

Onsite Construction T Level: Core

▲ Figure 8.12 A small contractor or sole trader can benefit from keeping accurate data from past projects

3 Data management and confidentiality

Data and information sources used in the construction industry are of great value, both operationally and financially. This means they must be stored securely and protected.

> **Improve your English**
>
> Write a short account explaining why data and information sources are valuable operationally and financially.

3.1 Data storage

Physical storage

In the past, data tended to be stored as paper documents and physical drawings. While hard-copy documentation is still used extensively, digital data storage is increasingly used.

There can also be a crossover between physical and digital data storage, if paper documents are scanned and converted to a digital medium for easier transmission to others in different locations. This can speed up communication of useful data that might be needed urgently.

> **Industry tip**
>
> Everyone involved in using and storing physical data must be conscientious in keeping documents and drawings secure and in good condition. Even if they do not contain sensitive details, they are valuable sources of information that should be taken care of.

Virtual storage

Virtual storage is the storage of data in a digital format, for example on a computer hard drive, on a portable flash drive or in a location that is remote from the user.

Remote storage is often referred to as 'cloud storage' or 'in the cloud'. Data are transmitted digitally through a network and then stored on a server. These storage facilities are often owned by specialist companies who use multiple data centres around the world.

Data files are stored remotely in a number of different ways:
▶ Public cloud: data are stored in the service provider's data centres across multiple regions or continents. Users pay the provider based on the amount of data stored.
▶ Private cloud: data are stored in an in-house storage facility, forming a dedicated environment that is protected from outside access by a **firewall**.
▶ Hybrid cloud: this is a mix of public and private cloud storage, offering businesses flexibility and more ways that data can be managed and made available to authorised users.

Data storage companies must provide safe and secure storage and have facilities in place to back-up data to prevent loss.

> **Key term**
>
> *Firewall:* a protective software program or hardware device that monitors, filters and may block data entering and leaving a network

Chapter 8 Information and data principles

▲ Figure 8.13 A server room in a data storage centre

> **Test yourself**
>
> What is meant by storing data 'in the cloud'?

3.2 Confidentiality

Some data must be kept confidential. For example, data for a construction project commissioned by the government or a bank may contain sensitive details which could threaten security if accessed by unauthorised parties.

Confidential data stored in a physical format must be locked away and protected in secure facilities. It may be necessary to consider protection from fire damage, in addition to prevention of unauthorised access.

▲ Figure 8.14 Confidential physical data should be stored securely

Measures to maintain the security of digital data include:
- using anti-virus software and keeping it up to date
- using a system of strong passwords so that access to data is restricted
- using **encryption** to convert data from a readable format into an encoded format
- installing **firmware** and software updates as soon as they become available (these often contain security patches and new security features)
- logging off (or turning off) computers and locking them if possible.

The term 'hacking' has become increasingly familiar, referring to theft or manipulation of critical data, usually with malicious intent. Hackers are focused on bypassing the security measures designed to keep them out of data processing and storage systems. Security measures must therefore be rigorously designed, carefully adhered to, and updated when necessary, in order to meet the growing challenge of these attacks. Increasingly sophisticated systems are being devised to maintain the security of digital data.

> **Key terms**
>
> **Encryption:** the process of converting data or information into a code to prevent unauthorised access
>
> **Firmware:** software that has been permanently installed in a device to make it work as the manufacturer intended

> **Improve your maths**
>
> Find out about the mathematical processes used to encrypt data.

The news often refers to 'cyberattacks' when describing the work of individuals or organisations who steal data or install damaging software known as **malware**. Malware can behave like a virus, in that it can easily be passed from computer to computer, or system to system, so that the 'infection' spreads.

A Trojan horse or Trojan is a type of malware that is often disguised as legitimate software, misleading the user as to its true intent. It tricks systems operators into allowing unauthorised access.

> **Key term**
>
> **Malware:** software that is specifically designed to disrupt, damage or allow unauthorised access to a computer system

> **Research**
>
> Search online for the names of some common Trojans.

> **Test yourself**
>
> Why is malware sometimes referred to as a Trojan?

Loss of data can be very damaging to the operational efficiency of construction companies that are managing complex project commitments. Recovery of lost or stolen data has developed into an important activity that uses specialist techniques to beat the abilities of hackers and data thieves.

A physical or digital data source may become obsolete or out of date and require removal from data storage. It may not be sufficient to simply archive the material since it could still contain sensitive information that should not be accessed by unauthorised persons.

An example could be documentation related to a government or financial institution contract (as mentioned earlier in this chapter).

It may be the case that many design details remain the same for similar future contracts but information specific to the completed contract may compromise a building's security if allowed to fall into the wrong hands.

Physical documents containing sensitive data can be shredded to prevent the information they contain being of use. Digitally stored information and data can be electronically 'shredded' using specialist software or the stored data can be **sanitised**.

> **Key term**
>
> **Sanitised:** the process of treating data held digitally to reduce the likelihood of retrieval and reconstruction to an acceptable level. Some forms of sanitisation will allow the subsequent reuse of media, while others render the media completely unusable

Legal requirements

Data protection in the UK is governed by the UK General Data Protection Regulation (UK GDPR), which came into effect on 1 January 2021. It should be considered alongside the Data Protection Act 2018. All companies that sell goods or services need to comply with these legal texts.

The aim of this legislation is to ensure that personal data are gathered legally, and that those who collect and hold data protect them from misuse and exploitation. A construction company that stores personal data about its workers should always ensure confidentiality.

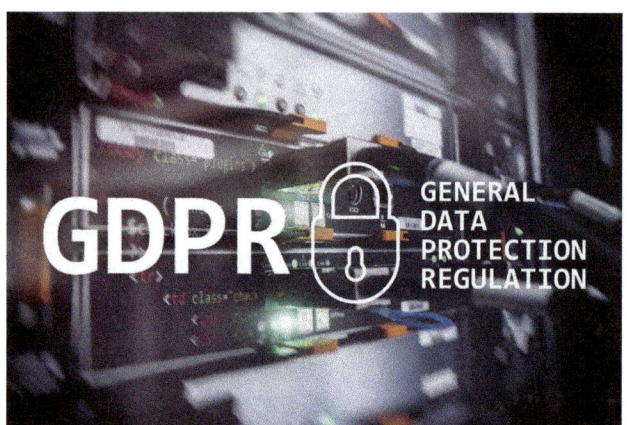

▲ Figure 8.15 GDPR protects the confidentiality of personal information

> **Industry tip**
>
> GDPR is not simply guidance – it is law. As such, there are legal penalties if it is not adhered to. If your role in construction involves dealing with anyone's personal information, never treat it casually.

> **Improve your English**
>
> Search online for the seven principles of GDPR and write a summary.

Chapter 8 Information and data principles

Assessment practice

Short answer

1. In what different forms can data be produced?
2. Write a definition of data interoperability.
3. Who can make use of data generated at the start of the building design process?
4. What do the letters EAM stand for?
5. Why might commissioning and certification of systems in a new building continue over the first year of occupancy?

Long answer

6. Describe a work task where metadata would contribute to efficiency when designing a building.
7. Explain how generalisation can make producing information from data analysis easier and more manageable.
8. Explain how BIM is structured to manage large volumes of data to allow collaboration between project personnel.
9. Explain the purpose and list the benefits of using data for post-occupancy evaluation (POE).
10. Explain how data that are no longer needed but may still contain sensitive information should be securely disposed of. Include physical and digital data in your explanation.

Project practice

A detached house has an integral garage which the owner wishes to convert into living space. The proposal is to remove the existing wall between the lounge and the garage to create a larger lounge area.

The existing garage door opening will be closed by building a cavity wall as specified in the working drawing, with a new window frame installed above the level of the new masonry.

Review the section in this chapter on product data, and then research and list manufacturers' product data (sometimes called 'technical data') for the following items which will be used in the conversion:
- cement
- face bricks
- thermal blocks
- concrete blocks
- stainless steel wall ties.

Include data for the following where appropriate:
- acceptable temperature limits when using (cement)
- weight and water absorption (brick)
- strength (concrete blocks)
- thermal conductivity (thermal blocks)
- spacing (wall ties).

Separately, list data that provide health and safety guidance where appropriate for each listed item.

▲ Figure 8.19 Working drawing

Chapter 9 Relationship management in construction

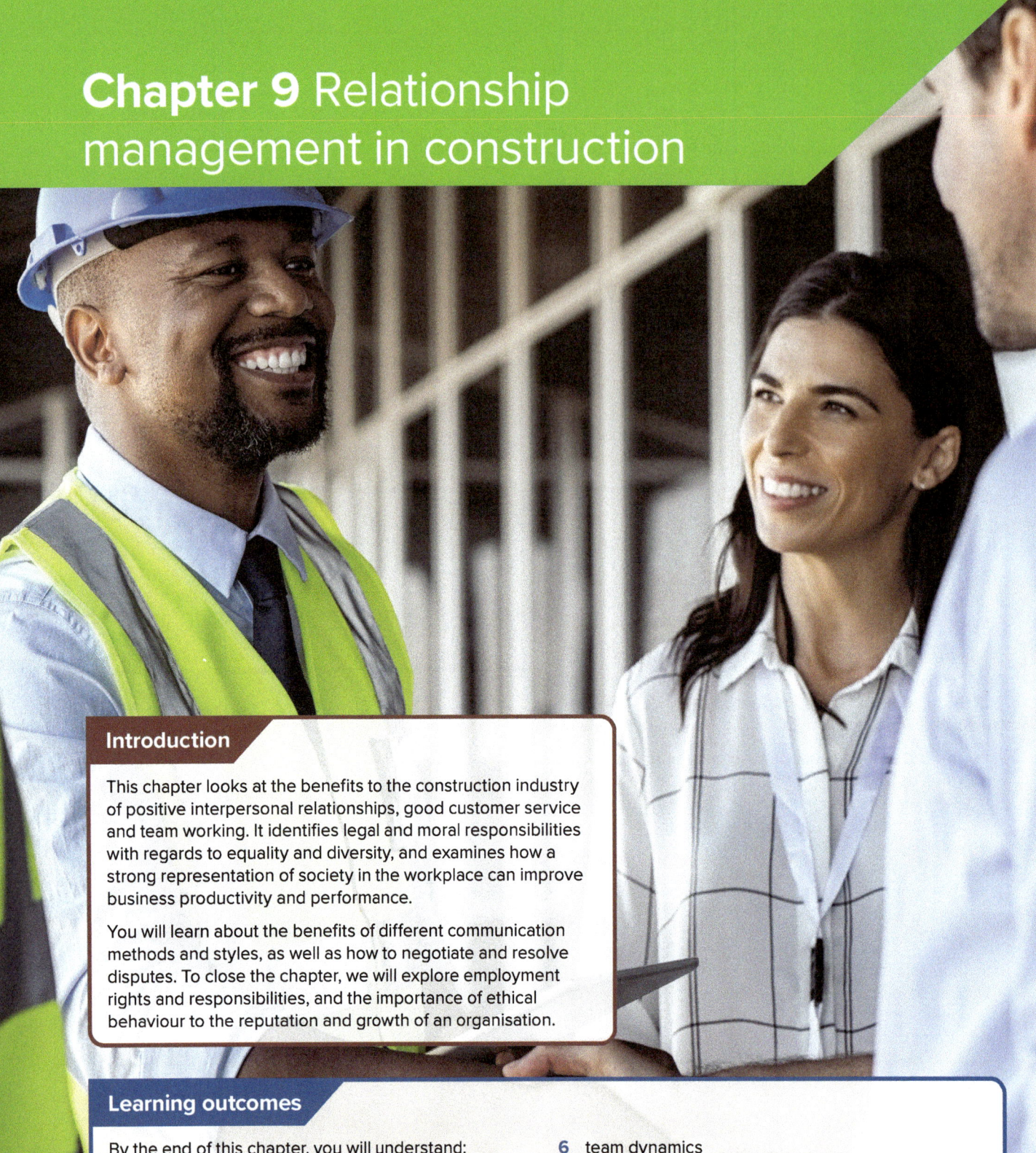

Introduction

This chapter looks at the benefits to the construction industry of positive interpersonal relationships, good customer service and team working. It identifies legal and moral responsibilities with regards to equality and diversity, and examines how a strong representation of society in the workplace can improve business productivity and performance.

You will learn about the benefits of different communication methods and styles, as well as how to negotiate and resolve disputes. To close the chapter, we will explore employment rights and responsibilities, and the importance of ethical behaviour to the reputation and growth of an organisation.

Learning outcomes

By the end of this chapter, you will understand:

1. stakeholders
2. roles, expectations and interrelationships
3. the importance of collaborative working to project delivery and reporting
4. customer service principles
5. the importance of team work to team and project performance
6. team dynamics
7. equality, diversity and representation
8. negotiation techniques
9. conflict-management techniques
10. methods and styles of communication
11. employment rights and responsibilities
12. ethics and ethical behaviour
13. sources of information.

Chapter 9 Relationship management in construction

1 Stakeholders

Any group or individual that has an invested interest in the long-term success of an organisation is referred to as a stakeholder. Stakeholders can affect and be affected by the achievement of an organisation's business objectives. Therefore, it is important that they can work together.

Internal stakeholders support or have concern for an organisation and benefit from their direct relationship with it, for example employers and employees. If an organisation did not have any internal stakeholders, it would simply cease to exist.

There is a risk that some internal stakeholders will have conflicting interests. For example, employees want to earn a good wage, whereas their employer may want to reduce outgoings on staff remuneration. To avoid this problem, successful construction businesses keep the number of stakeholders to a minimum and ensure they have a good balance between all of their interests in order to prevent conflict.

External stakeholders do not have a direct relationship with an organisation and are not employed by it, yet they can still have an indirect effect on it. For example, a building control officer enforcing building regulations will determine the standards that a contractor has to adhere to.

External stakeholders that do not benefit from a direct interest in an organisation are also known as secondary stakeholders.

Table 9.1 gives examples of internal and external stakeholders.

▼ Table 9.1 Types of stakeholders

Internal stakeholders	External (secondary) stakeholders
• Clients • Architects • Investors • Managing directors • Site managers and supervisors • Structural and civil engineers • Buyers • Contracts managers • Estimators • Trades forepersons • Quantity surveyors • Project managers • Subcontractors	• Suppliers • Government • Trade associations • Emergency services • Building control officers (BCOs) • Regulatory authorities • Communities and local residents • Health and Safety Executive (HSE) • Local councils • Lobby groups and activists, for example **environmental activists** • Trade unions • Professional bodies • End users in a construction project

Key term

Environmental activists: people who campaign for the protection of the natural environment

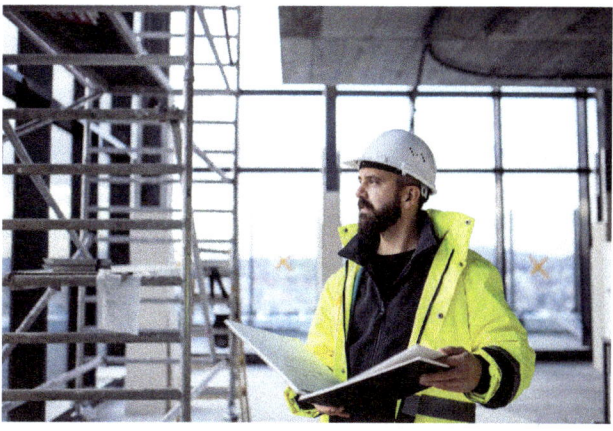

▲ Figure 9.1 A site manager is an internal stakeholder

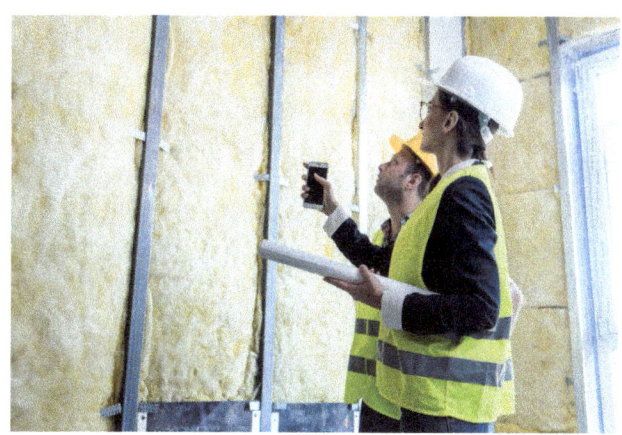

▲ Figure 9.2 The Health and Safety Executive (HSE) is an external stakeholder

Test yourself

Give a definition of 'stakeholder' in the construction industry.

Stakeholders will always be present in construction projects, and their influence plays a part in determining success or failure. For this reason, whenever possible, internal and external stakeholders should be selected to suit specific construction projects because of their skills, knowledge and experience working on similar jobs.

> **Research**
>
> Find out what shareholders are. Describe the main differences between shareholders and stakeholders.

> **Test yourself**
>
> How could an external stakeholder, such as a community, indirectly affect an organisation?

2 Roles, expectations and interrelationships

The roles, expectations and interrelationships of stakeholders throughout all stages of a construction project, from design to construction to handover and use, are covered in detail throughout this book:

- hierarchy of project management (see Chapters 3, 4)
- promoting good relationships across the project (see Chapters 1, 9, 11)
- cost-control measures (see Chapters 4, 6, 8, 11)
- time-management methods (see Chapter 3)
- Corporate Social Responsibilities (CRSs) (see Chapter 3)
- Affordable housing – Section 106 Agreement (see Chapter 3)
- follow-up and review (see Chapter 3).

> **Test yourself**
>
> Explain the benefit of using a hierarchy to structure an organisation in the construction industry.

3 The importance of collaborative working to project delivery and reporting

A **collaborative** approach to project delivery and reporting is essential to ensure work is completed on schedule, within budget and to the client's specification.

Reporting any foreseeable problems before they occur, and working with the project team to resolve them, is essential for meeting project aims and objectives. Collaborative working also ensures projects are completed to the minimum industry standards and building regulations.

Project teams and other stakeholders that communicate effectively at every stage of construction work are more likely to provide a safe working environment for employees and others that may be affected by work activities.

Figure 9.3 identifies various stages of construction work when members of a project team need to collaborate in order to achieve a successful project outcome.

Pre-design stage: preparing design drawings, obtaining planning permission

↓

Design stage: tender process, appointing the principal contractor/subcontractors, health and safety file, risk assessments, organising plant and equipment

↓

Construction: off-site manufacturing, e.g. wall panels and roof trusses; onsite construction

↓

Tracking/monitoring/controlling performance: progress meetings, health and safety meetings/tours, site inductions, toolbox talks

↓

Handover: snagging, handover to the client

↓

In use

▲ Figure 9.3 Stages of construction work

In Chapter 4, we looked at Building Information Modelling (BIM) and how this can be used to collaborate at various stages of a construction project, by identifying design problems and communicating efficiently with the project team. The construction management team can also use workflow software packages to plan and organise work. Information can then be shared easily with others and worked on as a **living document** to bring a project in on schedule.

Later in this chapter, we will look at the benefits for project delivery of members of the project team collaborating using other communication methods, including face-to-face meetings.

> **Key terms**
>
> **Collaborative:** involving people or groups of people working together with a single common interest or aim
>
> **Living document:** a document that can be changed, amended or updated when necessary for the benefit of the user

Chapter 9 Relationship management in construction

▲ Figure 9.4 A project management team discussing project progress

▲ Figure 9.5 A contractor meeting with a client

> **Test yourself**
>
> Why is a collaborative approach important for project delivery and reporting?

4 Customer service principles

Clients are responsible for initiating and financing construction projects. It is therefore important for contractors to maintain good working relationships with their clients, in order to keep them happy throughout a project and increase the chances of repeat business.

Contractors need to create a good first impression. This can be achieved through a portfolio of successfully completed contracts and positive **testimonials** from previous clients. Demonstrating good product knowledge is also extremely important, as the client is unlikely to be an expert and will be looking for guidance to steer them through the project and achieve their vision.

When a contractor works with a client for the first time, it is essential to establish a good level of trust. This can start during the planning stage of construction by maintaining good lines of communication with the client, such as responding promptly to any messages and involving the client in decision making. Listening to the views of clients at every stage of a project and treating them with empathy is also important in building their confidence.

Total trust between parties can take time to establish, but just a moment to ruin. However, if a contractor can demonstrate their ability to manage the client's project efficiently by meeting agreed timescales and working with honesty and integrity, then their professional bond will grow.

Dealing with clients and employees is just as important as managing a project itself, especially if they have a complaint. The way a contractor responds to negative feedback is important; they need to demonstrate that they are continuously looking for ways to improve, in order to keep clients satisfied and maintain a good reputation.

> **Key term**
>
> **Testimonials:** statements of recommendation produced by satisfied customers or clients that confirm the quality of a product or service

> **Improve your English**
>
> Write a sentence to explain the term 'empathy'.

> **Improve your English**
>
> Write a testimonial on a supplier's website for a product you have recently purchased.

> **Test yourself**
>
> Describe the customer service principles that should be applied at various stages of a construction project.

Onsite Construction T Level: Core

5 The importance of team work to team and project performance

When employees collaborate towards common goals without friction or conflict, they create a healthy work environment that improves staff morale. Employees who feel part of a team and are offered the chance to be creative and learn new skills develop positive 'can-do' attitudes that contribute to the long-term success of an organisation.

Teams perform best when individual members are motivated by satisfaction with their roles and everyone has accountability for the part they play.

▲ Figure 9.6 Employees working together

Team members will work well together if they can communicate openly, without fear of reprisal. Individuals should be comfortable both asking their peers for help when needed and providing feedback to the team when the opportunities arise. This will create an environment that improves trust between team members and ultimately leads to improvements in efficiency and productivity.

Conversely, conflict, tension, low engagement and lack of trust in a team will have a negative impact on project performance.

> **Test yourself**
>
> Explain the advantages of team work for team and project performance.

6 Team dynamics

Team dynamics refers to psychological processes and behaviours occurring in a team that influence its direction and performance.

People's personalities and behaviour are often uncontrollable and unpredictable, and they can have either a positive or negative influence on team dynamics. With the right strategies in place to address any issues early on, a good team leader can manage the relationships between team members, in order to keep lines of communication open and ensure active participation and co-operation.

When all members of a team work collectively, listen to and support each other, they are more likely to resolve challenges and achieve objectives. Signs of positive team dynamics are when members work together without conflict and trust the expertise, knowledge and abilities of individuals within the team.

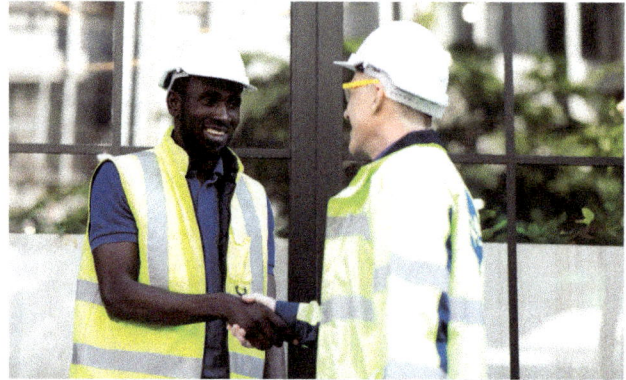

▲ Figure 9.7 Construction professionals demonstrating good team dynamics

Chapter 9 Relationship management in construction

Businesses rely on positive team dynamics to generate new ideas and improve performance. It is therefore important to identify poor team dynamics as soon as possible. Signs of negative team dynamics include:
- disruptive work behaviour
- failing to achieve positive results
- not being accountable
- poor decision making.

Poor team dynamics can cause individuals to feel vulnerable, making them less reliable and less able to adapt to the changing situations that may arise during building projects.

7 Equality, diversity and representation

Employers that embrace **equality**, diversity and inclusion develop an organisational culture that helps people to achieve their full potential based on their talent, not on factors such as their age, background, race, religion, gender, disability or sexual orientation. Employing a talented, diverse representation of society allows different views and ways of thinking to be shared within the organisation, to reflect the local culture and community.

Organisations that promote the values of equality, diversity and inclusion are often more sociable, productive and efficient, which improves profitability for their shareholders.

> **Key term**
>
> *Equality:* a state where all members of a society or group have the same status, rights and opportunities

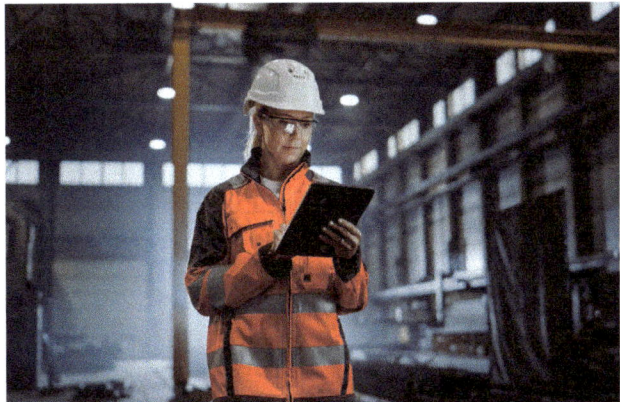
▲ Figure 9.8 A female construction worker

▲ Figure 9.9 People with disabilities should be supported to work in the construction industry

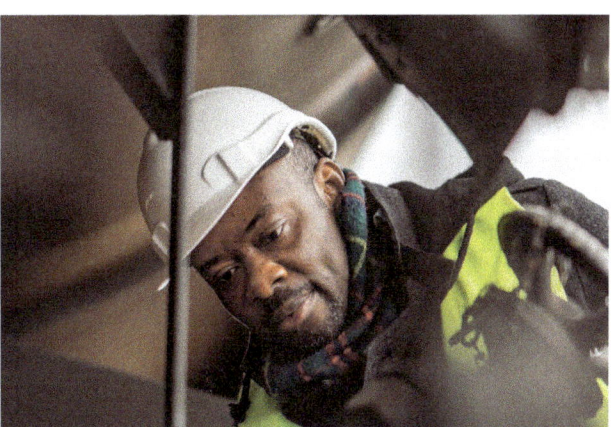
▲ Figure 9.10 The construction industry must aim for a truly inclusive and diverse working environment

To demonstrate its commitment to equality, diversity and inclusivity, an organisation usually has a written equality and diversity policy, which should be read and signed by all employees as part of their induction.

The policy should include:
- values
- legal duties

235

- processes which employees should follow to challenge discrimination or if they have a **grievance**, for example **whistleblowing**.

The policy also helps to identify any staff training needs and how to support them.

> **Key terms**
>
> **Grievance:** a feeling of having been treated unfairly
>
> **Whistleblowing:** the act of reporting information about wrongdoing

Employees who work for organisations that have a robust equality and diversity policy and recognise equal opportunities will feel more valued.

> **Improve your English**
>
> Read the equality and diversity policies for your employer and your training provider. Summarise the main differences in a short paragraph.

Discrimination is the unfair treatment of someone because of their characteristics. An employer could openly discriminate against certain groups or individuals while recruiting new employees or promoting existing employees within their organisation. For example, they should not advertise for specific age groups or genders, unless they are specifically needed to fulfil the job role.

Besides the moral reasons for equality and diversity, businesses also have legal duties as outlined below.

7.1 Equality Act 2010

This is the main piece of equality and diversity legislation that protects people from discrimination at work and in wider society. It outlines the different ways that it is unlawful to treat someone, and it strengthens protection in some situations.

Under the act, the following are protected characteristics:
- age
- disability
- gender reassignment
- marriage and civil partnership
- pregnancy and maternity
- race
- religion or belief
- sex
- sexual orientation.

> **Improve your maths**
>
> A construction site has 267 workers, with age ranges as follows:
> - 18–25 years – 25 workers
> - 26–35 years – 116 workers
> - 36–45 years – 53 workers
> - 46 years and over – 73 workers.
>
> What percentage of the workers are aged between 18 and 25 years?

7.2 Employment Rights Act 1996

This act legally protects the personal rights of employees and **workers**. It covers areas such as:
- contracts of employment
- protection of wages
- zero-hours workers
- Sunday working
- flexible working
- rest breaks
- study and training
- unfair dismissal
- maternity and parental leave
- redundancy payments.

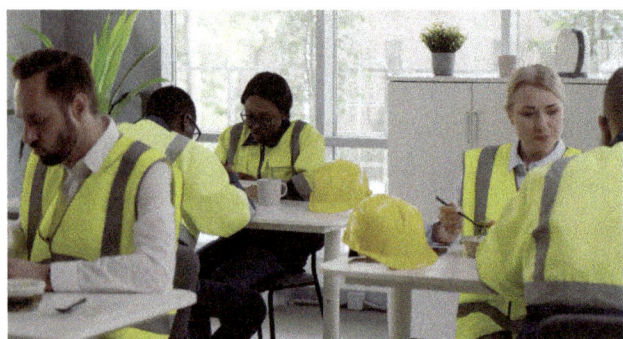

▲ Figure 9.11 Rest break on a construction site

> **Key term**
>
> **Workers:** people who do not have a permanent contract of employment with an employer but are contracted for work or services

7.3 Employment Act 2008

This act covers:
- the procedure for the resolution of employment disputes between employers and staff
- compensation for financial loss in cases of unlawful underpayment or non-payment of staff

- the enforcement of minimum wages
- the enforcement of offences under the Employment Agencies Act 1973
- the right of trade unions to expel or exclude members on the grounds of membership of a political party.

7.4 Human Rights Act 1998

The Human Rights Act 1998 incorporates the European Convention on Human Rights into UK law. It sets out fundamental rights and freedoms that everyone in the UK is entitled to, including:
- the right to life
- freedom from torture or inhuman or degrading treatment
- freedom from slavery
- the right to liberty and security
- the right to a fair trial in a court of law
- the right to respect for private and family life, home and correspondence
- freedom of thought, conscience and religion
- freedom of expression
- freedom of peaceful assembly and association
- the right to marry freely
- the right to an effective remedy in a national court
- freedom from discrimination
- the right to education
- the right to vote.

Under this act, people who are not treated fairly, equally and with dignity and respect can take legal action to defend their rights.

> **Case study**
>
> Ashley is employed as an electrical engineer by a small construction company with a workforce of 12 full-time employees. She is the only female member of staff. Since completing her apprenticeship when she was 19 years old, she has worked for her current employer for six months.
>
> Ashley has started to notice that she is being treated differently to the other employees, for example being given less-challenging jobs and being paid slightly less than another electrical engineer who started at the same time as her.
> - What employment rights does Ashley have?
> - Working in a small group, discuss Ashley's situation and explain the steps she should take to achieve a satisfactory outcome.
> - Do you think this is a case of workplace discrimination? Why?

> **Test yourself**
>
> Explain the benefits of equality and diversity in the workplace.

8 Negotiation techniques

At various stages of construction projects, different parties need to negotiate money, assets or resources in the best interests of the organisation or as an individual.

Negotiation skills are important in the construction industry, for example when buyers acquire land. Negotiations between buyers and landowners must be managed carefully due to the large sums of money involved. A poorly negotiated deal to purchase land will impact on profit margins when projects are completed and sold.

Once land has been acquired, planning permission may be needed before building work can start. This involves the project planning team working closely with the local planning department, to develop and negotiate initial concepts into a mutually agreeable proposal. If the initial planning application is refused, there is usually an opportunity to appeal the decision at a hearing, where negotiation skills will be tested once again.

▲ Figure 9.12 Local authority planning committee

In situations where a client already owns a building plot, they will usually select a principal contractor through the tender process and award the contract after carefully negotiating the terms. For more information about principal contractors and the tender process, see Chapter 4.

Different negotiation techniques are needed during the construction phase of building work, for example when changes occur to the original contract agreement (change orders) or time extensions may be needed to avoid penalty clauses or to resolve disputes with subcontractors.

8.1 Distributive negotiation

Distributive negotiation is used to **haggle** over a common single interest at stake, known as a fixed sum.

A fixed sum is best described as a pie that parties are battling over for a bigger slice, with exchange offers back and forth. If one party gains more of the pie through the distributive negotiation, then the other party loses a percentage.

When preparing a bargaining strategy for a negotiation, both parties should have a preconceived goal. This is the point at which they would walk away without a deal, known as the **reservation point**.

Due to the nature of this method of negotiation, it is not possible for both parties to have a whole pie each. However, a mutual agreement can be reached where one party has a smaller slice of the pie, providing they have accepted a deal no less than the reservation point.

> **Key terms**
>
> **Haggle:** to negotiate for the best terms of an agreement or financial arrangement
>
> **Reservation point:** the highest price a buyer is willing to pay for an item and the lowest price a seller will accept for the item

> **Improve your maths**
>
> A developer negotiates to buy a plot of land for £470,000. The plot measures 23 m × 29 m. The developer then decides to sell a part of the plot that measures 29 m × 9 m, and adds a further 15 per cent to the rate at which he bought it.
>
> How much will the developer sell the plot for?

8.2 Win–lose approach

This is probably the most common negotiation method used to settle disputes between two parties. However, an agreement is more difficult to reach, because one side has to compromise in order for the other to experience a positive outcome.

If the win–lose approach is used to negotiate the distribution of resources, bitter disputes could cause conflict between the parties and may damage their future relationship, especially if negotiations break down and one party walks away.

8.3 Lose–lose approach

Sometimes during business negotiations, all concerned parties end up worse off and not achieving their desired result, for example unavoidable financial cuts within an organisation due to low annual profits.

In these situations, all participants should try to minimise their losses as much as possible, and to make sure they are fair. Even though participants are no better off using the lose–lose approach, their relationship remains intact because the loss has been evenly distributed.

8.4 Compromise approach

Negotiations between professionals to find an acceptable middle ground can be time consuming. When negotiators are unable to reach a mutual agreement and have nothing more to negotiate with, they often make **concessions** in order to meet the needs of the other party or to get something else they want.

The compromise approach is used to settle disputes quickly by one party settling for less than it may have hoped for, in order to reduce strained negotiations and maintain or fix a relationship. The risk is that one party gives up much more than it should, resulting in financial loss.

> **Key term**
>
> **Concessions:** something granted in response to a demand

8.5 Integrative negotiation (integrative bargaining)

This type of negotiation takes place between parties with common interests, in order to collaborate in finding a mutually beneficial solution. It is used when multiple issues have to be agreed.

Negotiators often make trade-offs across the issues being discussed and add further issues into the negotiations to create additional value for all parties, so that a successful outcome can be achieved.

8.6 Win–win approach

In this approach, negotiators with shared interests work together to find resolutions they are both satisfied with, rather than seeking to fulfil self-interests. This avoids disagreements and helps to maintain strong relationships between parties, while achieving a fair outcome for both sides.

9 Conflict-management techniques

Working in a fast-paced industry such as construction can be very rewarding. However, there can sometimes be conflict between the management team, subcontractors, suppliers and client.

Disputes can occur for several reasons, such as:
- ambiguous contract terms
- breaches of contract conditions
- late supply of building materials, resources or equipment
- breaches of site rules, for example failing to wear mandatory PPE
- programme delays.

People working at all levels in the construction industry will experience the pressure of having to meet tight deadlines while working long hours. This can have an impact on their physical, physiological and mental health, and is also a contributing factor when conflicts occur.

Initial grievances may be verbal exchanges, although these could escalate to formal written correspondence, for example emails and letters. Wherever possible, conflict should be avoided, because litigation can be a time-consuming and costly process; it may also damage professional working relationships between the development team and the client.

If a dispute occurs, each side will compete for their best interests using one of a number of conflict-management techniques, in order to force the other side to concede or reach an amicable agreement.

There are no benefits to conflict, so it should be avoided to prevent friction between staff, a negative working environment and delays to a project. Building Information Modelling (BIM) (see Chapter 4) can be used as a preventative measure by identifying issues that could provoke conflict between parties before they become a problem. When changes or alterations occur to a project design or specification, BIM can be used to update all concerned parties.

A dispute can quickly heighten emotions, unless a conflict-management technique is used. One way to resolve a problem is for parties to have an informal discussion in a neutral environment, where both sides have an opportunity to air their views without fear of backlash. If an amicable agreement cannot be reached however, the dispute will have to be resolved using mediation, conciliation or arbitration.

9.1 Mediation

This process involves appointing a mediator – an impartial third party who uses negotiation and communication techniques to encourage participation from both sides to find common ground and resolve their dispute.

One side cannot force another to use mediation; however, agreeing to follow this route can avoid the financial expense of going to court.

9.2 Conciliation

This is similar to mediation. An unbiased conciliator is appointed to meet with the parties separately and together, to weigh up both sides of an argument. They will then make a proposal based on the relative merits of each side, to find an amicable solution to the matter and bring it to a close as quickly as possible.

9.3 Arbitration

During this process, an appointed arbitrator (or tribunal hearing) resolves disputes between parties. When both sides of an argument have been heard, the arbitrator makes a legally binding decision, known as the award. If either party disagrees with the decision, they can take the case to court.

> **Test yourself**
>
> Why is it important to be able to use different negotiation techniques when working in the construction industry?

> **Improve your English**
>
> For the last two months, a supplier has been late in supplying building materials to your construction projects. Write a formal letter to the supplier to try to resolve the matter.

Onsite Construction T Level: Core

10 Methods and styles of communication

Communication is the process of exchanging information from one place, person or group to another.

The success of any project depends on accurate and effective lines of communication throughout the hierarchy of the construction team. Inaccurate, ambiguous, incomplete or confusing information can lead to delays, missed deadlines, initiated penalty clauses and additional project costs.

10.1 Methods of communication

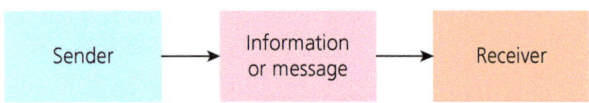

▲ Figure 9.13 Three-part communication process

Information can be exchanged verbally, non-verbally or in the form of visualisations (graphics). While there are advantages and disadvantages to each of these methods, information should always be clear, comprehensible and in a format that is easy to understand.

▲ Figure 9.14 Graphical communication – a safety sign

Complex information received by two or more people can sometimes be interpreted slightly differently by each recipient; just because information has been sent and received does not mean it has been understood as the sender intended. To confirm the recipient has fully understood the information, feedback needs to be given to the sender. However, this is not always possible using some forms of written or graphical communication.

The number of methods used to communicate in the construction industry has grown substantially in recent years, with the development of new digital platforms and technology.

> **Health and safety**
>
> Under the Construction (Design and Management) (CDM) Regulations 2015, employers have a legal duty to ensure information is easy to understand to protect the health, safety and welfare of people from work activities. (See Chapter 1 for more information on these regulations.)

In this section, we will look at some of the most commonly used methods of communication, their advantages, barriers to their effective use and problems that can occur.

Verbal communication

Face-to-face communication and indirect communication, such as talking on the phone or using walkie-talkies, are the simplest, quickest and probably most frequently used methods of transferring information from one person to another. Working in the construction industry, you will interact with your colleagues each day to build good working relationships, trust and confidence.

When team members engage positively and have respect for one another, they are generally happier, which in turn creates a better working environment. **Top-down** positive communication is a sign of a strong management team, which can result in employees having increased confidence in their employers, being more productive and having less desire to leave their jobs.

Negative forms of communication, for example in terms of pitch and tone of voice, cause conflict in the workplace, damage staff morale and increase stress levels. They often create **passive-aggressive behaviour** in individuals and a toxic culture, impacting on the business's performance and causing an increase in staff turnover.

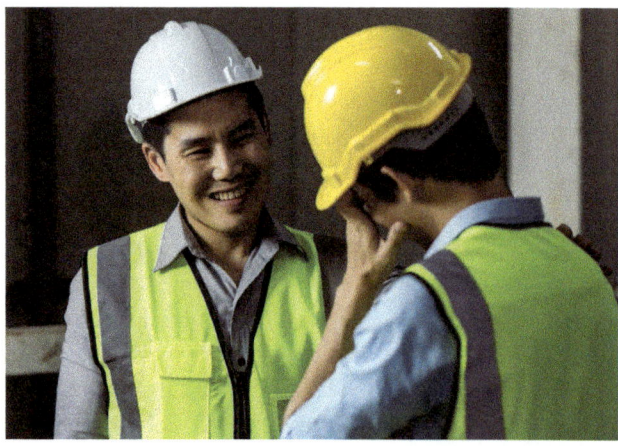

▲ Figure 9.15 Construction workers communicating verbally

240

Chapter 9 Relationship management in construction

> **Key terms**
>
> **Top-down:** proceeding from the most senior to the least senior
>
> **Passive-aggressive behaviour:** indirectly or subtly expressing negative feelings

> **Test yourself**
>
> Explain the term 'arbitration' and describe a situation in the workplace when this may be used.

Table 9.2 outlines strengths and weaknesses of using verbal communication.

▼ Table 9.2 Strengths and weaknesses of using verbal communication

Strengths	Weaknesses
• It is quick and simple. • There is no reliance on technology, for example an internet connection. • Two-way conversation is possible. • It is personal. • It is direct. • It allows the recipient to confirm that they have understood the information (give feedback). • Non-verbal impressions can also be used with verbal communication, for example raised voice, gestures or body language. • The sender can use both open and closed questioning (open questioning allows someone to give a free-form answer; closed questioning requires a response from a limited set of answers such as 'yes' or 'no').	• There is no written record. • Information can be forgotten by the recipient. • Information can be misunderstood if the sender has a strong accent or regional dialect. • Language barriers may exist between people who speak different languages. • Information can be misheard with background noises. • The recipient may have a hearing impairment. • The message could be **ambiguous**.

> **Key term**
>
> **Ambiguous:** unclear and difficult to understand

> **Industry tip**
>
> Listening is as important as speaking, as this is one way that we learn. While undertaking your industry placement as part of your T-Level course, your employer will assess both of these skills.

> **Health and safety**
>
> The Health and Safety Executive (HSE) reported in 2021 that one in four people living in the UK has had mental health problems. Work-related stress can aggravate pre-existing mental health issues, such as anxiety and depression. A report by the Office for National Statistics (ONS) stated that more than 1400 construction workers took their own lives between 2011 and 2015, over three times the national average.
>
> Talking with co-workers informally has many benefits, including providing support and improving wellbeing and general mental health.

Non-verbal communication

People can communicate with each other without the need to speak, using written information, body language, gestures and even the way they dress.

Wherever possible, formal communication in a work environment is written down, because it provides a permanent record and is easier to distribute to others without misinterpretation.

Written information in some forms is a legal requirement, for example building contracts and policies. When information is recorded, it can be referred to or even used as evidence if a dispute or grievance occurs between employers, employees or clients.

▲ Figure 9.16 Non-verbal communication – a contract of employment

▼ Table 9.3 Strengths and weaknesses of written communication

Strengths	Weaknesses
• It can be referred back to. • There is a permanent record of the communication. • The same information can be distributed easily without diluting it. • The sender does not have to meet the recipient to pass on the information. • It can be used to communicate if the recipient has a hearing impairment.	• There may be no opportunity to feed back quickly. • It is not possible to clarify immediately whether information has been received and understood. • It is impersonal. • It is indirect. • A recipient may have dyslexia or other reading difficulties. • There may be language barriers. • It takes time to write. • It is slow to distribute. • A recipient may have impaired vision. • The message could be ambiguous.

Besides written communication, non-verbal communication can involve body language and behaviours such as eye contact and facial expressions. However, these methods can be easily misinterpreted and there is no permanent record.

> **Industry tip**
>
> Good communication skills are essential in most job roles in the construction industry, and they are a desirable quality for employers. The first communication you have with an employer could be a job application, so it is vital that it is well written without any spelling or grammatical errors, in order to create a good impression. Before submitting your CV and job application to an employer, always get someone to **proofread** them.

> **Key term**
>
> **Proofread:** to check a piece of written communication for errors in spelling, grammar, punctuation and accuracy before it is shared or published

> **Industry tip**
>
> The way you dress, act and speak during an interview is vital in creating a positive first impression with a potential employer. Even though you may be applying for a job working in a particularly dirty environment, you should still dress appropriately for an interview. Remember, you only get one chance to make a first impression!

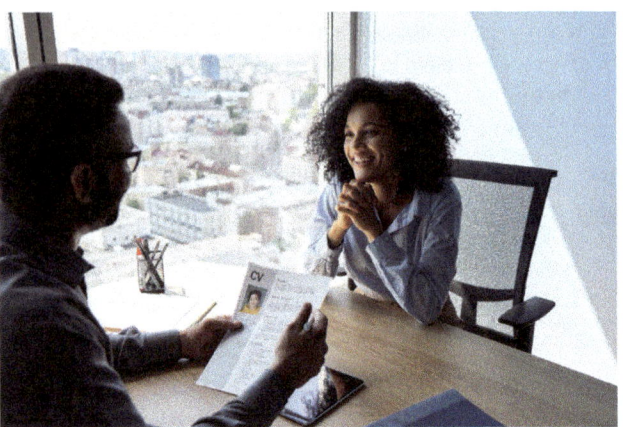

▲ Figure 9.17 A job interview

Visualisations (graphics)

Information displayed in a graphical form, such as posters, safety signs and notices, is often used in the construction industry to reinforce information already communicated by some other method. For example, during a site induction you may have been told to wear a range of personal protective equipment (PPE); mandatory safety signage displayed around the site may also illustrate this requirement.

Technical information is often communicated using detailed visualisations, either in hard-copy format or using digital technology.

> **Industry tip**
>
> Posters and information boards should be updated regularly to keep them current, and to ensure they are not disregarded by the intended recipients.

Chapter 9 Relationship management in construction

▼ Table 9.4 Strengths and weaknesses of using visualisations

Strengths	Weaknesses
• The message can be repeated to a wide audience. • There is no language barrier. • The message is clear and consistent. • It is quick to interpret a simple message. • There is a written record. • They are eye-catching. • They are jargon-free. • Symbols and pictograms are often standardised in the UK construction industry.	• The recipient may have a visual impairment. • There is no confirmation that they have been acknowledged. • They are easily dismissed. • The recipient may have to receive training to understand the symbols, pictures or pictograms. • They can be lost, removed or defaced if on display. • The recipient cannot ask questions. • There is no immediate feedback. • Only simple messages can be conveyed.

Table 9.5 outlines various methods of communication and how they might be used in a typical construction project.

▼ Table 9.5 Methods of communication

Type of communication	Use in a typical construction project
Face to face	• Job interviews • Site inductions • Toolbox talks • Informal discussions with work colleagues • Site meetings • **Appraisals**
Email	• Communications with clients and other stakeholders • Quotations and estimates • Exchanging of building contracts
Letter	• Formal exchanges of information between contractors, duty holders and clients, for example building contracts • Business letters • Formal exchanges of information with the local authority, HSE and professionals involved in a construction project • Notes • Memorandums (memos)

Telephone (land line)	• Internal communication between staff within a business:
	  • Better signal strength compared with mobile phones in remote locations
	  • Less likely to run out of battery charge
	  • Gives the impression of a professional business when communicating with clients

Mobile phone	• Communication between members of the site management team on a construction site (providing they have a good network signal) • Accessing the internet for emails, livestream video calls and text messaging • Recording videos or taking photographs to communicate information quickly from the construction site to the office or other stakeholders • Contacting the emergency services

Walkie-talkies	• Shortwave radio contact between members of the construction team on building sites:
	  • durable and moisture resistant
	  • quick and direct
	  • ideal for use in remote locations where phone signals may be weak

Drawn information	• Communicating information from an office to construction sites, for example as site plans, elevations, section drawings or detail drawings • Scheduling and planning • Sharing technical information with clients, the planning department and subcontractors • Forming part of a tender package • Supporting oral communication on construction sites in the form of sketches

Onsite Construction T Level: Core

▼ Table 9.5 Methods of communication (continued)

Type of communication	Use in a typical construction project

Tablets, notebooks and laptops	• Video-conferencing site meetings
	• Site inductions
	• Letter writing
	• Snagging lists
	• Surveying
	• Requisitions (Orders)
	• Auditing
	• Scheduling
	• Digital project management:
	  • augmented reality (AR)
	  • virtual reality (VR)
	  • Building Information Modelling (BIM)
	  • controlling drones to broadcast livestreams in real time

Signs and notices	• Safety signage displayed around the workplace, for example fire escape • Labelling on packaging • Site notice boards • Improvement and prohibition notices served by the HSE
Videos	• Site inductions • Toolbox talks • Marketing • Site inspections • Monitoring progress and health and safety standards

Key term

Appraisals: scheduled routine meetings between an employee and their employer to review their work performance against their job description

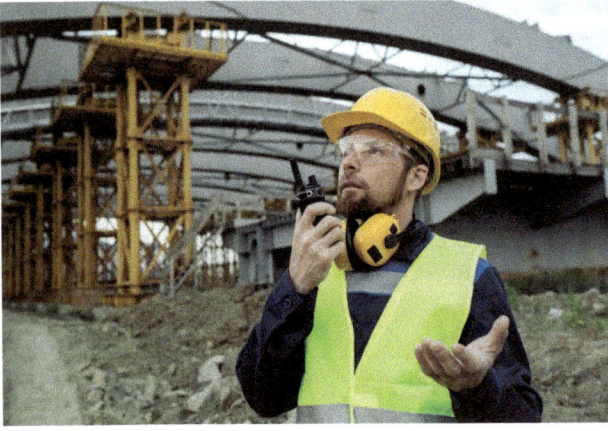

▲ Figure 9.18 Walkie-talkies are ideal for use in remote locations where phone signals may be weak

10.2 Communication styles

Communication styles can be either formal or informal:
▶ Formal communication is used by organisations to pass information through prescribed official channels, following an organisational structure.
▶ Informal communication is much quicker, because there are no rules restricting which direction or lines of communication have to be used; therefore, it is 'free flowing'.

The number of non-English-speaking workers, and those who speak English as a second language, is on the increase in the UK construction industry. All employers have a duty of care to communicate effectively with everyone under their control, especially when it concerns their health and safety. To overcome these problems, employers may use a translator or a **bilingual** supervisor or co-worker to interpret key information. Where this is not possible, they may have to translate written information to use for site inductions, toolbox talks and other training materials. In situations where language is a barrier, simple pictorial illustrations are also a useful tool to support employers communicating with their employers.

Chapter 9 Relationship management in construction

> **Improve your English**
>
> Make a short toolbox talk video on a topic of your choice, to develop your speaking and presenting skills.

> **Key term**
>
> **Bilingual:** fluent in two languages

> **Case study**
>
> Lucas has been a self-employed roofer for the past ten years. Last summer, he was subcontracted to tile the roofs on 15 new houses on a development. Unfortunately, he had an accident while working on one of the roofs, and sustained a back injury that has prevented him from continuing to trade as a roofer.
>
> During the investigation into the accident, Lucas said that he attended a site induction, but he had difficulties reading and understanding the RAMS (risk assessments and method statements).
> ▶ Whose fault is it that Lucas could not understand the RAMS?
> ▶ Work in a small group to consider an alternative site induction for Lucas.

11 Employment rights and responsibilities

Every employee in the UK has legal rights and responsibilities under the Employment Rights Act 1996. Employment law controls employees' rights at work and relationships between employees, employers, trade unions and the government.

As soon as someone accepts a job offer from an employer, they technically have a contract of employment, so before starting work it is important that they understand what they are entitled to and what to expect from their employer. A contract of employment does not have to be in writing, although most employers prefer to record the following terms of the agreement to avoid confusion or disputes:
▶ employment rights
▶ employee's duties
▶ responsibilities – employer to employee
▶ responsibilities – employee to employer
▶ employment conditions.

11.1 Employment rights

Wages

Minimum wage

Almost every person with a legal right to work in the UK has employment rights, such as being entitled to the national minimum wage and living wage. The minimum wage a person should receive depends on their age and whether they are an apprentice:
▶ Workers of at least school-leaving age are entitled to the national minimum wage. In England, you can leave school on the last Friday in June, if you will be 16 by the end of the summer holidays. You will have the following options until you are 18:
  – Spend 20 hours or more a week working or volunteering, while in part-time training or education.
  – Stay in full-time education.
  – Start a traineeship or apprenticeship.
▶ Workers over the age of 23 are entitled to the national living wage.

▲ Figure 9.19 An apprentice tradesperson

The rates of the National Minimum Wage and National Living Wage are reviewed on 1 April every year. To check the current rates, visit **www.gov.uk** and search for national minimum wage or national living wage.

Apprentices who have completed the first year of their apprenticeship and are aged 19 or older are entitled to the minimum wage for their age.

If an employee is absent from work for at least four days in a row and has earned at least £120 per week, their employer must pay **statutory sick pay** (SSP) for up to 28 weeks. Employers can pay more than the minimum SSP; this is usually agreed in the contract of employment before starting work.

The law states that employers must give their employees **payslips** as proof of their earnings, tax paid and any other deductions, for example pension contributions.

> **Key terms**
>
> **Statutory sick pay:** payment made by employers to employees when they are too ill to work
>
> **Payslips:** written documents provided by an employer that contain personal information about the employee and specific details about the amount of money that they have earned for a given period (for example, a week or month), as well as their total earnings, tax deductions, National Insurance contributions and pension payments made for that financial year

▲ Figure 9.20 A payslip

> **Research**
>
> Find out who is not entitled to the national minimum wage or national living wage.

Pension

Everyone who has made National Insurance contributions through their earnings is entitled to a government **state pension**, which can be claimed on reaching state pension age (although this is currently under review). The amount a person receives depends on their National Insurance record. A state pension alone is unlikely to provide enough money to live on when someone retires; therefore they will also have either a workplace pension (set up by their employer) or a private pension (set up by the individual).

Since 2012, all employers must automatically enrol their employees into a workplace pension scheme and make contributions towards it, provided they are aged 22 or above and earn over the low-income threshold set by the government.

All eligible employees make their pension contributions through deductions taken directly from their income. If an employee prefers not to have a workplace pension, they can opt out of the scheme. However, this is best done within the first month of employment to prevent payments going into a pension fund that can only be accessed when they retire.

> **Key term**
>
> **State pension:** a regular sum of money paid by the state to people of or above official retirement age

Time off

Almost all employees have the right to paid leave for public duties and responsibilities, such as jury service, under the Working Time (Amendment) Regulations 2007. People who work a five-day week are also entitled to at least 28 days' paid leave each year, which is equivalent to 5.6 weeks of holiday. Part-time employees have the same entitlement as full-time staff, although this is dependent on the number of days that they work, for example four days per week amounts to 4 × 5.6 weeks = 22.4 days per year.

Employees who are parents of children under the age of 18 may also be eligible for unpaid leave to look after their child's welfare, providing they have been with a company for more than one year.

Employees over the age of 18 are entitled to three different types of unpaid rest breaks:

▶ rest breaks at work – one uninterrupted 20-minute break if they work more than six hours a day
▶ daily rest – at least 11 hours between shifts
▶ weekly rest – 24 hours' uninterrupted break each week, or 48 hours each fortnight.

Employees are entitled to 52 weeks' statutory maternity leave, made up of 26 weeks' ordinary maternity leave and 26 weeks' additional maternity leave. Statutory maternity pay (SMP) is paid for up to 39 weeks and includes:

▶ 90 per cent of the employee's average weekly earnings (before tax) for the first six weeks

Chapter 9 Relationship management in construction

▶ either a fixed sum set by the government or 90 per cent of average weekly earnings (whichever is lower) for the next 33 weeks.

When an employee's partner is having a baby, adopting a child or having a baby through surrogacy, they might be eligible for one or two weeks' leave or paternity pay. The weekly rate for paternity pay is 90 per cent of average weekly earnings. Employees may not be entitled to both paternity leave and pay, but they may be eligible for shared parental leave and pay.

There are different eligibility criteria for birth, surrogate and adoptive parents to get shared parental leave and pay; further information can be found at www.gov.uk/shared-parental-leave-and-pay.

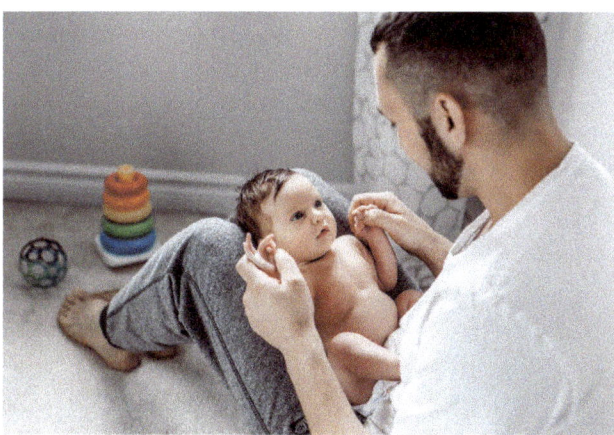

▲ Figure 9.21 Parental leave

Improve your maths

A 19-year-old carpenter earns the national minimum wage:
▶ One week, he worked 47 hours in total.
▶ His employer pays him 'time and a half' for every hour after 39 standard-time hours.
▶ 37 per cent of his total pay is deducted for tax, National Insurance and pension.
▶ The threshold for PAYE (pay as you earn) is £242 per week, or £1048 per month.

How much has the carpenter earned after deductions?

Research

Find out the employee leave and pay entitlements for paternity and adoption.

Research

Find out which type of employees would not be permitted time off work for jury service.

Health and safety

Rest breaks from the workplace are extremely important to protect people from physical and psychological harm.

Equal rights

Under the Equality Act 2010, people at work and in wider society are legally protected from direct or indirect discrimination, victimisation and harassment. See section 7.1 of this chapter for details of protected characteristics.

Health, safety and welfare

The Health and Safety at Work etc. Act 1974 (HASAWA) protects the health, safety and welfare of:
▶ people at work
▶ those who could be affected by work activities.

An employer's legal duties under this legislation are explained in Chapter 1.

Access to representation in times of grievance

Occasionally, an employee will raise a problem, complaint or concern with their employer that cannot be resolved amicably without intervention.

A grievance is usually raised with an employer following internal procedures laid out in the contract of employment. This often involves notifying an employer in writing at first, and then following this up with a meeting between the employer, a representative from the organisation's **human resources department** (if appropriate), the employee and a companion of the employee's choosing. Employees can represent themselves, although they may exercise their legal right to take a representative from their trade union or a colleague for support.

Key term

Human resources department: department of an organisation that deals with recruiting, administrating and training staff

Redundancy

Employers and employees must adhere to the terms of the employment contract until it ends, when the employee is given notice or dismissed.

If an employee who has worked for the same employer for two or more years is made redundant, they are normally entitled to statutory redundancy pay. The amount of pay received depends on the employee's age at the time of leaving and their length of service (although this is capped at 20 years).

11.2 Employment responsibilities

Employers have responsibilities to their employees to make sure they are treated fairly, by maintaining good practices and complying with employment law.

Employers' responsibilities towards their employees include:
- protecting their health, safety and welfare, for example by providing personal protective equipment, toilets and rest facilities
- providing a contract of employment
- informing and consulting when necessary
- being an inclusive employer and not discriminating while recruiting, employing or promoting staff
- paying the minimum wage, sick pay, maternity pay, holiday pay and other entitlements
- allowing staff to return to the same job after a leave of absence, for example after maternity leave or a **sabbatical**
- following the Working Time Regulations 1998
- abiding by the terms and conditions of the contract of employment
- considering requests from staff for **flexible working**
- auto-enrolling employees into a workplace pension.

> **Key terms**
>
> **Sabbatical:** an extended period of unpaid leave from work, taken in agreement with an employer, often used for holidays, travelling or pursuing interests
>
> **Flexible working:** a working arrangement that allows an employee to choose when and where they work, in order to improve their work–life balance, reduce their stress levels and provide better job satisfaction

Employees also have responsibilities to their employer, including working to the terms and conditions agreed in their employment contract and company handbook.

A company handbook, also known as a company policy, provides employees with job-related information, for example working conditions, health and safety, standards, procedures and reasonable behaviour expectations. It also provides employees with details about the rules that everyone responsible for using personal data has to follow, in order to protect the confidentiality of people at work and to comply with the UK General Data Protection Regulation (UK GDPR) and the Data Protection Act 2018 (see section 3.2 of Chapter 8).

Company handbooks often contain a social media policy and code of conduct. These documents explain rules which employees should follow outside of their working hours, to ensure their behaviour does not impact negatively on the employer or damage their brand or reputation. Improper or illegal behaviour by an employee can sometimes prevent them performing their job role as outlined in their job description, and may lead to disciplinary action or dismissal by their employer. Furthermore, they could be prosecuted by external enforcement bodies, for example the police, if they have committed a criminal offence.

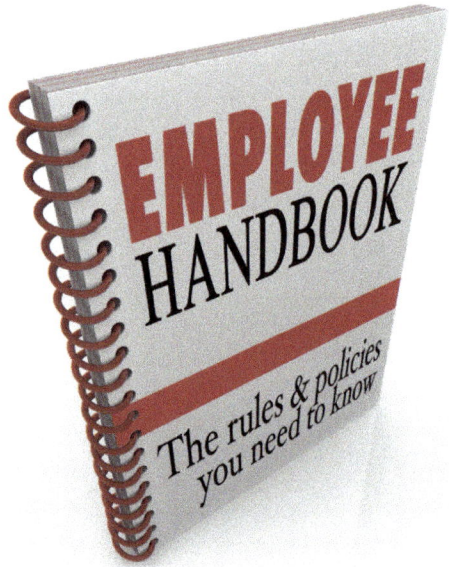

▲ Figure 9.22 A company policy document

> **Test yourself**
>
> Describe the most likely outcome for an employee following a minor breach and a serious breach of their contract of employment.

> **Test yourself**
>
> List the employment rights of an employee.

12 Ethics and ethical behaviour

In principle, people should treat others the way they would like to be treated themselves. This is essential to the long-term success of a business and the route to making a better environment in our workplaces and the world around us.

When you think about **ethics**, you should consider how the actions you take could affect others. For example, a business may choose not to provide PPE for employees as a way to increase its profits; however, this is considered **unethical** because of the risk to people's health.

Larger businesses often develop a system of policies and practices to ensure they operate in a fair, moral and legal way; this is known as a code of ethics. Businesses with a consistent set of values and robust moral code develop a strong ethical culture, which means they are less likely to face the consequences of unethical behaviour. Unethical behaviour in business is damaging and can have a negative impact on an organisation's future prospects.

▲ Figure 9.23 As well as being a legal responsibility, providing PPE is ethical behaviour

> **Key terms**
>
> **Ethics:** moral values that govern a person's behaviour towards others
>
> **Unethical:** morally wrong or unacceptable
>
> **Defamation:** the act of damaging someone's good reputation through a false written or verbal statement, also known as libel (written) or slander (spoken)

Table 9.6 provides examples of ethical and unethical behaviour in business.

▼ Table 9.6 Ethical and unethical behaviour in business

Ethical behaviour	Unethical behaviour
• Adherence to health and safety legislation and guidance • Equality and diversity • Fulfilling the terms of building and employment contracts • Honesty, integrity, commitment and loyalty • Corporate social responsibility • Environmental protection • Accurate business statements • Protection of personal or sensitive data • Fair employment rights and responsibilities • Adherence to laws • Respect • Accountability for actions • Submission of accurate and true annual accounts to HMRC	• Ignoring health and safety legislation and guidance • Discrimination • Failure to honour commitments • Corruption • Harassment, for example sexual harassment • Disregard for corporate social responsibility • Wilful damage to the environment, for example inappropriate disposal of waste or destruction of protected trees • False business claims • Disclosure of sensitive personal information • Mistreatment/exploitation of employees, for example paying below the minimum living wage or offering poor working conditions • Sabotage • Disrespect • **Defamation** of a competitor • Tax avoidance and manipulating accounts • Theft • Bribery • Money laundering • Deception

Businesses that demonstrate ethical traits and encourage and reward good attitudes are often more productive and successful, as they develop trust between the management, the employees and the public. When a business with power is seen to be acting responsibly and unselfishly, it can also improve its reputation.

> **Test yourself**
>
> Describe a situation where ethical behaviour is applied in the workplace.

> **Key term**
>
> **Networking:** an activity where businesses and people with a common interest meet to share information and develop contacts

13 Sources of information

Organisations need to create opportunities to meet new customers and develop relationships with contacts and partners for new business ventures.

Networking is a low-cost process used by many organisations to make initial introductions with likeminded people, share information and form long-lasting business relationships. Construction businesses can network on construction sites, attend trade events or exhibitions, join networking groups or use social media. Distributing business cards is another way to expand a networking community, although this method is often slower and restricted to a small demographic compared with the use of digital media.

The internet is used by most construction businesses in the UK to market their product or service, raise their profile and develop people's interest for future sales opportunities. A professionally made and managed website is essential as a platform to advertise and promote a business to the public, customers and other stakeholders.

Businesses can use social media to engage quickly with a wide demographic. It allows them to share ideas, knowledge and good practice with other businesses, through posts, blogs, podcasts, articles and videos.

Businesses that use social media often provide customers with the opportunity to write reviews, as a way of giving unfiltered honest feedback, although it may not always be positive or constructive. Some businesses also use social media to their advantage by redirecting traffic from posts to their main website, as a marketing strategy to increase sales.

Social media must be controlled carefully by an organisation. It can quickly and easily damage both reputation and brand if errors are made or negative feedback is received and publicised. Due to the fast pace of social media, businesses have to continually monitor and update their content to ensure it remains current and has not been abused with unwanted or inappropriate posts. There is also a risk that confidential information could be accessed by hackers or leaked online, unless a good cyber security system is used to protect it.

> **Assessment practice**
>
> **Short answer**
>
> 1. What term describes a group or individual that has an invested interest in the long-term success of an organisation?
> 2. Why is it important to maintain a good working relationship with customers?
> 3. What is the main piece of legislation that prevents discrimination in the workplace?
> 4. Where would an employee find information on working conditions, health and safety, standards, procedures and reasonable behaviour expectations?
> 5. What must all employers automatically enrol their employees into and make contributions towards, providing the employees are aged 22 or above and earn over the low-income threshold?
>
> **Long answer**
>
> 6. Describe the negotiation techniques used in the construction industry.
> 7. Describe a situation involving unethical behaviour in a business.
> 8. Explain the term 'networking'.
> 9. Describe a situation when arbitration may be appropriate to use in the workplace.
> 10. Explain the advantages of non-verbal communication.

Project practice

Liam has just started a new job as a solar panel installer. In the first week, he has been asked by his employer to install six photovoltaic (PV) panels with two other employees.

Liam and his work colleagues attended a site induction on the new development under construction, where they were informed about the site rules and mandatory PPE that has to be worn at all times on the site.

While installing the first PV panel, Liam was challenged by the site manager for not wearing his hard hat. Liam said that it was uncomfortable, and too hot to be wearing on a roof in the middle of the summer. Later that day, Liam was caught again by the site manager not wearing his hard hat as instructed and was asked to leave the site. During an investigation into the matter by his employer, Liam claimed he was being discriminated against.

▶ Describe the conflict-management technique used by the site manager when he challenged Liam on the second occasion.
▶ Work in a small group to research Liam's employment rights and responsibilities, and write a brief report to explain his behaviour on the day and whether this was acceptable.
▶ In your opinion, was Liam discriminated against by the site manager?

Chapter 10 Digital technology in construction

Introduction

The expanding and evolving use of digital technology in the construction industry has led to improvements in management and production methods, for example:
- better efficiency in project planning
- a more streamlined design approach
- enhanced collaboration between the building team and stakeholders
- greater innovation in construction methods and design.

While Chapter 8 introduced the term 'digital' and outlined a limited number of applications for digital technology, this chapter considers the use of digital technology in construction in greater detail, identifying digital systems and processes that are employed to achieve specific results.

Learning outcomes

By the end of this chapter, you will understand:
1. the internet of things (IoT)
2. digital engineering techniques
3. opportunities for the use of technology.

Chapter 10 Digital technology in construction

1 Internet of things (IoT)

The term 'internet of things' (IoT) refers to the system of applying unique digital identifiers to physical objects, such as buildings, which can be connected to a digital network. Communication data can flow between the connected objects, allowing interactivity and collaboration to achieve specific goals.

Objects connected through a digital network can be referred to as 'smart' objects, when the processing of information and data happens without human intervention. Communication can take place directly from machine to machine (M2M). This means that systems management and decision making can be automated in operational areas of a smart building, such as:

- monitoring and optimising systems performance
- detecting systems inefficiencies
- alerting remote operators to potential problems
- diagnosing possible causes of problems.

▲ Figure 10.1 In the IoT, communication can take place from machine to machine

The potential for using the IoT in construction applications is vast. Designing buildings and infrastructure elements to incorporate interconnected and unified sensors and monitoring equipment allows data analysis, which can translate into:

- reduction in energy use
- improvements in manufacturing efficiency
- improvements in safety
- streamlining of materials delivery and supply-chain activity.

Smart technology can help building designers to identify optimal construction materials and methods for future projects, leading to the construction of buildings that rely less on highly skilled workers, use fewer resources and produce less carbon.

The interconnected nature of the IoT allows data analysis which can provide performance information on individual components and materials in a range of structures. As a result, manufacturers can improve and refine manufactured resources.

▲ Figure 10.2 The interconnected nature of the IoT allows for many applications

Smart technology is evolving quickly, and the term 'artificial intelligence' (AI) is becoming more familiar. AI describes digital systems that go beyond simply following programmed instructions. They use methods of analysis that identify patterns and correlations, in order to draw appropriate conclusions more quickly than humans are able to.

> **Industry tip**
>
> AI is used in many devices we take for granted. For example, face and fingerprint recognition software is often used in mobile phones. This application of AI is also often used to prevent unauthorised access to construction sites, using a **biometric** turnstile.

> **Key term**
>
> **Biometric:** involving the detection and analysis of individuals' unique physical characteristics as a means of verifying identity

253

Onsite Construction T Level: Core

> **Test yourself**
>
> How can AI be used to control who enters a construction site?

1.1 Uses of digital technology

The development and implementation of digital technology, in both the construction industry and society generally, has led to new methods of working, with technological tools being refined constantly.

Let us look at some uses of digital technology in the operation of completed buildings and during the construction phase.

Digital technology used in completed buildings

Completed buildings that are designed to integrate with the IoT can produce large quantities of useful data.

Sensor technology and data management systems embedded in a structure can record and archive data, as well as produce live data streams, for analysis and evaluation. This allows a clear understanding of the performance of an occupied building and the environment it provides for its occupants.

▲ Figure 10.3 Controlled access to a construction site using a fingerprint reader

Data analysis using AI holds the promise of real-time **autonomous** decision making when operating buildings systems. It also offers the ability to compare different features, options, costings and performance factors during the design of a new building.

Smart learning that uses digital technology and the networking capability of the IoT is likely to become an integral feature of future design processes and operating systems in the construction industry.

> **Key term**
>
> **Autonomous:** functioning independently without external control

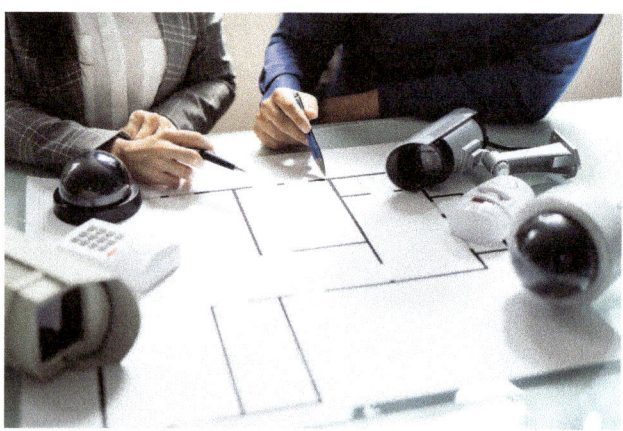

▲ Figure 10.4 Sensor technology embedded in a structure can provide valuable data

Occupier comfort

In buildings such as offices, smart control systems can be used to monitor and adjust levels of relative humidity, air temperature, ventilation, dust and even air pressure within individual rooms, in order to ensure a comfortable and productive working environment.

Chapter 10 Digital technology in construction

> **Research**
>
> Search online for companies that offer control systems for buildings. Write down the range of features in a building that can be controlled by digital technology.

Analysis of data over time can be used to refine the operational **parameters** of control systems, in turn creating and maintaining consistently comfortable working and living conditions.

Experts in AI and smart technology suggest that control systems in the future may be able to sense in real time whether a building's occupants are tired, too hot or too cold, and automatically make adjustments to heating and ventilation to enhance wellbeing.

> **Key term**
>
> *Parameters:* limits which define the scope of a system, process or activity

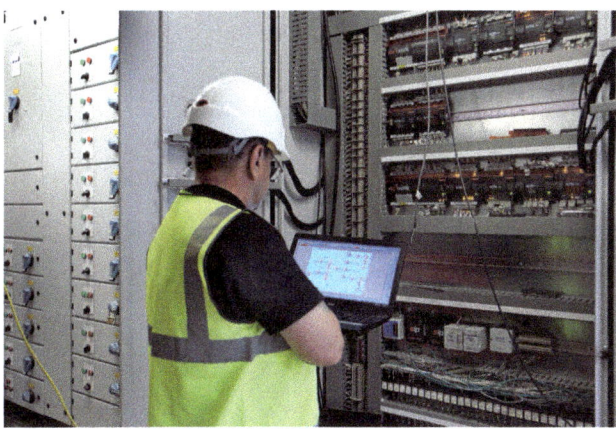

▲ Figure 10.5 Analysis of data can be used to set operational parameters of control systems

> **Improve your English**
>
> Increasing numbers of smart products are being marketed for home use. Find out what is available and write an account of how a smart home could be set up for the comfort and convenience of the occupants.

> **Test yourself**
>
> How can the analysis of appropriate data help to make a building comfortable to work in?

Power management

Digital technology can support energy efficiency through the management of electrical power. Demand for electrical power in an occupied building will fluctuate depending on the number of occupants active at certain times. Intelligent systems can gather and compare information over time to detect changes in energy use and potentially predict periodic peaks and troughs in demand.

Through the IoT, energy suppliers could request that consumers command their internal systems to reduce energy consumption when increased general demand puts stress on supply capacity. This model of real-time management of demand is likely to play an increasingly important part in improving efficient energy usage.

Sustainable electricity generation for individual properties, for example in the form of solar panels and wind turbines, is increasing. However, these energy sources may not provide sufficient amounts of energy for larger buildings, which may still need to rely on external energy supplies.

Renewable energy sources such as solar and wind power provide variable energy outputs dependent on weather conditions, so it is necessary to balance the use of onsite and off-site energy sources in a building according to the demand from occupants.

▲ Figure 10.6 Digital technology allows sophisticated control of electrical power

Systems using AI and the IoT have the potential to make these decisions in real time, using a constant feed of up-to-date information. In order to optimise energy use and minimise the production of carbon, sophisticated control systems could make decisions to use onsite energy generation, store energy in battery facilities, export energy to the National Grid, use only external energy sources, or combine some of these options.

> ### Research
>
> By 2030, all new cars in the UK will be fully electric (or at least heavily electrified) instead of powered by petrol and diesel.
>
> Find out how many electric car charging points there are in the UK. Could AI be used to manage this part of the infrastructure?

Maintenance

Benefits of digital technology include the vast quantities of data that can be processed and the speed of data transfer. AI systems can compare a huge range of scenarios based on live data streams, from which warning signs of equipment failure can be identified. This allows predictive maintenance to be a reality.

A building is an asset which must be managed and maintained efficiently in order to preserve its value. The development of AI systems that can detect potential faults and anomalies in equipment means that maintenance interventions can be made before that equipment fails.

Equipment failures cause disruption that can be costly and even dangerous to the building's occupants, so digital technology is a valuable tool in asset management and maintenance of the built environment.

> ### Test yourself
>
> How can digital technology be used to maintain a building effectively?

Digital technology used during the construction of buildings

While some types of digital technology have been used in construction for many years, especially during the design process, new technologies are constantly being developed to improve site safety and efficiency and make construction more sustainable.

Design

Computer-aided design (CAD) has become a standard design tool when developing the concept for a construction project (see later in this chapter for further details).

> ### Research
>
> Research the range of CAD software available online. How many different programs can you find? Write down the cost of two CAD software packages.

> ### Health and safety
>
> Using digital technology can mean spending a long time focused on a screen and sitting in one position. Be aware of the possible damage to your health – poor posture can cause back and neck problems, and excessive screen time can affect your eyesight.

Digital systems can speed up the design process considerably, leading to greater productivity. Building design ideas created using specialist software can be presented as a three-dimensional (3D) model, allowing examination and refinement of the design concept.

▲ Figure 10.7 Digital tools can produce 3D models of a structure

Digital tools are able to generate accurate materials and components lists from 3D models or 2D drawings. These can be linked digitally to materials costs databases, in order to produce an up-to-date costing for a project.

Speeding up the process of designing and costing a project, coupled with the ability to update or amend project details quickly, streamlines development and improves productivity.

On site

Once project design details have been finalised and costings completed, digital technology has a significant role to play on the construction site. Smart equipment, such as surveying tools that use satellite positioning data for setting out buildings, can link with remotely stored project data through the IoT. This type of

Chapter 10 Digital technology in construction

equipment is becoming increasingly sophisticated and can speed up site preparation considerably.

A similar smart approach can be used to avoid damage to existing buried pipes and cables when excavating a new site. Mechanical excavators can be equipped with satellite positioning equipment in order to locate pipes and cables, and digital depth controls can assist operators in limiting the depth of excavations to safe levels.

> **Industry tip**
>
> Experienced site supervisors will judge the safest way to work around existing services on site. Sometimes the safest way to expose pipes and cables is to dig by hand.

▲ Figure 10.8 It is vital to avoid damage to buried cables and pipes on site

Smart technology can also support the efficient operation of machines on site. For example, removing too much material when shaping the ground will require some material to be replaced later, which is uneconomical. Smart equipment in earth-moving and excavation machines can ensure pinpoint accuracy in levelling the ground and creating gradients and slopes when contouring the terrain.

Digital tracking systems can also be installed in machinery. **Geofencing** uses digital technology such as radio frequency identification (RFID) or GPS to pinpoint the location of construction equipment and create a virtual boundary, ensuring that it remains within a specified area.

> **Key term**
>
> **Geofencing:** using technology to pinpoint the location of equipment and create a virtual boundary

> **Improve your English**
>
> Search online for 'AI excavators'. Write a report about how construction machines are being developed to operate without drivers. (Hint: Japan is leading the way.)

Materials and resources

Delivery of materials and components to site must be reliable and consistent, in order to avoid project delays and increased costs. However, storage of materials on site is often limited due to congestion and a lack of suitable storage space.

A strategy developed in the car manufacturing industry known as 'just in time' (JIT) can be used in construction activities to improve efficiency, reduce waste and eliminate the need for extensive onsite storage facilities.

> **Improve your English**
>
> Find out the history of the JIT strategy. Write a short report about how the idea came about and where it was first applied.

A JIT strategy requires delivery of the right materials, in the right order, in the right amount, at the right time. For example, concrete is usually mixed off site to the required specification. It must therefore be delivered at the right time and in the right quantity to allow it to be poured at the required location before setting occurs.

There are many software packages designed to facilitate JIT systems.

▲ Figure 10.9 JIT systems can be used in congested locations where access is difficult

> **Industry tip**
>
> The timing and sequencing of deliveries is critical for a JIT system. Goods must therefore be handled carefully to avoid damage. Damaged goods may not be replaced quickly, leading to disruption of production schedules.

Digital technology can be used to track and monitor materials and components during manufacture, transport and delivery. This is useful during complex projects, where a large catalogue of materials and components can be controlled using processing systems.

> **Test yourself**
>
> What is a JIT system?

2 Digital engineering techniques

Digital engineering uses a digital skillset to create and edit data as part of a design process. Engineering principles are applied in a virtual environment, allowing designers to explore a range of design possibilities and develop innovative solutions to design problems.

> **Improve your English**
>
> Find out what is meant by the term 'engineering principles' and write a definition in your own words.

In this section, we will discuss three main techniques applied in the field of digital engineering:
- simulation
- animation
- modelling.

2.1 Simulation

Simulation uses digital methods to mimic the behaviour of real systems and processes. It helps to answer important questions about a proposed construction project, such as:
- Will the building design be energy efficient?
- Which construction methods are best suited to the design concept?
- What is a realistic timescale for the construction of the building?

By using a simulation model, 'what if' analysis becomes possible, so that the design and selected construction processes can be checked for feasibility. For example, it can be confirmed that the intended building plot will accommodate the design, allowing access to deliver and store materials without causing undue disruption to surrounding properties and their occupants.

The sequencing of project stages can be simulated, so that construction processes follow a smooth path without disruptive clashes. Any necessary changes can be made before work commences on site, avoiding errors and expensive delays.

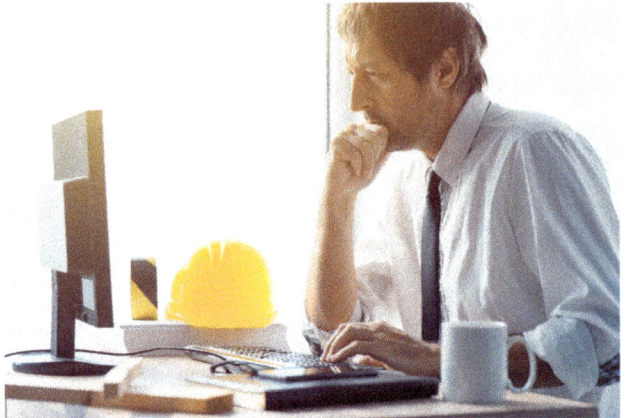

▲ Figure 10.10 Simulation software can assist in planning and design

Simulation can also be used for structural analysis. Performance data harvested from past projects can be used to model the behaviour of materials and components in the planned project, to ensure the proposed design is capable of carrying the stresses and loadings that will be imposed on it.

> **Test yourself**
>
> Suggest some questions about a new building project that simulation can help to answer.

When retrofit or restoration work is planned, simulation can be used to ensure the existing structure can accommodate new materials or components without overloading or stressing load-bearing elements.

Chapter 10 Digital technology in construction

▲ Figure 10.11 Digital simulation can be used to confirm structural stability during restoration work

Research

Search online for 'before and after' images of construction restoration projects. Copy and paste the images into a Word document and write a summary of each project.

- visually assess the effects of lighting and the proportions of room sizes
- gauge the number of occupants who can use the internal spaces
- evaluate the access to and movement of occupants in shared spaces and corridors to ensure safety and comfort.

▲ Figure 10.12 Digital technology can be used to create a walk-through of a proposed building to assess lighting effects and room proportions

2.2 Animation

Visualising a completed building and imagining how it will function can be difficult when simply viewing a 2D drawing or diagram. While a 3D digital model can deliver an impressive view of a project, the viewer's experience can be greatly enhanced if that model is animated to help them focus on key elements.

Digital animation is a powerful tool that can be used for refining conceptual details of a project. It can provide an engaging and lifelike view of a building, often by creating a walk-through to show room areas and features. This has many benefits, such as the ability to:

Improve your English

Search online for an example of a construction animation. Watch the whole animation then write a detailed description of what it showed you. (If necessary, you can watch the animation a number of times.)

Test yourself

What does the term 'walk-through' mean in the context of animation?

Case study

Following the Grenfell Tower fire in 2017, a contract has been awarded to replace dangerous external cladding on a group of multi-storey residential buildings. The buildings will be occupied throughout the project, and traffic in the streets around the apartment blocks must be disrupted as little as possible. Storage for the replacement cladding panels is limited, so a JIT delivery system will be used, with delivery vehicles routed to specific offloading points around the area.

In pairs or small groups, discuss how creating an animation of the organisational requirements for the project would be beneficial.

Think about:
- traffic flow
- personnel movements
- equipment movements
- materials handling
- sequence of work
- protection of occupants.

259

2.3 Modelling

So far, this section has considered how simulation and animation use digital modelling to assist in designing buildings and planning construction work. However, digital modelling can also be used throughout the life cycle of buildings.

For example, sophisticated digital modelling is used to survey existing buildings, in order to collect data that can be used for:
- assessing maintenance and repair requirements
- planning restoration projects
- carrying out alterations and extensions
- making decisions on demolition methods at the end of a building's useful life.

A laser scanner can be used to create 3D imagery of a building's complex geometry or survey surface areas that are not easily accessible. From its setup position, it can digitally record the precise distance of densely grouped points at rapid speed. This process is often referred to as a 'point cloud survey'.

> **Industry tip**
>
> Point cloud surveying may also be referred to as 'light detection and ranging' or LIDAR – a combination of the words light and radar.

▲ Figure 10.13 A laser scanner

The point cloud data generated by laser scanning can be integrated into Building Information Modelling (BIM) software or CAD systems to digitally create detailed 3D models of a structure or features of the built environment for a range of uses. Laser scanners can be mounted on drones to capture the exact contours of landscapes, road layouts, railway routes and even entire towns.

▲ Figure 10.14 Laser scanners can be mounted on drones for surveying

A major advantage of using digital technology in this way is that 2D or 3D digital models and representations of structures can be easily amended and instantly shared by being transmitted to other users anywhere in the world.

> **Test yourself**
>
> State one use for point cloud surveying.

The scanned data can be utilised in **immersive** technologies, such as virtual reality (VR). A VR user is visually completely shut off from the outside world by wearing a head-mounted display (HMD). Whatever input the user sees through the HMD becomes their 'reality', allowing them to experience a digitally generated scene as if they were part of it.

> **Key term**
>
> *Immersive:* creating a 3D image which appears to surround the viewer

The HMD senses user movement, so that when the user looks up, down or from side to side, the scene they see tracks their range of movement over 360 degrees, to give the sense of looking around a space within the virtual building.

> **Health and safety**
>
> There are obvious safety considerations if a user cannot see their real-life surroundings while moving through a virtual environment. If you use a VR system, plan the way you use it with this in mind.

Chapter 10 Digital technology in construction

> **Research**
>
> VR systems have become quite common. Research and make a list of construction and non-construction uses for VR systems.

> **Industry tip**
>
> Other types of immersive technology are often mistakenly grouped under the title 'virtual reality'. Augmented reality (AR) and mixed reality (MR) are different systems that can be applied in various construction-related contexts. For more on AR, see Chapter 4.

Multiple users can enter this virtual world to collaborate on design decisions, making the process more creative and efficient. Often, a client will have definite ideas about how they want their project to be completed but they may not be able to express their ideas clearly to designers. Using immersive technology, the designer can accompany the client using the virtual representation of the building to confirm that the proposed design fully matches the client's expectations.

As well as producing dynamic visual representations of buildings and other structures, the data in digital models can also be used to create static illustrations or artistic impressions of a project for use in conventional non-digital documentation, such as a client brochure or report.

3 Opportunities for the use of technology

Digital technology is widely used in many industries to achieve specific user-defined results. Manufacturing, research, transport, medicine and a host of other sectors routinely make use of digital technology in order to operate and grow. Systems and processes using digital data are often taken for granted in our daily life, for example in items such as mobile phones and computers.

Table 10.1 details some of the benefits of using digital technology in construction, especially in relation to planning and organising a project and monitoring the construction phase.

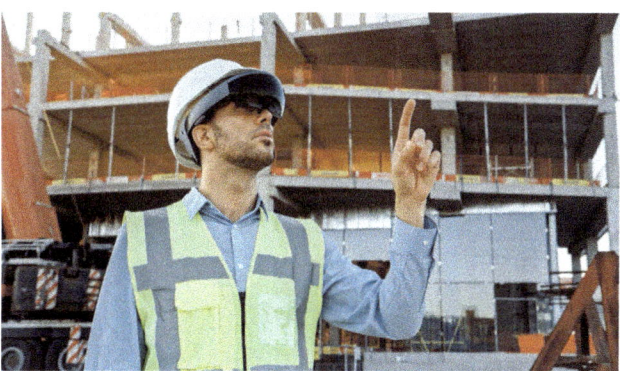

▲ Figure 10.15 An architect using an MR system

▼ Table 10.1 Benefits of technology in construction

Benefit	Application
Accuracy	Digital technology allows large volumes of data to be processed with high degrees of accuracy.
	Data concerning measurements, costings, projections and evaluations can be cross-referenced and linked in complex ways to produce rich and valuable assessment data.
	This can inform reliable decision making and forecasting during the planning of a project and the monitoring of progress during the construction phase. Accuracy during the design and construction phase of a project can be supported by using Geophysical Surveys (Geo surveying) to accurately map the shape and topography of a site. Technologically advanced equipment such as laser scanners and drones (mentioned previously) can be used for this purpose.
Accessibility	Digital data can be stored and retrieved quickly and easily.
	Data can be accessed by anyone who is authorised to use relevant networking systems, regardless of their location. This encourages collaboration and allows data to be instantly updated or modified.
Efficiency	Digital networks can be permanently open to authorised users, allowing efficient working in both local teams and global partnerships across different time zones.
	Whether working in teams or in isolation, working simultaneously or sequentially, effective work patterns can be created and refined to respond to current needs in maintaining efficiency.

▼ Table 10.1 Benefits of technology in construction (continued)

Benefit	Application
Risk reduction	Digital simulation of construction operations can be used to identify areas of operational risk. Analysis of accident and injury data allows the identification of emerging trends, possibly due to inappropriate behaviour or incorrect work practices. These can be mitigated by the rapid introduction of safer work methods – risk assessments and method statements can be matched more closely with actual working patterns and conditions. Risks associated with project startup on site can be reduced by using advanced technology such as Ground Penetrating Radar (GPR). This can perform non-intrusive surveys to locate underground services or potential hazards from past mining activity and natural or man-made buried structures.

> **Test yourself**
>
> What are the benefits of using technology in construction?

There will no doubt be intriguing opportunities in the future to take advantage of the power of digital equipment, computer modelling and smart technologies.

3.1 Robotics

Industries such as car manufacturing use robotics with digital control systems to assemble components repetitively.

There are some repetitive activities in construction, where industrial robots are already being used or are being developed, for example in off-site construction. This involves the manufacture of parts or sections of a building away from the site location. The manufactured items are then transported to site, where they are assembled to complete the building. Off-site construction is often referred to as 'prefabrication' and may involve modular construction methods. For more information on off-site construction, see Chapter 7.

▲ Figure 10.16 Robotics can be used for repetitive construction operations

When manufacturing sections of a building, the use of robots has many benefits, including:
- consistent quality standards
- a regulated work rate that allows for reliable scheduling of materials deliveries to the factory
- accuracy and efficiency (so less waste)
- enhanced safety standards, as there is reduced human involvement.

▲ Figure 10.17 Factory-produced modules being installed on site

> **Research**
>
> Research off-site construction companies and find out the types of projects modular construction has been used for.

Other types of robots that could be used in the construction industry include:
- inspection robots that can reach areas that are difficult or dangerous to access
- maintenance robots that can clean and repair buildings safely
- 3D-printing robots that 'print' components or whole buildings
- demolition robots that safely dismantle unstable or fragile structures.

Chapter 10 Digital technology in construction

▲ Figure 10.18 3D-printing robots can 'print' components or entire buildings

Test yourself

How can robotics be used in modular construction?

3.2 CAD/CAM

CAD/CAM refers to software that is a combination of two processes – computer-aided design (CAD) and computer-aided manufacturing (CAM).

As mentioned above, CAD is a digital tool used to create 2D and 3D representations and drawings during the design process, which allows simulation, testing and refining of ideas to arrive at a satisfactory finished design.

CAM uses digital **geometric** data to control manufacturing machinery, such as computer numerical control (CNC) machines which have motorised tool positioning and manoeuvring capabilities controlled by pre-programmed computers.

Key term

Geometric: consisting of defined angles, patterns and shapes

The combination of CAD and CAM results in a faster production process that provides dimensional control and ensures product consistency. By using automated systems that can work around the clock, the need for skilled production workers is reduced. However, these systems still require input from trained personnel who set up and program the machines and monitor and intervene when problems arise.

▲ Figure 10.19 A CNC cutting machine being set up to perform repetitive tasks

Digital processes such as CAD, CAM and BIM can be integrated, in order to save time and achieve complex outcomes without the possibility of human error in interpreting drawings and written instructions.

For example, a 2D or 3D CAD drawing of a building consists simply of lines to represent its shape and the component parts within it. If the CAD data is linked to BIM, the combinations of simple lines can be allocated grouped geometric patterns, features and dimensions.

This can allow automatic generation of a range of accurate views of a component, along with details of the materials it is made from, so that a CAM machine can manufacture it.

Onsite Construction T Level: Core

Assessment practice

Short answer

1. What does M2M stand for?
2. How can the IoT improve energy efficiency in a building?
3. State two benefits (besides energy efficiency) that can result from integrating buildings into the IoT.
4. What alternative name is sometimes given to point cloud surveying?
5. How can robotics be used in a construction setting?

Long answer

6. Describe how AI systems go beyond simply following programmed instructions.
7. Explain how AI systems could be used to improve worker productivity in an office building.
8. Describe the range of benefits achievable by using CAD when designing buildings.
9. Explain how an item of equipment or machinery used on site can be described as 'smart'.
10. Describe the difference between simulation and animation when used for analysis of buildings.

Project practice

An old office block in a city centre is to be extended to allow for company expansion. It is a familiar landmark in the area because of its attractive period architectural features. A railway line and a river run close to the building.

The existing parking area next to the building will become the location of the office extension. Parking for existing and new office staff will in future be provided by a new underground car park below the extension.

You have been commissioned to conduct a survey of the building plot and the existing structure to establish the feasibility of the proposed extension project.

- Write a report to compare the advantages and disadvantages of conducting the survey visually on foot or using a drone to create a point cloud survey.
- Make a list of potential problems that must be considered regarding the proposed underground car park. (Hint: think about the railway and river nearby.)
- Consider technological methods that could be used to record details of the architectural features of the existing landmark building, with a view to matching them in the new extension.

Chapter 11 Construction commercial/business principles

Introduction

This chapter looks at different business structures in the construction industry and identifies legal obligations towards registration and taxation. We will compare the benefits and financial risks for proprietors when trading as a particular business type and learn how they can protect themselves personally from any liability.

We will then examine the roles of values, aims and objectives in achieving business goals and establish how the success of a business can be measured.

Learning outcomes

By the end of this chapter, you will understand:
1. business structures
2. business objectives
3. business values
4. principles of corporate social responsibility
5. principles of entrepreneurship and innovation
6. measuring success
7. project management
8. quality management.

1 Business structures

1.1 Business types

There are many different types of business that operate in the UK construction industry, ranging from individuals trading on their own to national public limited companies employing thousands of people across the country.

Some business structures are not designed for everyone because of the size of the organisation and its legal status. However, as a business grows it may need to change to a more appropriate business model.

The advantages and disadvantages of each business type have to be considered carefully, because there could be serious legal implications if trading unlawfully.

In this section, we will look at the following types of business:

- sole traders, for example a self-employed tradesperson
- partnerships, for example an architects' practice
- limited companies (private, Ltd, or public, plc), for example construction companies
- small and medium-sized enterprises (SMEs), for example construction companies
- not-for-profit organisations, for example Her Majesty's Revenue and Customs (HMRC) or the Health and Safety Executive (HSE)
- community interest companies (CICs), for example local community development trusts
- franchises, for example building maintenance businesses.

Sole traders

As soon as someone registers with HMRC as self-employed, they are effectively a sole trader. As a **proprietor**, they own and have full control of their entire business or enterprise, although they can employ other people to work for them.

A sole-trader business can operate under the name of its owner. However, if the proprietor would prefer to use a different business name, they must register it at companies house to prevent another company trading with the same name.

> **Key term**
>
> **Proprietor:** an individual who owns a business

▲ Figure 11.1 A sole trader

Operating as a sole trader is one of the simplest business structures to run, because there are fewer tax responsibilities (such as filing accounts with Companies House) and therefore less administration. However, a sole trader is still responsible for the day-to-day management of their company and:

- accurately recording expenses, sales and profits
- running payroll
- filing annual self-assessment tax returns with HMRC
- paying National Insurance contributions
- paying value added tax (VAT) if annual turnover exceeds, or is likely to exceed, £85,000.

> **Research**
>
> Find out about value added tax (VAT) and explain the advantages and disadvantages for a sole trader.

> **Improve your maths**
>
> An electrician who is employed as a sole trader has an annual turnover of £92,000 (inclusive of VAT) and is paid a salary of £42,000.
>
> Calculate how much tax they will have to pay to HMRC (excluding any tax savings) if the rates of tax are:
>
> - VAT 20 per cent
> - tax and National Insurance 27 per cent.
>
> Note: VAT is paid on whole turnover, but tax and National Insurance are paid on salary.

To help manage the financial aspects of running a business, many sole traders appoint an accountant. However, the sole trader is still legally responsible for the information submitted to HMRC and Companies House each year.

Chapter 11 Construction commercial/business principles

A sole trader is personally liable for any losses or debts (for example, to the bank, HMRC or suppliers) that the company may incur if things go wrong. **Creditors** can legally recover any money owed by the **debtor** from their business or their personal assets such as their house. If the debt is considerable and the individual is unable to pay, they could be made personally bankrupt. This could result in further difficulties, for example being unable to obtain any type of credit such as a mortgage, credit card or finance on a vehicle because they are deemed to be a high risk.

> **Key terms**
>
> **Creditors:** individuals or organisations that are owed money because they have provided goods, services or a monetary loan; HMRC would also be described as a creditor where a tax duty is owed
>
> **Debtor:** an individual or organisation that owes money

Partnerships

Partnerships are businesses owned by two or more individuals. Partners share the costs, duties and risks of managing a business together, although they may delegate certain responsibilities to their employees.

Each partner's share of the profits is based on the **partnership agreement** between all stakeholders and can be determined by looking at the amount of money they invested in the business and/or the level of involvement they have in the day-to-day running of the business. It is possible to have 'sleeping partners' who have no involvement in managing the business but who contribute financially.

Each partner is personally responsible for paying tax and National Insurance contributions, based on their share of the profits, and must therefore register for self-assessment tax returns with HMRC.

A nominated person must be chosen to manage the partnership's business accounts and file its annual tax returns with HMRC.

> **Key term**
>
> **Partnership agreement:** a legally binding contract that sets out terms and conditions for each partner in the business

▲ Figure 11.2 A nominated person filing a tax return

There are three different types of business partnership:
- ordinary partnership
- limited partnership
- limited liability partnership (LLP).

There are no special responsibilities for an ordinary partnership other than those described above.

Limited partnerships have to be registered with Companies House. They must have at least one general partner and one **limited partner**:
- General partners are responsible for managing and controlling the business and are liable for any debts that the business cannot repay, in the same way as a sole trader. They can apply for the limited partnership to act as an authorised contractual scheme (ACS), which means that the partners co-own the assets but only pay tax on their share of the profits and are exempt from paying **corporation tax**.
- Limited partners are not responsible for the management of the business and are only liable for any debts up to the amount they have contributed.

A partner cannot be both general and limited at the same time.

> **Key terms**
>
> **Limited partner:** a part-owner of a business whose financial liabilities cannot exceed their investment
>
> **Corporation tax:** a duty paid annually to HMRC based on a company's profits

Onsite Construction T Level: Core

> **Improve your English**
>
> Explain the term 'sleeping partner'.

> **Research**
>
> How long does a business legally have to keep tax records? What are the consequences of not keeping tax records in order?

A limited liability partnership (LLP) can be created (**incorporated**) with two or more partners. While it can have any number of ordinary (non-designated) members, it must have at least two designated members.

Designated members have responsibilities to manage the business, for example:
- registering the business for self-assessment with HMRC (individual LLP members must also register separately)
- registering the business for VAT, if annual turnover is expected to be over £85,000
- keeping accounting records
- completing self-assessment and VAT returns
- preparing and signing annual accounts and filing them with Companies House
- sending an annual confirmation statement to Companies House.

A formal LLP agreement should be written to outline:
- how members can join or leave the LLP
- the roles and responsibilities of members
- how decisions are made
- how profits are shared among members.

Having a legally binding agreement in place when the business is first set up avoids potential disputes or complications between partners at a later stage of trading.

LLPs are sometimes preferred to other types of business because they clearly define their partners as individuals when it comes to earnings, unlike corporations. They also protect the personal assets of partners as a separate legal entity from the liabilities of the business.

As with any partnership, if there are only two members and one leaves, the LLP may have to be dissolved.

> **Test yourself**
>
> Why are LLPs sometimes preferred to other types of business?

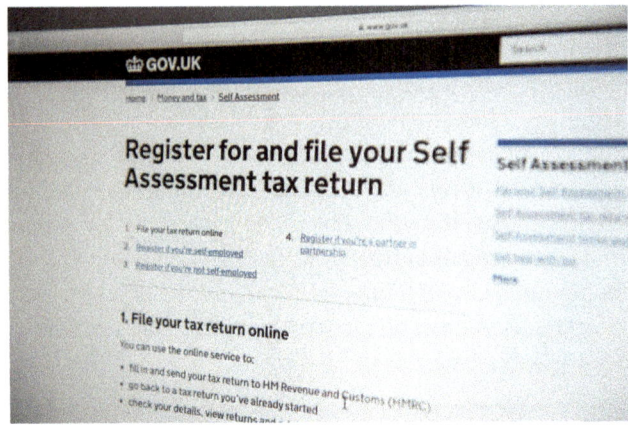

▲ Figure 11.3 Self-assessment tax return

> **Test yourself**
>
> What factors could cause a business partnership to fail? Explain how these could be avoided before starting an enterprise.

> **Key term**
>
> *Incorporated:* legally registered as a limited company

Limited companies

Businesses often change their legal status to limited when they become more established, because it gives clients more confidence that the business is a credible organisation. Other benefits of limited companies include:
- protection of owners' personal assets from any business debts
- owners given complete control of their business
- no requirement to pay National Insurance
- lower rates of tax
- more tax relief than sole traders.

A disadvantage of being a limited company is that information about the directors, shareholders, registered office and yearly financial statements is on public record. These details are visible not only to the general public, but also to the business's potential competitors.

Private limited companies (Ltd)

A private limited company (Ltd) is a business that is legally registered with Companies House. It is referred to as incorporated, meaning that it has been made into a corporation that is a separate legal entity, with its own rights and obligations. The main benefit of a limited company is that the owners and shareholders have limited liability, so their personal assets are protected against any business debts up to the value of the money that they have invested.

As with any business, a limited company and its owners (known as directors) must be registered with HMRC and complete annual tax returns, regardless of whether any **duty** is owed. Annual business accounts must be filed with HMRC, Companies House and shareholders at the end of each **financial year**.

Directors of the business are paid in the same way as its employees, through the pay as you earn (PAYE) system, and they do not have the flexibility to withdraw money from the company without restrictions or tax consequences. This means that tax and National Insurance are calculated and deducted from salaries, rather than the directors and employees having to complete annual self-assessment returns with HMRC.

At the end of each financial year, the board of directors considers whether to distribute any profits among the shareholders or reinvest the money in the business. Payments made to shareholders are known as dividends; these can vary in size, depending on the amount of investment shares that have been purchased.

The limited company is responsible for paying PAYE duty to HMRC either monthly or quarterly, and any corporation tax annually. There are strict deadlines for filing tax returns and making any payments due, with potential penalties or fines if they are not met. Therefore, most businesses employ an accountant to complete their accounts, or a larger company may have an in-house accounts team.

▲ Figure 11.4 A limited company

> **Key terms**
>
> **Duty:** a type of tax charged on the value of goods and services
>
> **Financial year:** a 12-month period over which a business' accounts are calculated, as opposed to a calendar year (January to December); also known as a fiscal year

> **Research**
>
> Research the purpose of National Insurance contributions made through personal income.

Public limited companies (plcs)

A public limited company (plc) is similar to a private limited company. The main difference is that money can be raised for the business through investors (for example, the public) buying shares on the stock exchange.

This type of business benefits from many tax and National Insurance advantages up to a certain value of its profits, including:
- paying less corporation tax
- better dividend payments for shareholders
- pensions
- reinvesting money back into the business.

For a company to change its legal status to plc, it must have at least two shareholders and a minimum of £50,000 worth of shares to be issued. As with a private limited company, a plc must be registered with HMRC and Companies House and file tax returns.

Small and medium-sized enterprises (SMEs)

In the UK, small and medium-sized enterprises (SMEs) are defined as follows:
- small – employs on average no more than 50 people and has an annual turnover of £10.2 million or less
- medium-sized – employs on average no more than 250 people and has an annual turnover of £36 million or less.

In 2020, 99.9 per cent of all businesses in the UK were SMEs, which represents the employment of the majority of the working population. These businesses play an important role in the country's economy because of the jobs and tax revenue that they generate. To encourage further growth and innovation, and to help keep these enterprises in business, the government regularly offers various incentives, such as better access to loans and favourable tax treatment.

Not-for-profit organisations

As the name suggests, these are charitable businesses that do not make a financial profit. They often seek to provide a public service or social benefits for individuals or communities in need.

These types of business are usually run by a board of directors that use similar management techniques to for-profit organisations. However, any money earned is used to cover operational costs and reinvested back into the organisation to pursue its objectives or goals.

The type of management structure adopted often depends on whether the organisation is limited, a legal charity or an unincorporated association with voting members.

> **Research**
>
> Find out what unincorporated associations are and how they operate:
> ▶ What is their usual purpose?
> ▶ Do they have to register with Companies House?
> ▶ How much does it cost to set them up?
> ▶ What happens if they make a profit?

A not-for-profit organisation can be started by anyone without any money, although they are often funded through charitable donations, grants or self-funding. In many cases, they also have tax-exemption status. However, they are still obligated to deduct tax from their employees' wages and complete annual company tax returns with HMRC.

A successful not-for-profit organisation will demonstrate ethical standards and practices, such as trustworthiness, openness and accountability. This instils public and government confidence in the organisation, resulting in greater financial investments in the future.

> **Research**
>
> Research not-for-profit construction organisations in the UK and explain the motives for running their businesses.

Community interest companies (CICs)

The aim of community interest companies (CICs) is to provide a benefit for the community or trade with a social purpose. Unlike not-for-profit organisations, they are expected to make a profit, which is predominantly reinvested in the company or used for the community they are set up to serve. However, returns to company owners and investors are allowed, as long as they are balanced and reasonable, and a dividend cap must be put in place. A CIC is primarily about benefiting the community and not making a private profit.

Franchises

Successful and profitable businesses may look to grow further by creating a franchise.

A franchise allows a business to expand quickly with lower **capital** outlay by selling the rights to the business name, logo or trademark and business model to self-employed **entrepreneurs**, referred to as franchisees. By operating under the banner of an established business with a proven track record, a franchisee can seek to replicate the success of the bigger brand.

It is sometimes more beneficial to buy a long-term franchise than start a business from scratch with a new name, because of the business system (processes, suppliers and other resources), existing customer loyalty and support provided by the **franchisor**. New businesses often take years to establish themselves with a strong customer base before making a profit, whereas they are more likely to be profitable from the outset if they have invested in a franchise.

Business support provided for a franchisee often includes:
▶ marketing (promoting the business)
▶ recruitment of personnel (human resources)
▶ provision of training and equipment
▶ provision of resources, for example machinery or vans
▶ assistance from head office
▶ collaborative networking with other franchisees
▶ dealing with suppliers
▶ financial planning
▶ accounting.

Chapter 11 Construction commercial/business principles

▲ Figure 11.5 Business support training

> **Key terms**
>
> **Capital:** the amount of funds or **liquid assets** owned by a business
>
> **Liquid assets:** things that a business owns that can be sold quickly and easily for cash without any loss in value
>
> **Entrepreneurs:** individuals who start up their own businesses, taking on financial risks in the hope of making a profit
>
> **Franchisor:** the original owner of a franchise

Once an agreement has been reached and the contract has been signed, the franchisee must pay any initial fee for the rights to the business. In addition, the franchisor will be paid a regular share of the profits by the franchisee, known as royalties. Royalties can be calculated from a percentage of the business turnover, a mark-up on goods supplied or a flat fee.

For a franchise business to work properly, it has to maintain its standards and brand control. The business model must therefore be transferable, regardless of geographical location. The business activity should also be something that another business can do or be trained to do successfully.

Evidence suggests that franchises often perform well because of the personal financial investment made by their owners and the support network provided. Once established, the franchisee has the option to buy further investments or sell their business to a buyer authorised by the franchisor.

A disadvantage of this type of business is that the franchisee does not have any independence compared with running their own business. They have to adhere to the franchise business plan, operational plan for day-to-day running, systems, policies and procedures. Regardless of how well a franchise is managed, there is always a risk that the organisation can be tarnished by other franchisees, which could impact on future success.

> **Test yourself**
>
> Which term describes an individual who starts up their own business, taking on financial risks in the hope of making a profit?

2 Business objectives

Objectives can be defined as the incremental steps a business needs to take in order to achieve its overall aims, which are often closely aligned to its **business plan**.

Businesses working in different sectors vary in size and scale; therefore their aims and objectives must be individually tailored. For example, an entrepreneur who has recently started a new business may initially have an aim to financially survive the first couple of years of trading, whereas an established organisation may have an ambition to increase its annual turnover or profit.

> **Key term**
>
> **Business plan:** a written document that defines a business's goals and the strategies and timeframes to achieve them

> **Improve your English**
>
> You have decided to start your own business as a sole trader in your chosen profession and you are looking for investment to get you started, for example to buy a van and tools.
>
> Design headed paper with your business details and write a letter to a bank, requesting a loan and outlining your business plan.

> **Key term**
>
> **Market share:** the percentage of total sales or output that a business has in a specified market; sometimes referred to as market leadership

Objectives should be challenging, to give a business purpose and direction. Once they have been achieved, they should be updated to improve the business and move it one step closer to achieving its main goal.

The main categories of business objective used to measure performance of an organisation in the construction industry are:
- financial
- social
- organisational culture
- innovation
- quality
- sustainability
- compliance.

2.1 Financial objectives

Entrepreneurs often prefer the independence associated with having control of their own organisation, rather than being employed by someone else. The freedom to run a business and make important decisions with the potential of determining successful outcomes can be personally satisfying, especially when the owner is challenged beyond their comfort zone.

The financial objective for private organisations is usually to develop their business with innovative ideas, products or services in order to maximise profits. As the business grows, other opportunities can be created to adjust or set new goals in order to achieve a bigger **market share** in a particular geographical location.

As mentioned above, the aims and objectives of businesses can be very different. For example, the financial objective of not-for-profit organisations is not to create wealth but to make enough money to:
- cover their costs
- establish reserves
- fund activities that benefit the community
- reinvest in the business.

2.2 Social objectives

Private organisations have social objectives to provide quality products and services that are useful to customers.

Most private organisations serve society by employing competent people (skilled and unskilled) at different levels. Organisations with strategic social objectives to treat their employees fairly and inclusively will create meaningful jobs, which enables their businesses to grow and in turn generate further employment opportunities. These types of job often provide opportunities for continuing professional development (CPD) and promotion, as well as offering other staff benefits. Besides the fair wages that employees should receive from their employers, they may also benefit from a workplace pension or welfare schemes such as private healthcare.

Not-for-profit organisations, such as charities and social enterprises, work for the benefit of society by providing a public service (such as housing, healthcare and education) or helping people, in order to create positive social change.

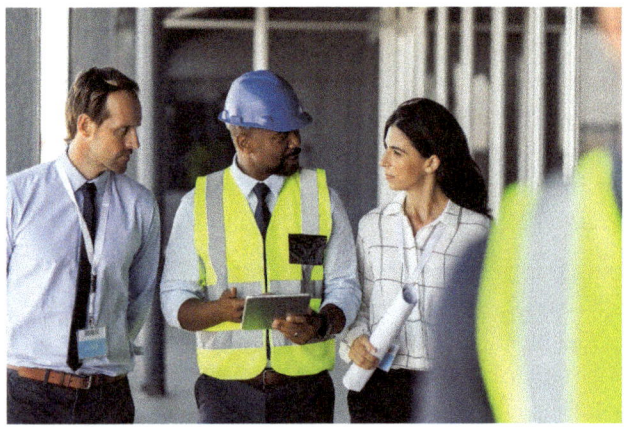

▲ Figure 11.6 Private organisations seek to provide quality products and services that are useful to customers

2.3 Organisational culture objectives

Organisational culture can be described as a collection of core beliefs, ethical values, expectations and behaviours that define the corporate personality of an organisation. It is embodied in the way an organisation conducts business, treats its customers,

Chapter 11 Construction commercial/business principles

manages **workflow** and interacts with its employees. It is essential that all members of an organisation understand its shared cultural objectives and promote them while performing their roles and responsibilities.

Creating a strong organisational culture is vital for the success of a company. Not only does it define a company's internal and external identity, but it also helps to build a strong, unified team of employees who enjoy greater wellbeing. In turn, this will reduce staff turnover and assist with recruiting high-calibre employees.

An organisation that recognises and rewards the success of its employees is likely to transform them into advocates for the business, so that they continue to promote and contribute to its consistent cultural behaviours.

> **Key term**
>
> **Workflow:** a sequence of activities needed to complete a work task

> **Test yourself**
>
> Why is it important to have a clearly defined organisational culture?

2.4 Innovation objectives

Innovation objectives involve greater levels of uncertainty and risk than general business objectives. They should be challenging but obtainable, by agreeing steps to achieve them with all the business's shareholders.

The primary innovation goals for business improvement can be defined by:
▶ setting aims and developing a strategy for how these goals can be achieved
▶ understanding the primary challenges that may prevent targets being met
▶ understanding data, costs and risk
▶ being able to implement innovative new ideas in the organisation
▶ increasing employee retention
▶ engaging with employees and promoting innovation and new ideas
▶ identifying and rewarding talent.

A business's innovation objective may be to launch a new product or service, outperform its competitors or become a market leader. However, in order to accomplish these long-term goals, it may first have to achieve the following typical objectives of:
▶ optimising existing processes or procedures
▶ increasing business productivity and efficiency
▶ improving the quality of service or existing products
▶ being more competitive
▶ meeting customers' needs
▶ improving the skills and knowledge of the workforce
▶ developing performance
▶ refining sustainability within the organisation.

Innovative construction methods and technology, such as self-driving vehicles, virtual reality (VR), computer-controlled manufacturing robots, 3D printers and drones, are explained in Chapter 7.

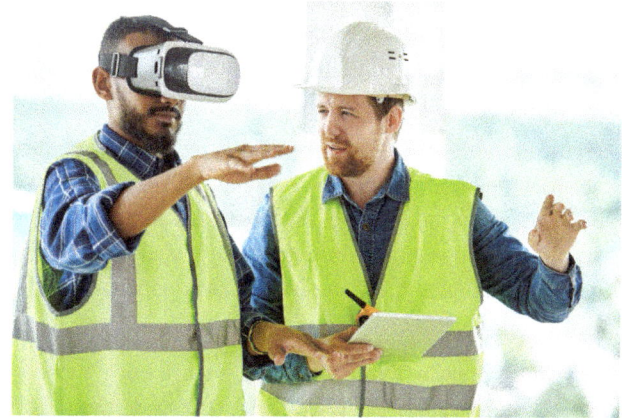

▲ Figure 11.7 Virtual reality is a good example of innovation in construction design and development

2.5 Quality objectives

The minimum requirements for behaviour, work and materials in the UK construction industry are specified in health and safety legislation, British Standards and building regulations. However, businesses often aim to exceed these standards, in order to ensure the welfare of their employees and offer an improved service to their clients.

Construction businesses can subscribe to a number of voluntary assessment and certification schemes, to ensure best-practice standards that often far exceed those required by legislation.

The Building Research Establishment's Home Quality Mark (HQM) is an independently assessed certification scheme for new homes that focuses on the expectations and needs of occupants and protection of the environment. It assesses the quality and sustainability of the home itself, its surroundings and the construction or renovation in the following key areas, set out in the 'Home Quality Mark ONE' technical manual:

- 'Our surroundings', for example recreational space, local amenities and sustainable transport options
- 'My home', for example sound insulation, ventilation, water efficiency, energy and cost
- 'Delivery', for example construction energy and water use, commissioning and testing, and aftercare.

Where the HQM scheme recognises that the standards of a new home are significantly higher than the minimum requirements, it awards a star rating from 1 to 5. A 1-star rating means that a home meets key baselines beyond minimum standards; a 5-star rating means that a home is outstanding and far exceeds minimum standards.

Attaining the HQM instils confidence in buyers, tenants, investors, developers and insurers that higher levels of build quality have been met and there are significant benefits to living in the home, for example a lower risk of defects usually identified in new-build properties.

The International Organization for Standardization (ISO) is an international standard-setting body with headquarters in Geneva, Switzerland. Its membership comprises 165 national standards bodies from around the world. They work together to develop and publish market-relevant commercial, industrial and technical standards.

The main goal of the ISO is to facilitate trade. It offers solutions to global challenges and supports innovation by providing guidelines to streamline processes and improve quality and safety across a range of businesses and products.

There are over 23,000 ISO standards that have been voluntarily developed by industry experts by consensus, covering all aspects of manufacturing and technology. ISO identifies its certified standards with a unique reference number, for example:

- ISO 45001 – Occupational health and safety management systems
- ISO 9001 – Quality management systems
- ISO 14001 – Environment management systems.

Access to general ISO information can be obtained free online. However, the full guidelines must be purchased from the organisation.

The ISO is not a governing body and does not have authority to enforce any regulations and laws. It is also not involved in the certification of its standards; this is carried out by external certification bodies. Therefore, businesses can only become certified to specific standards, for example 'ISO 9001:2015 certified' and not 'ISO certified'. To become certified, a business will need to adhere to the requirements in the standards and may therefore use these as quality objectives.

> **Improve your English**
>
> Write a short sentence to explain the meaning of the word 'consensus'.

> **Test yourself**
>
> Explain the purpose of the ISO.

2.6 Sustainability and compliance objectives

Construction organisations often embed sustainability into their business objectives for moral reasons, for example to protect the environment by maintaining **ecological balance**, conserving natural resources and preventing pollution. The way we design and construct buildings plays an important part in achieving these aims, from energy-efficient construction to eco-friendly use of building materials.

Building regulations and other external regulatory control measures determine mandatory obligations, for example Part L of the Building Regulations 2010 outlines requirements for conservation of fuel and power. However, businesses may create their own internal control measures beyond these standards and build them into their business objectives.

Besides the moral reasons mentioned above, businesses may set further goals to improve their sustainability in order to reduce waste, increase profits and attract the interest of future clients with similar interests in protecting the environment.

> **Key term**
>
> *Ecological balance:* where living organisms, such as plants, animals and humans, co-exist in a sustainable environment

Chapter 11 Construction commercial/business principles

▲ Figure 11.8 Protecting the natural environment from development activities

3 Business values

Earlier in this chapter, we acknowledged that organisational culture stems from a company's beliefs and values, sometimes referred to as its **philosophy**.

Transparent business values play an essential role in any organisation. They can:
- create a sense of purpose and commitment
- improve **cohesion** of the workforce
- drive an organisation forward, helping it to reach the goals set by the management team
- help motivate employees by building trust and security
- help develop relationships with partners, stakeholders and customers
- influence sales, customer service and marketing strategies
- demonstrate the business's culture outside of the organisation, which may attract new talent.

It is important that potential new employees share the same values as the organisation; this will determine whether they enjoy their job, engage with the company and work productively.

> **Key terms**
>
> **Philosophy:** values and beliefs that act as guiding principles for behaviour
>
> **Cohesion:** a state of working together in unity

The fundamental business values for construction organisations are:
- financial stability
- customer service
- care for life
- ethics and transparency
- codes of conduct
- collaborative working.

3.1 Financial stability

Financial stability means a business having sufficient funds to pay overheads, repay any loans and still make a profit to prepare for risk management in times of potential **economic downturn**.

A business with steady revenue that does not rely too much on debt should be making a healthy profit through sales while its expenses remain the same. This should enable growth of the business.

> **Key term**
>
> **Economic downturn:** when the economy has stopped growing and is on the decline, resulting in reduced financial turnover

3.2 Customer service

When a customer first approaches a construction business with a concept for a project, they often have expectations of what the business can provide and the level of service it should offer. To build and maintain good customer service, the business must:
- pay close attention to the customer at all times
- be helpful – make the customer feel at ease exchanging information and ideas
- be polite and courteous when communicating verbally or in written form
- listen actively to the customer by responding to all of their needs
- solve problems – offer practical solutions with design, product and scheduling ideas
- provide support to meet the customer's needs and expectations
- be timely – provide the product or service in a mutually acceptable timeframe
- be proactive – anticipate problems before they occur
- demonstrate skills and knowledge – have a strong and up-to-date understanding of the industry and be able to produce work to the highest standards
- know the business product or service offered – understand the business's limitations

- ensure the customer's needs are met – measure the success of customer service by making sure the customer's initial expectations have been met or exceeded
- offer value for money – be able to provide a satisfactory product or service for a reasonable price
- personalise support for each customer.

Providing good customer service is vital to the success of all construction businesses, to ensure a good reputation. Customers who experience good customer service are more likely to provide repeat business and recommend a company to other potential customers.

3.3 Care for life

The health, safety and welfare of all employees and others affected by work activities, for example the general public, should be one of the core values of all businesses.

Besides the legal obligations that businesses have to adhere to, they should also consider the mental and physical wellbeing of their staff and the impact that their job roles may have on them. Businesses that value their employees will make sure they have a good work–life balance and actively promote health and wellbeing through a number of different benefits, such as a tax-deductible cycle-to-work scheme or mentoring.

3.4 Ethics and transparency

Business ethics can be described as business practices and policies when faced with arguably controversial subjects, for example corporate social responsibility. Some business ethics are embedded in legislation; for example employers have a duty to treat employees fairly and with respect and not to discriminate. However, ethics often goes beyond legal responsibilities to include a moral code for the business and its employees.

Businesses with positive values and behaviours gain public trust and approval, which will enable them to grow. There are negative implications for businesses that do not have business ethics, such as damage to reputation, difficulties retaining employees and loss of work.

Business transparency is about open and honest communication across all levels within an organisation and the sharing of information both internally and externally.

For a business to be completely transparent, it must provide access to all the information needed, not just what it is willing to share. Some businesses may not want information such as salaries, revenue, future ambitions and hiring policies to be available to the public, although this level of transparency can provide potential investors and customers with an insight into the organisation and the level of service they can expect.

When employees feel financially, physically and emotionally safe within an organisation, they are more likely to communicate openly and honestly with their employers, without fear of losing their job. In turn, when employees feel they are being listened to, they are more motivated and productive.

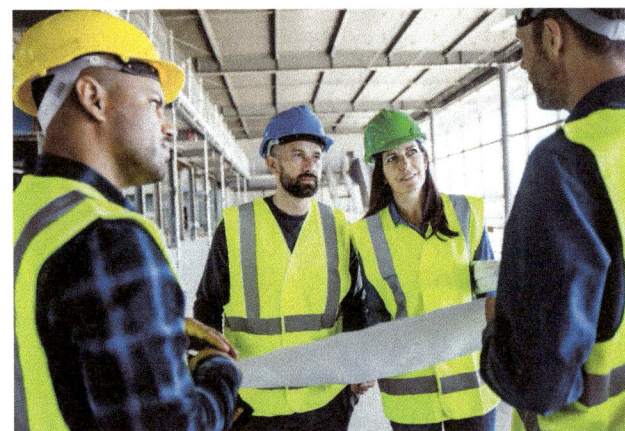

▲ Figure 11.10 Employees who can communicate openly and honestly with their employers are more motivated and productive

▲ Figure 11.9 Cycle-to-work scheme

Chapter 11 Construction commercial/business principles

3.5 Codes of conduct

A code of conduct is essentially a set of rules written by an employer for their employees, in order to protect their business and its reputation. It should explain an organisation's values and principles and link them with standards of professional behaviour.

A code of conduct typically includes standards of behaviour regarding equality and diversity, protection of sensitive data, and disclosure of any criminal convictions that an employee may have prior to employment or during their service.

All employees must read and understand their employer's code of conduct before starting work; they must also appreciate the consequences of not following the rules. Any breaches of the code of conduct could result in disciplinary action being taken, or in some cases termination of employment.

3.6 Collaborative working

Internal collaborative working enables a diverse range of employees with specific skills or traits to work together to problem solve and achieve business objectives.

Collaborative working with another organisation is referred to as networking. Organisations of different sizes can collaborate in a non-hierarchical way either in person or remotely using a variety of different technologies, for example video conferencing and BIM.

Besides being more competitive, there are several other advantages of working with another organisation, including:
- sharing responsibilities
- sharing knowledge and expertise
- sharing materials, resources and facilities
- developing skills and innovation
- improving motivation
- increasing productivity and sales
- accelerating business growth
- being able to tender for larger contracts
- achieving aims that may have been out of reach if working independently.

Provided that employees understand and share the same business values, and their employer honours them when making important decisions, a strong business can be created with a positive company image.

> **Test yourself**
>
> List the benefits of networking with another organisation.

> **Research**
>
> Research your industry placement's key business values and explain how it is working to achieve these.

> **Case study**
>
> You and your business partner have successfully been trading for a couple of years in the construction industry and have decided to grow the enterprise further by employing some staff.
>
> Complete the following tasks in preparation for your new employees:
> - Outline your business values.
> - Write a brief code of conduct.

4 Principles of corporate social responsibility

Construction and maintenance of the built environment can have a negative impact on the natural environment and local communities, through pollution and disruption. Employees in the construction industry can also experience short- and long-term physical and mental health issues due to a number of factors, for example working to time constraints in a competitive market.

Corporate social responsibility (CSR) is the commitment of an organisation to carry out its business activities in a socially and environmentally responsible way.

▲ Figure 11.11 Disruption to a community due to construction activities

CSR is not a legal requirement. However, developing a CSR strategy that will have a positive impact on the community and wider society, and integrating it into an organisation's values, makes good business sense, because it often contributes to risk management and legal compliance. It also affects how stakeholders such as clients and investors view an organisation, and may impact on their decision to work with or support it.

It is important for construction businesses to establish links with local schools, community groups and charities, with a view to being as transparent as possible about their aims to both minimise negative impacts on the community and actively make positive impacts.

Table 11.1 provides examples of how CSR is applied in the construction industry.

▼ Table 11.1 CSR in the construction industry

Principles of CSR	Application in the construction industry
Economic	• Career management • Use of local operatives, trades and suppliers • Positive impact • Supporting housing initiatives • Boosting the economy by increasing local employment and paying above the minimum living wage
Legal	• Ensuring inclusivity (for example, providing affordable homes)
Ethical	• In design (community led, inclusive, meets local needs), for example providing functional community spaces, landscaping, cycle paths, community lighting for safety • Buildings are well designed to suit occupants' lifestyles • Responsible purchasing • Ensuring equality and diversity in the workforce by creating opportunities for minority and under-represented groups
Environmental	• Local sustainable materials • Sustainable initiatives, for example energy-efficiency initiatives • Reducing the environmental impact of construction work • Using ethically and sustainably sourced building materials through the entire supply chain, for example Forest Stewardship Council-certified timber • Using recycled materials • Minimising waste • Constructing greener buildings using low-impact materials • Working more efficiently to reduce carbon footprint • Reducing the use of fossil fuels • Reducing air, water and ground pollution • Making provision to improve the biodiversity of the local area
Philanthropy	• Improving quality of life and wellbeing • Supporting training and development programmes, for example work experience, apprenticeships and internships • Encouraging flexible working • Improving working conditions for employees • Supporting local community groups and charitable activities with time and resources

> **Key term**
>
> **Philanthropy:** the activity of donating time, experience, skill or money to help the poor and other charitable causes, and to promote the welfare of others

Chapter 11 Construction commercial/business principles

▲ Figure 11.12 Supporting training and development programmes such as apprenticeships is an important part of CSR

Besides the moral obligations businesses have to protect the environment and people from construction activities, there are also legal duties that may align with their CSR commitments. Failure to follow legal duties can result in prosecution and damaged public relations.

Further information on CSR can be found in Chapter 3.

> **Test yourself**
>
> List five different ways that a construction business can implement CSR principles of sustainability in their objectives.

5 Principles of entrepreneurship and innovation

As mentioned above, an entrepreneur is an individual who starts up their own business, taking on financial risk in the hope of making a profit. To be successful, a business owner must know their market well, including its strengths and weaknesses, in order to provide a service or product for which there is a demand or that has a unique selling point (USP).

5.1 Creating a viable product or service

Market research is a valuable tool used by many businesses to analyse industry and **demographic data**, in order to gather information about consumer needs and preferences.

Typical data that a business may want to analyse in a particular location or region could include:

- age
- gender
- marital status
- education
- employment
- income.

The data gathered will help to create two different groups of people: those with no interest in the business and a possible customer base.

Once a viable business opportunity has been identified, the business' mission should be established – what does the business hope to achieve? A business' goals could be purely economic or financial, although some businesses are created for social or environmental reasons.

> **Research**
>
> Research the demographic data in your local area to answer the following questions:
> - What is the average age?
> - How many people are living in social rented accommodation?
> - How many people have access to a car or van?

> **Key term**
>
> **Demographic data:** statistical data about a population in a particular location or region

Entrepreneurs who are willing to take a risk, and have the vision to solve problems with an innovative product or service, are likely to have a successful enterprise. Besides developing the initial business concept, they are also the driving force behind an organisation, and must therefore have other qualities in order to grow their business. They must be confident in their product or service and motivated by the **business model**, not just by the idea of generating a lot of money.

Developing innovative technologies, products and processes is extremely important to the future of the construction sector, in order to:

- create employment
- improve productivity
- provide value for money
- improve the way we live
- improve product service
- increase growth and profit
- make advancements in the industry.

279

Onsite Construction T Level: Core

> **Key term**
>
> **Business model:** a plan usually created by a business owner which describes the strategy or framework that an organisation will use to operate and includes the identification of products/services, revenue sources and customer base

5.2 Capital

To get a new business venture off the ground often requires an investment known as working capital. This is a sum of money remaining after all the business's debts have been covered. It does not amount to the value of the company in terms of assets or customers' unpaid bills.

The money needed to start a business could come from the owner/s, investors, shareholders or a business loan. Business loans from banks are often secured against the owner's personal assets or those of the business, depending on the type of organisation. (See section 1 of this chapter for details on different types of organisations.)

▲ Figure 11.13 Securing a business loan from an investor

▲ Figure 11.14 Principles of entrepreneurship

5.3 Growth and marketing

For a business to thrive, potential customers need to be made aware of the product or service being sold, the people selling it, the price and the place where they can buy it. Therefore, some of the money invested in a business should be allocated to promoting the product or service through marketing. Active advertising will communicate with a bigger audience more quickly than waiting for the business to grow through word of mouth alone.

Before a customer decides to invest their money in a product or service, they may want evidence that it is as good as it sounds. As a business begins to grow, it can build customer confidence by sharing evidence of its success through showcasing satisfied customers or completed projects

> **Research**
>
> Research different methods of marketing and explain the impact each one may have on a construction business.

5.4 Priorities

Entrepreneurs with a concept for a new business must have a list of priorities to get their venture off the ground, become financially stable as soon as possible, and grow in the future. Even with a strong initial investment, a business will need to earn money quickly to repay any loans it may have for resources and equipment and to cover the cost of overheads and wages.

Below are some of the general priorities an entrepreneur should have when they start a business:
- Exploit a concept or product and develop it to add value.
- Research the market.
- Produce a business plan, including business aims and objectives.
- Seek investment.
- Produce business documentation, for example a contract of employment, code of conduct and health and safety policy.
- Acquire premises, such as an office, workshop or factory.
- Acquire resources, such as machinery, equipment and staff.
- Register the business.
- Arrange suitable insurance for the business.
- Market/advertise the business, for example by networking.

Chapter 11 Construction commercial/business principles

- Ensure the business's income is more than its outgoings.
- Regularly measure performance against benchmarks (covered later in this chapter).
- Set improvement strategies to meet targets.
- Keep accurate business accounts.
- Submit annual tax returns and pay any duty due.

It is important to acknowledge that just as many entrepreneurs fail as succeed. However, there are many more who never progress beyond their initial idea because of the fear of failure. Entrepreneurs with a passion about something important to them, who have patience, confidence and tenacity, are most likely to achieve their goals.

6 Measuring success

6.1 Benchmarking

Successful organisations regularly evaluate their performance and make improvements, either to tackle areas of weakness or to demonstrate excellence.

Many new businesses fail within the first few years of trading because they cannot manage their cash flow effectively. The concepts for the business may be sound, and there may be no shortage of work, but if the business is poorly managed then the money goes out quicker than it comes in.

The term '**benchmarking**' means measuring an organisation's internal and external performance against pre-determined industry standards, competitors or completed projects. It is a powerful management tool that can increase both productivity and profits with minimum input (time and money) and maximum output (benefits). Any business targets set by an organisation must be realistic and achievable, if the business is to accomplish its aims.

> **Key term**
>
> **Benchmarking:** measuring an organisation's internal and external performance against pre-determined industry standards, competitors or completed projects

6.2 SMART objectives

SMART is a common acronym used in the built environment and construction industry for target setting. Smart objectives must be:

- **S**pecific – clear, unambiguous and understood by those who are expected to achieve them
- **M**easurable – in terms of time, productivity or cost
- **A**chievable – in terms of the size of the organisation, resources available and budget
- **R**elevant – appropriate to what is being done
- **T**ime-bound – in terms of a start and finish date (objectives with no deadline for completion are usually dismissed and never achieved).

6.3 Key performance indicators

Benchmarking is a recognised system that involves identifying problem areas in a business that needs improvement, for example:
- project completion times
- planned budgets against actual costs
- profit margins.

Areas that can be measured in terms of a value are known as key performance indicators (KPIs). Measuring business performance against KPIs helps to establish if objectives have been met or whether new strategic targets need to be planned. If a business is not honest with the findings of KPIs, fails to learn from its mistakes and starts the cycle again, then the process of benchmarking is pointless.

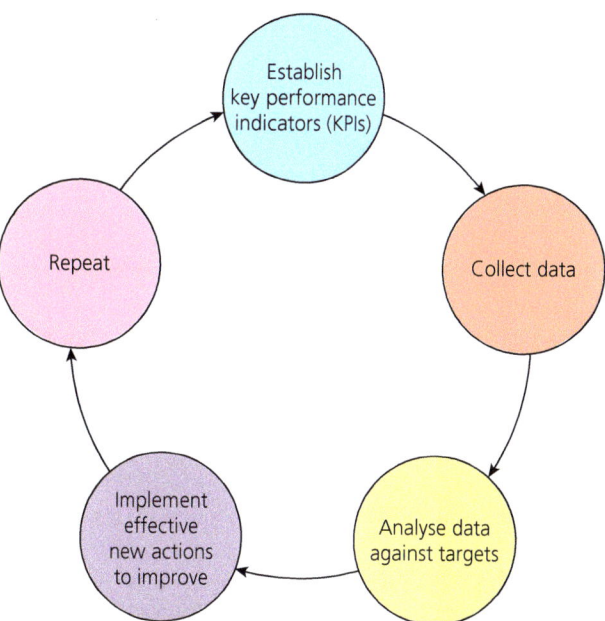

▲ Figure 11.15 The process of benchmarking

Every business selects its own KPIs based on what is important to it. For example, they could be used to measure the performance of a construction project based on data from a similar completed project. By looking at the results, a business can establish what

281

worked well, as well as what worked less well, and find areas for improvement (for example, when scheduling and sequencing work).

> **Health and safety**
>
> Using a business's accident and near-miss statistics over a given period is an accurate way to determine benchmarks and strategies to improve health, safety and welfare in a workplace. These benchmarks can be used to measure future performance of a business.

Gathering information for benchmarking within an organisation is much easier than trying to measure the success of a business against its competitors, because of difficulties in accessing all of the data needed. Internal benchmarking is often more valuable than external benchmarking, because it is focused and can be tailored to meet specific business objectives.

External benchmarking is used by businesses and other organisations such as the government to focus on the industry as a whole. Construction industry KPIs can be gathered from the results of national surveys of clients and construction professionals and performance data based on thousands of projects across the UK. Glenigan and Constructing Excellence publish annual sets of KPIs with the Department for Business, Energy and Industrial Strategy, to support businesses and drive improvement.

> **Improve your maths**
>
> There are various financial KPIs that can be used to measure the success of a business. Explain the difference between:
> - gross profit margin
> - net profit
> - turnover
> - working capital.

> **Test yourself**
>
> What acronym is commonly used when target setting? Explain the meaning of each letter.

> **Research**
>
> Look at the latest annual KPI survey published by Glenigan and Constructing Excellence (www.constructingexcellence.org.uk/kpi-reports).
>
> What were the results of the contractor and client satisfaction surveys?
>
> Contractor satisfaction survey:
> - Provision of information – ? per cent
> - Overall performance – ? per cent
> - Payment – ? per cent.
>
> Client satisfaction survey:
> - Service – ? per cent
> - Value for money – ? per cent
> - Product – ? per cent.
>
> Are these results better or worse than the previous year? Explain why you think the results of the surveys may be different to last year.

7 Project management

Throughout this book we have looked at the principles of project management, including the importance of:
- effective planning
- setting clear goals and objectives
- defining roles and areas of responsibility
- setting realistic milestones
- constraints on cost and time.

To ensure all the business objectives are measurable and achievable, the SMART technique (see section 6.2) can be applied.

> **Research**
>
> Research PRINCE2 (PRojects IN Controlled Environments) and explain how this can be used in project management.

8 Quality management

There are a number of quality management systems and techniques used in business to maintain the standard or quality of the work in a consistent manner, to track progress and to measure success. These include the following:

Chapter 11 Construction commercial/business principles

- self-assessment – a process of evaluating an organisation against a model for continuous improvement
- internal audit – completed by the organisation itself to measure the effectiveness of its quality systems. The aim of the audit is to identify any areas of weakness and develop a strategy for further improvement
- external audit – a quality management system evaluation completed by an external auditor to ensure that systems, processes and documentation are appropriate for the business. This technique is often the best and most beneficial because the results are impartial
- quality control – an internal process often used by organisations to maintain or improve standards. This system may involve the sampling, testing or reviewing of products, processes or documentation
- quality improvement – a systematic approach to measuring the success of a business against benchmarks and set new goals for continuous improvement
- ISO 9001:2015 (see section 2.5 and Chapter 7).

Assessment practice

Short answer

1. Which type of business sells its products and brand name to other businesses hoping to replicate its success?
2. What is another name for the fiscal year?
3. Which term describes collaborative working outside of an organisation?
4. Which term describes the measurement of an organisation's internal and external performance against pre-determined industry standards, competitors or completed projects?
5. Which term describes the incremental steps a business needs to take in order to achieve its overall aims, which are often closely aligned to its business plan?

Long answer

6. Describe one of the business types.
7. Explain the main disadvantage of being a sole trader.
8. Outline the main objectives of the International Organization for Standardization (ISO).
9. Explain the purpose of an organisation's corporate social responsibilities (CSR).
10. Explain what is meant by key performance indicators (KPIs) and why they are used.

Project practice

You have been working as a self-employed tradesperson for three years and have successfully networked with other sole traders on a few construction projects. One of these sole traders has suggested going into partnership. However, you have some concerns about growing the business.

Both partners are expected to invest £12,000 each, in order to purchase some innovative construction equipment to maximise efficiency and profits. You do not have the money to invest in the business; therefore you will have to take out a personal loan from your bank.

The projected turnover for the partnership for the next financial year is expected to exceed £90,000.
- Explain the risks of going into partnership, taking out the loan and increasing your turnover.
- Explain how you could reduce these risks and still expand your business.
- Present your concerns and suggestions in a digital format to one of your peers, and discuss any suggestions for improvement they may have.

About the exams and employer-set project

The T Level in Onsite Construction is made up of two components:
- A core component (300), which all students will complete.
- An occupational specialist component (or components), which is a trade-specific pathway that you will select with your lecturer.

This textbook covers the core component only, although the content you have covered and the skills you have learned will provide the foundation for your development in the occupational specialist component.

You can find out more about the qualification and how you will be assessed in the specification, which is available on the City & Guilds website.

Assessment

The core component is assessed in two ways:
- Papers 1 and 2, which are traditional written exam papers.
- An employer-set project, which is an extended real-world project that allows you to demonstrate the skills you have learned across the core component.

Core component exams

The two exams will test your knowledge and understanding of the content covered in the core component.

The 'Assessment practice' learning features, which appear at the end of each chapter, have been designed to help you apply your skills and knowledge in a similar style to what you can expect in the two exams.

Paper 1

This exam will include a range of short-answer questions, structured questions and extended-response questions.

The exam is 2 hours long and worth 35 per cent of the final grade for the core component.

The paper will cover:
- Health and safety in construction (Chapter 1)
- Construction design principles (Chapter 3)
- Construction and the built environment industry (Chapter 4)
- Sustainability principles (Chapter 5)
- Building technology principles (Chapter 7)

Paper 2

This exam will also include a range of short-answer questions, structured questions and extended-response questions.

The exam is 2 hours long and worth 35 per cent of the final grade for the core component.

The paper will cover:
- Construction science principles (Chapter 2)
- Measurement principles (Chapter 6)
- Information and data principles (Chapter 8)
- Relationship management in construction (Chapter 9)
- Digital technology in construction (Chapter 10)
- Construction commercial/business principles (Chapter 11)

Exam hints and tips
- Always read the instructions carefully. Think about what the command/keyword is asking for. These could include: identify, select, state, describe, explain, discuss, analyse, evaluate, justify.
- Concentrate on one question at a time and ask yourself the following:
  - Do I understand what the question is about?
  - How many marks is the question worth? You should try to work to one minute per mark.
  - How many parts are there to the question?
  - Can I provide a well-constructed answer?
  - How am I going to answer the question?
  - Do I need to include examples?
  - Do I need to relate my answer to a particular context?
  - Do I need to use technical terminology?
- It is important that the person marking your paper can not only read your handwriting but can also understand what it is you are trying to tell them. If they cannot read it or understand it, they cannot award you marks!

About the exams and employer-set project

- If you make a mistake, cross it out neatly and then start again. There may be extra pages at the back of the exam paper, or you can ask for extra paper. If you use the extra pages or paper, then you must make it clear where your answer can be found.
- When you have finished answering the questions, and if you have time, go back over your answers.
- Read carefully what you have written and ask yourself:
    - Have I answered the question?
    - Have I answered all parts of the question?
    - Have I met the demands of the command/keyword, for example, have I explained?
    - Have I used the correct technical terminology?

Employer-set project

This is an externally-set and externally-marked project. The scenario, which will be written by City & Guilds in collaboration with employers, will allow you to use the skills, techniques, concepts and knowledge that you have learned across the core component. You will receive the scenario from your lecturer, along with detailed guidance.

The employer-set project is worth 30 per cent of the final grade for the core component.

The 'Project practice' learning features, which appear at the end of each chapter, have been designed to help you apply your skills and knowledge in a similar style to what you can expect in the employer-set project.

In addition to the core component content, the project also links to the core skills:
- Core skill A: Applying a logical approach to problem solving
- Core skill B: Primary research
- Core skill C: Communication
- Core skill D: Working collaboratively with other team members and stakeholders

The learning features in this book allow you to develop these skills.

Answers

Assessment practice

Chapter 1

Short answer

1. The Health and Safety Executive (HSE)
2. Mandatory, safe condition, prohibition, warning, fire fighting
3. Reporting of Injuries, Diseases and Dangerous Occurrences Regulations (RIDDOR) 2013
4. Tablets and medicines
5. During the site induction

Long answer

6. Under this legislation, employers must calculate the amount of vibration that employees may be exposed to at work. At a specific level (referred to as the 'exposure action value'), employers must introduce technical and organisational measures to reduce the risk of personal injury to an acceptable level.
7. A safe system of work is a formal set of procedures that must be followed when hazards cannot be eliminated completely.
8. Toolbox talks are short training sessions arranged at regular intervals at a place of work to discuss health and safety issues; they give safety reminders and inform personnel about new hazards that may have recently arisen. These are usually delivered to small groups of workers, in an area of the workplace where they should not be disturbed. These talks usually cover a single aspect of health and safety, such as good housekeeping.
9. A confined space is a workplace which may be substantially but not always entirely enclosed, where there is a foreseeable serious risk of injury because of the conditions or from hazardous substances. Excavations, loft spaces, sewers or wells could be described as confined spaces, because they are enclosed with restricted access and egress.
10. Working in a confined space should be avoided wherever possible, and work should be completed in another way without entering the space. If this cannot be done and there is still a significant risk of injury, then the work must be properly risk assessed, planned and organised, with appropriate control measures in place before it starts. Everyone involved in working in a confined space must be competent and specifically trained to undertake their tasks.

Chapter 2

Short answer

1. – kg/m^3
   – Radius is half the diameter and is 0.42 m so $\pi \times 0.42^2 = 0.55\ m^2$
2. Answers may include
   – Aluminium
   – Copper
   – Silver

   Brass or bronze would not be acceptable as these are alloys and not pure metals.
3. Force of load is:
   $180 \times 9.81 = 1765.8\ N$
   Force needed is:
   $f \times d = f \times d$
   So:
   $\frac{1765.8 \times 0.5}{3} = 294.3\ N$
4. As the load is a weight, gravity is not required. The time must be in seconds so 1.5 minutes = 90 seconds, so:
   $\frac{5820 \times 28}{90} = 1810.67\ W$
5. Firstly, the current needs to be determined:
   $I = \frac{20}{58} = 0.34\ A$
   So:
   $P = 20 \times 0.34 = 6.8\ W$
6. Stored hydro is where stored water, in a lake or reservoir, is released when needed to turn turbines. This can produce electricity almost immediately. Water is then pumped back into the lake or reservoir when the system is not required.
7. The RCD monitors the current flow in the circuit Line and Neutral. If the two values are balanced, the RCD will not trip. If a fault occurs and some current leaks to earth, less returns in the Neutral and the values become unbalanced. If the value of unbalanced current exceeds the residual current rating of the device, it will trip disconnecting the circuit.
8. If somebody, for example at work, is in a position where they are subjected to glare from a lamp, where they can see the full harsh light from the lamp, they will likely squint due to the harsh bright light. Exposure can cause discomfort and possibly

Answers

headaches. Glare can be avoided by deflecting the bright light by diffusers or louvres that soften and scatter the light.

9 Attenuation is the reduction of sound reverberation. This can be achieved in several ways including:
   – Using soft furnishings/fabrics
   – Using baffles such as hanging special panels made of soft material/fabrics from the ceiling
   – Using soft materials or fabrics on the walls or carpeting the floor.

10 Wall area is:
   $3 \times 3 \times 2 = 18\ m^2$
   $2 \times 3 \times 2 = 12\ m^2$
   Less door:
   $2 \times 0.9 = 1.8\ m^2$
   So:
   $18 + 12 - 1.8 = 28.2\ m^2$
   So heat loss through wall is:
   $28.2 \times 0.4 \times (28-(-5)) = 372.24\ W$

Chapter 3

Short answer

1 Conservation of fuel and power: Approved Document L
2 Permitted development rights
3 Any three of the following:
   – excavation of the foundation
   – laying of foundation concrete
   – installation of damp-proof course (DPC) and damp-proof membranes (DPM)
   – laying of drains
   – completion of the roof structure
   – completion of first-fix installations (before plastering or drylining)
   – testing of drains
   – completion of the project.
4 The RIBA Plan of Work
5 Any one of the following:
   – panelised – timber frame
   – steel-frame construction
   – structural insulated panels (SIPs)
   – volumetric (pod/modular).

Long answer

6 A greenfield site is an undisturbed piece of land that has never been developed (for example with roads, buildings or other structures). However, a brownfield site may have an existing structure (above or below ground), or may have been previously developed and the structure or building removed. Developing a brownfield site can be more expensive compared with a greenfield site, because of the costs involved in preparing the site and the removal of any contaminated soil. Brownfield sites are often preferred for use by local planning departments as they have less impact on the natural environment.

7 Vernacular construction is where the design of houses is sympathetic or particular to a region, relying on locally sourced materials and traditional skills that have developed over generations. Given the distinct regional characteristics of such buildings, it is problematic to design and specify this form of construction anywhere other than where it is usually found. It would be difficult to achieve the same standards of workmanship without employing labour and sourcing materials from further away.

8 A BIM designer works with and advises clients on the implementation of Building Information Modelling (BIM).

9 CDM Regulations, budget, site analysis, planning, site of special scientific interest (SSSI), animals or infestations, protected site

10 Step 1: Pre-planning application
   Step 2: Full planning application
   Step 3: Consultation process
   Step 4: Decision-making process/outcome
   Step 5: Appeal (if necessary)

Chapter 4

Short answer

1 When annual turnover exceeds £85,000
2 Any one of the following:
   – government
   – public limited company (PLC)
   – commercial
   – private.
3 Any one of the following:
   – transportation engineering
   – structural engineering
   – environmental engineering
   – geotechnical engineering
   – hydraulic engineering
   – construction engineering.
4 A snagging list
5 Any one of the following:
   – political
   – economic
   – social
   – technological
   – legal
   – environmental.

Long answer

6 Self-employed subcontractors are responsible for every part of running a business, for example estimating, invoicing, ordering and accounts. They do not receive some of the benefits that employed people enjoy, for example holiday pay (including bank holidays), sick pay, maternity or paternity leave or pay.

7 Building Information Modelling (BIM) uses digital technology to share construction documentation and provide a platform for effective and efficient collaboration between designers and the construction team at every stage of a building project. It is adaptable to suit the size and complexity of each project and allows technical information to be shared throughout the management and construction teams.

8 Clients sometimes withhold a percentage of money due to the contractor at each stage of the building work; this is known as a retention. The exact percentage of money to be withheld has to be agreed between the client and contractor before work starts and is usually 3–5 per cent. The retention acts as financial security for the client, to make sure the contractor finishes the building work and resolves any snagging within a reasonable amount of time after completion of the work; this is known as the defects liability period. If the contractor does not return to complete the work within the period agreed in the contract, the client has reasonable grounds to use the money to instruct other contractors to undertake the outstanding work. On the other hand, if the contractor does complete the project and resolve all of the defects identified within the defects liability period, the client must release the outstanding retention payment without delay.

9 Answer to include:
   - protecting clients, customers and the public
   - keeping up to date with the latest regulation changes, product developments and technological advancements
   - developing product knowledge
   - working more efficiently
   - improving knowledge and skills
   - enhancing the company image
   - career progression.

10 A tender package may include the following documents:
   - letter inviting the contractor to submit a bid for the work
   - outline of the proposal
   - form of tender and timeline to return the completed bid
   - form of contract and conditions (including the process for payments and interim valuations)
   - programme of work
   - design drawings
   - specifications
   - site-specific information or issues
   - preliminaries
   - special planning-permission requirements
   - bill of quantities (cost framework)
   - tender return document.

Chapter 5

Short answer

1 Any two of the following:
   - Building Research Establishment Environmental Assessment Method (BREEAM)
   - Leadership in Energy and Environmental Design (LEED)
   - Timber Research and Development Association (TRADA)
   - WELL Building Standard
   - PAS 2035
   - PAS 2038.

2
   - building management systems
   - automated controls
   - smart controls
   - smart meters.

3 Any three of the following (or any other appropriate answer):
   - self-healing concrete
   - green roofs
   - smart glass
   - grey-water recycling
   - reed beds
   - soakaways
   - smart cement.

4 The Domestic Building Services Compliance Guide

5 Grade I: buildings of exceptional interest; Grade II*: particularly important buildings of more than special interest; Grade II: buildings of special interest, warranting every effort to preserve them

Long answer

6 When planning and delivering a construction project, sustainability is achieved by:
   - using renewable and recyclable resources

- sourcing materials locally
- protecting resources
- reusing and refurbishing materials
- reducing energy consumption and waste
- creating a healthy and eco-friendly environment
- protecting the natural environment.

7 Social sustainability focuses on wellbeing and quality of life. It involves recognising the needs of everyone impacted by construction projects, from design to demolition. This includes construction workers, local communities, project supply chains and users of the building.

8 Passivhaus, or 'Passive house' in English, is an energy performance standard intended primarily for new buildings. It ensures buildings are so well constructed, insulated and ventilated that they require little energy for heating or cooling.

9 Grey water is waste water that has been used for washing and is generated from hand basins, washing machines, showers and baths. Rather than sending it down the drain, it can be reused for watering plants and flushing toilets, thereby reducing mains water usage.

Rainwater harvesting involves collecting rainwater from roofs, filtering it and storing it for reuse, thereby saving energy, lowering carbon emissions and reducing mains water consumption.

Chapter 6

Short answer

1 The volume of topsoil is calculated in cubic metres (m^3).
2 Answer to include:
   - The work should be completed within an agreed budget and timescale.
   - The design brief should be wholly fulfilled.
3 A number system based on the number ten, tenth parts and powers of ten
4 Any two of the following:
   - setting out the outline of a building
   - positioning internal walls, doors and windows
   - establishing floor and roof heights.
5 New Rules of Measurement

Long answer

6 Accurate calculation of quantities of materials and components allows for assessment of the time needed to use and install those materials, contributing to the creation of a realistic work programme. Completion targets can then be set to support efficiency and productivity and satisfy the client's requirements for timely handover of the building.

7 Tolerances allow for acceptable variations in the strength of materials, the performance of a heating or ventilation system, temperature ranges in which materials can be used and many other situations. Exceeding allowable tolerances for a given construction task will mean that the work does not meet the specification. The function of the structure could be compromised and other elements of the building may not be able to be installed properly.

8 If a building that is not dimensionally accurate or not built to square is handed over to the client, significant problems can emerge when the installation of items such as floor tiles is attempted, or fittings and modular components such as kitchen units are positioned. Necessary remedial work would be potentially very costly to the contractor and disruptive to the client.

9 If the height of floors is not measured accurately during installation, a staircase will be placed at the wrong angle of slope, and it may not sit properly where the top of the stairs meets the upper floor.

10 U-value expresses the rate of transfer of heat through a structure (or more correctly through one square metre of a structure), taking account of the difference between internal and external temperatures. It is expressed in watts per metre squared kelvin (W/m^2K). The lower the U-value, the lower the rate of heat transfer.

Measured resistance to heat transfer is expressed as an R-value. The higher the R-value, the more resistance a material provides to heat transfer.

Chapter 7

Short answer

1 Approved Document B
2 Gas Safe
3 Pad foundations
4 50 mm
5 Any one of the following:
   - masonry screws/bolts
   - screws and plugs
   - chemical fixings
   - masonry nails
   - cartridge fixings.

Long answer

6 Off-site construction involves manufacturing modules in factories and then transporting them to site where they are assembled to form major elements of a building. It offers many benefits:
 - Automated systems improve accuracy and quality control.
 - Large numbers of components can be manufactured quickly and efficiently with reduced waste.
 - By carrying out a large proportion of the construction work in a controlled factory environment, the delays associated with adverse weather conditions on site are avoided.
 - Up to 95 per cent of a building can be constructed in a factory, therefore reducing the number of trades needed on site during the assembly and final finishing of the project.
 - There are cost savings from mass production in factories.

7 First fix is the phase of work completed after the structure has been erected and before plastering commences. Once the various trades have completed their first-fix work and the plastering/drylining is finished, second-fix work can commence. Second fixing often involves the installation of items and equipment that could have easily been damaged or affected by earlier stages of construction work.

8 Benefits of autonomous vehicles include:
 - reduced costs by improving productivity and site progress, because they are able to operate 24 hours a day without a driver
 - fewer carbon emissions than vehicles driven by humans, due to less erratic movements and more efficient use
 - better safety and accuracy.

9 Components of a solid floor:
 - compacted hardcore – provides a firm level base to bear the weight of the floor materials above
 - sand blinding – protects the polythene damp-proof membrane (DPM)
 - DPM – prevents rising damp and weed growth
 - concrete slab – provides most of the structural integrity
 - insulation – prevents heat loss through the floor
 - screed – provides a flat, level surface (also used to cover underfloor heating pipes).

10 Answer should include:
 - increased risk of harm to workers and others
 - product failure
 - premature wear
 - invalidation of warranty or product guarantee
 - minimum building standards not met therefore unable to be signed off by building control
 - breaches of health and safety legislation.

Chapter 8

Short answer

1 Numbers, measurements, words and descriptions
2 The ability of data systems to exchange and use information
3 Quantity surveyors preparing costings; contractors and subcontractors preparing tenders; systems designers planning building services requirements
4 Enterprise asset management
5 Some systems are dependent on different weather conditions and must be checked over time.

Long answer

6 When a set of data is used to describe or analyse other data, this is referred to as metadata. It speeds up the analysis of information required for technical design processes and could inform an architect regarding multiple intricate design elements of a building.

7 Generalisation is essentially a process of 'pulling back' from the mass of raw data to gain a broader, more general view and reduce the extent of analysis required. It avoids the need to process large volumes of complex data, which could be time consuming and confusing.

8 BIM allows authorised users to access a range of important information at all stages of construction and beyond. It uses digitally processed information to analyse design elements of a building, including 3D modelling.

Using BIM, complex design ideas can be transformed into a medium that is easier for all personnel to work with. It allows collaboration between every designer, engineer and contractor working on a project, providing comprehensive information about different workflows.

The Common Data Environment (CDE) is a single central source of information used within a BIM system. Relevant documents and data are brought together in a shared digital environment that can be accessed by all authorised personnel collaborating on the project.

9 When a building is handed over from the contractor to the client for occupation and brought into use, data can be harvested to answer important questions, such as:
   - How successful was the delivery of the project by the contractor?
   - Was the project delivered and handed over on time – if not, why not?
   - Does the building meet its design brief fully now it is in use?
   - What changes could be made to improve performance?

   This data can be useful to contractors who build repeat structures or specialise in specific project types.

10 When calculating outcomes, data can be used to analyse past or current activities to model what could happen in the future. An example might be calculating the outcome of using timber of specific dimensions to produce a load-bearing element of a building, such as a suspended floor. Consistent data has been recorded over many years which can be applied to producing reliable structural calculations, ensuring that the proposed design is safe and capable of bearing the required loads. Other examples of data might be performance charts for heating systems.

   Cost calculations draw on data that is collated, analysed, modelled and interpreted by the Building Cost Information Service (BCIS) of the Royal Institution of Chartered Surveyors (RICS). All sizes of construction company depend on accurate data to produce costings that will result in profitability for the contractor and a fair price for the client.

Chapter 9

Short answer

1 Stakeholder
2 To ensure a good company reputation and increase the chances of repeat business
3 Equality Act 2010
4 Company handbook/policy
5 Workplace pension scheme

Long answer

6 Answer to include the following:
   - Distributive negotiation is used to haggle over a common single interest at stake, known as a fixed sum. A fixed sum is best described as a pie that parties are battling over for a bigger slice, with exchange offers back and forth.
   - The win–lose approach is probably the most common negotiation method used to settle disputes between two parties. An agreement is more difficult to reach, because one side has to compromise in order for the other to experience a positive outcome.
   - The lose–lose approach – following negotiations, all concerned parties end up worse off and not achieving their desired result. All participants should try to minimise their losses and to make sure they are fair.
   - The compromise approach is used to settle disputes quickly by one party settling for less than it may have hoped for, in order to reduce strained negotiations and maintain or fix a relationship.
   - Integrative negotiation is used by parties with common interests, in order to collaborate in finding a mutually beneficial solution.
   - The win–win approach – negotiators with shared interests work together to find resolutions they are both satisfied with, rather than seeking to fulfil self-interests. This avoids disagreements and helps to maintain strong relationships between parties, while achieving a fair outcome for both sides.

7 An example of unethical behaviour in a business: during the development of a new construction site, a contractor wilfully damages protected trees (TPO – Tree Preservation Order) with a digger and contaminates a local stream with harmful construction waste. The polluted watercourse damages the natural environment of fish and other animals.

8 Networking is a low-cost process used by many organisations to make initial introductions with likeminded people, share information and form long-lasting business relationships.

9 Arbitration may be used as a conflict-management technique in the workplace, if an employee feels they have been unfairly dismissed from their job. This process would involve the appointment of an unbiased conciliator to meet with both parties

separately and together, to weigh up both sides of an argument. They will then make a proposal based on the relative merits of each side, to find an amicable solution to the matter and bring it to a close as quickly as possible.

10 Strengths of non-verbal communication:
   - It can be referred back to.
   - There is a permanent record of the communication.
   - The same information can be distributed easily without diluting it.
   - The sender does not have to meet the recipient to pass on the information.
   - It can be used to communicate if the recipient has a hearing impairment.

Chapter 10

Short answer

1 Machine to machine
2 By monitoring energy use and matching it to demand
3 Any two of the following:
   - improvements in manufacturing efficiency
   - improvements in safety
   - streamlining of materials delivery and supply-chain activity.
4 Light detection and ranging (LIDAR)
5 For performing repetitive tasks, such as construction of building modules, and tasks that might be hazardous to operatives

Long answer

6 AI uses methods of analysis that identify patterns and correlations, in order to draw appropriate conclusions more quickly than humans are able to.
7 Future AI systems may be able to sense in real time whether a building's occupants are tired, too hot or too cold, and automatically make adjustments to heating and ventilation to enhance comfort and wellbeing.
8 Digital CAD systems can speed up the design process considerably, leading to greater productivity. Building design ideas created using specialist software can be presented as a three-dimensional (3D) model, allowing stakeholders to examine and refine the design concept.

Digital tools are able to generate accurate materials and components lists from 3D models or 2D drawings. These can be linked digitally to materials costs databases, in order to produce an up-to-date costing for a project.

Speeding up the process of designing and costing a project, coupled with the ability to update or amend project details quickly, streamlines development and improves productivity.

Since CAD drawings are produced digitally, they can easily be stored, retrieved, shared and copied.

9 Equipment and machinery can be fitted with digital sensor technology to operate as part of the IoT. The range of movement and operational area of excavators and other machines can be controlled automatically, and programming can be updated to match new instructions from remote control locations.

10 Simulation is used to create a two- or three-dimensional model, which can provide technical data on structural and systems performance. Animation provides an engaging and lifelike view of a building, allowing a viewer to travel through a structure to assess the various elements visually.

Chapter 11

Short answer

1 Franchise
2 A financial year
3 Networking
4 Benchmarking
5 Objectives

Long answer

6 Business types include:
   - sole traders – must be registered with HMRC as a sole trader. As a proprietor, they have full control of their entire business or enterprise, although they could employ other people to work for them. A sole trader is responsible for the day-to-day management of their company, including filing self-assessment tax returns with HMRC and paying National Insurance contributions. Sole traders are personally liable for any losses or debts that the company may incur if things go wrong.
   - partnerships – are owned by two or more individuals. Partners share the costs, duties and risks of managing a business together, although they may delegate certain responsibilities. Each partner is personally responsible for paying tax and National Insurance contributions, based on their share of the profits. There are three different types of business partnership: ordinary partnership, limited partnership and limited liability partnership (LLP).

Answers

- limited companies – private (Ltd) or public (plc) – must be registered with HMRC and Companies House and file tax returns. The owners and shareholders have limited liability, so their personal assets are protected against any business debts up to the value of the money that they have invested. In a plc, money can be raised for the business through investors (for example the public) buying shares on the stock exchange.
- small and medium-sized enterprises (SMEs) – in the UK, small and medium-sized enterprises (SMEs) are defined as follows:
  - Small – employs on average no more than 50 people and has an annual turnover of £10.2 million or less
  - Medium-sized – employs on average no more than 250 people and has an annual turnover of £36 million or less
- not-for-profit organisations – are charitable businesses that do not make a financial profit. They often seek to provide a public service or social benefits for individuals or communities in need.
- community interest companies (CICs) – aim to provide a benefit for the community or trade with a social purpose. Returns to the company owners and investors are allowed, as long as they are balanced and reasonable, and a dividend cap must be in place.
- franchises – allow a business to expand quickly with lower capital outlay by selling the rights to the business model to self-employed entrepreneurs, referred to as 'franchisees'.

7 A sole trader is personally liable for any losses or debts that the company may incur if things go wrong. Creditors can legally recover any money owed by the debtor from their business or their personal assets such as their house. If the debt is considerable and the individual is unable to pay, they could be made personally bankrupt. This could result in further difficulties, for example being unable to obtain any type of credit such as a mortgage, credit card or finance on a vehicle because they are deemed to be a high risk.

8 The main goal of the ISO is to facilitate trade. It offers solutions to global challenges and supports innovation by providing guidelines to streamline processes and improve quality and safety across a range of businesses and products.

9 The purpose of an organisation's corporate social responsibilities (CSR) is to actively make a positive contribution to the community and wider society, and to minimise any negative impacts caused by their business. Having a CSR strategy integrated into its values also affects how stakeholders (such as clients and investors) view an organisation and whether or not they decide to work with or support it.

10 Areas that can be measured in terms of a value are known as key performance indicators (KPIs). Measuring business performance against KPIs helps to establish if objectives have been met or whether new strategic targets need to be planned. If a business is not honest with the findings of KPIs, fails to learn from its mistakes and starts the cycle again, then the process of benchmarking is pointless.

Glossary

Accident book: a formal document used to record details of accidents that occur in the workplace, whether to an employee or visitor

Active monitoring: monitoring people, procedures, premises and plant in the workplace in order to identify potential hazards before an accident or incident occurs, so that control measures can be taken to prevent harm

Alternator: an electrical generator that converts mechanical energy into electrical energy

Ambiguous: unclear and difficult to understand

Appraisals: scheduled routine meetings between an employee and their employer to review their work performance against their job description

Arson: the criminal act of deliberately setting fire to property

Asbestos: a naturally occurring mineral that used to be mixed with other construction materials to create insulation, pipe lagging and flooring. It has been banned in the UK since 1999 because it is extremely hazardous to health; however workers may still be exposed to it in older buildings

Autonomous vehicles: vehicles able to operate without human intervention

Basements: habitable rooms or spaces constructed below ground level

Benchmarking: measuring an organisation's internal and external performance against pre-determined industry standards, competitors or completed projects

Bilingual: fluent in two languages

Bimetallic strip: a temperature-sensitive component comprising two different metals bound together; when heated, each metal expands at a different rate, causing the strip to bend and activate a switch

Biometric: involving the detection and analysis of individuals' unique physical characteristics as a means of verifying identity

Bonded: the arrangement of staggered joints

Building Information Modelling (BIM): the use of digital technology to share construction documentation and provide a platform for collaboration

Built to square: where all the corners of a square or rectangular building are accurately set at 90°

Bump caps: a type of PPE designed to protect the user's head when there is a low risk of bumping it; it is not designed to take an impact from an object falling from height

Business model: a plan usually created by a business owner which describes the strategy or framework that an organisation will use to operate and includes the identification of products/services, revenue sources and customer base

Business plan: a written document that defines a business's goals and the strategies and timeframes to achieve them

Calibration: comparison of a measurement device against a traceable reference

Capital: the amount of funds or liquid assets owned by a business

Carbon: a chemical element that can be released into the atmosphere when fossil fuels are burned

Catenary wires: strong wires which are tied at each end and used to support other objects, such as cables which may stretch or break under their own weight when hung between two buildings

Cavity walls: external, load-bearing, structural walls consisting of two individual leaves (skins) of masonry with a gap (cavity) between them

CE marking: a mark on a product that identifies it has been designed and manufactured to meet EU safety, health or environmental requirements; CE is an abbreviation of a French term *'conformité européenne'*, meaning 'European conformity'

Chases: cuts in a masonry or plaster wall to conceal pipes or cables; once fitted, they are covered with plaster or similar

Chemically inert: will not react with chemicals

Chronic: continuing for a long time

Civil engineering: a profession involving the design, construction and maintenance of infrastructure that supports human activities, for example roads, bridges, airports and railways

Glossary

Climate change: a large-scale, long-term change in the Earth's weather patterns and average temperatures

Cob: a blend of subsoil (clay or earth), sand and straw mixed with water to make an organic material historically used to construct walls for homes and agricultural buildings in Devon, Cornwall and Wales

Collaborative: involving people or groups of people working together with a single common interest or aim

Commercial: relating to buying and selling

Commissioning: the process of ensuring that a building system is performing or working as it has been designed

Common law: legislation made in the civil courts rather than statute law that is made in Parliament

Common rafters: rafters that run from a ridge board to the wall plate at 90 degrees on plan

Companies House: a government body that registers and stores information on all the limited companies in the UK and makes it available to the public

Computer-aided design (CAD): using computer software to develop designs for buildings and structures

Concessions: something granted in response to a demand

Conductors: materials that have atoms less densely packed together and allow electron flow

Contingency: provision for an unforeseen circumstance; financial contingencies are often considered when planning for a construction project

Conventions: agreed, consistent standards and rules

Corporate social responsibility: the commitment of an organisation to carry out its business activities in a socially and environmentally responsible way

Corporation: a business owned by its shareholders, who often appoint a board of directors to manage the day-to-day running of its activities; the business is a legal entity and the shareholders have no personal liability for its actions and finances

Credit agreement: a legal contract made between a person or party borrowing money and a lender; it states the terms and conditions of the acceptance of credit, including how the debt will be repaid

Creditors: individuals or organisations that are owed money because they have provided goods, services or a monetary loan; HMRC would also be described as a creditor where a tax duty is owed

Data interoperability: the ability of data systems to exchange and use information

Defamation: the act of damaging someone's good reputation through a false written or verbal statement, also known as libel (written) or slander (spoken)

Demographic data: statistical data about a population in a particular location or region

Department for Business, Energy and Industrial Strategy (BEIS): the department of the UK government responsible for the UK's business, energy and industrial strategy

Design brief: a working document which specifies what a client wants; it makes clear all the design requirements of a project

Dissipated: energy consumed by converting to heat energy

Duty holders: people with a legal responsibility under health and safety law

Dynamic: characterised by frequent change or motion

Eaves: the part of a roof that overhangs the internal skin of the external walls

Ecological balance: where living organisms, such as plants, animals and humans, co-exist in a sustainable environment

Economic downturn: when the economy has stopped growing and is on the decline, resulting in reduced financial turnover

Efficiency: the ratio of output power compared to input power expressed as a percentage

Effluent: liquid waste or wastewater

Egress: an exit or way out

Elevation: a view of the front, back or sides of a building

Encasing services: creating a framework of materials to cover pipes, cables and other services in a building; also referred to as boxing-in

Encryption: the process of converting data or information into a code to prevent unauthorised access

Environmental activists: people who campaign for the protection of the natural environment

Equality: a state where all members of a society or group have the same status, rights and opportunities

Ethics: moral values that govern a person's behaviour towards others

Feasibility: how easy or difficult it is to do something

Ferrous metal: a metal that contains iron and is magnetic

Firewall: a protective software program or hardware device that monitors, filters and may block data entering and leaving a network

First fix: a phase of construction work completed before plastering

Fulcrum: the pivot point of a lever

Gas Safe: a professional organisation that controls the health and safety of work completed on gas systems in the UK

Generalisation: (in data processing) creating layers of summarised information from mass data

Geofencing: using technology to pinpoint the location of equipment and create a virtual boundary

Geometric: consisting of defined angles, patterns and shapes

Geotechnical engineering: a field of civil engineering that deals with the behaviour of earth materials such as soil and rock

Graining: a method used by decorators to create woodgrain effects on different surfaces

Grey water: water that has not been purified for the purpose of drinking, for example recycled water from a sink

Grievance: a feeling of having been treated unfairly

Gypsum: a natural mineral, often used in building products such as plaster and plasterboard

Haggle: to negotiate for the best terms of an agreement or financial arrangement

Heat: does not, in BSE terms, mean hot, it simply means heat energy, which can be hot or cold

Human resources department: department of an organisation that deals with recruiting, administrating and training staff

Immersive: creating a 3D image which appears to surround the viewer

Incorporated: legally registered as a limited company

Infrastructure: the basic systems and services required for the proper functioning of society

Joules: the unit of measurement for energy. Where energy is expressed as mechanical energy, it is known as work

Kinetic energy: energy derived from motion

Kinetic lifting: the act of manual handling

Liable: legally responsible

Life cycle assessment: assessing the total environmental impact of a building, considering all stages of the life of the products and processes used during its construction

Limited partner: a part-owner of a business whose financial liabilities cannot exceed their investment

Linear measurement: the distance between two given points along a line

Local Authority Building Control: local authority department responsible for inspecting building work against building regulations and signing-off completed projects

Lone workers: those who work by themselves without close or direct supervision by their employer or work colleagues; they may work in a remote location or in a workplace with members of the public

Low-emissivity (low-E) glass: a type of thermally efficient glass that is covered with a microscopic coating on its surface to minimise the amount of infrared and ultraviolet light that can pass through it to keep a building warm in the winter and cool in the summer

Low-rise buildings: buildings with up to four storeys

Luminaires: complete electric lighting units, including the casing, lamp and any internal controlling devices or electronic equipment (known as control gear or drivers)

Malware: software that is specifically designed to disrupt, damage or gain unauthorised access to a computer system

Market share: the percentage of total sales or output that a business has in a specified market; sometimes referred to as market leadership

Metric units: decimal units of measurement based on the metre and the kilogram

Minerals: a solid, naturally occurring, inorganic substance

Mobile bowsers: a wheeled trailer fitted with a tank for carrying oil

Networking: an activity where businesses and people with a common interest meet to share information and develop contacts

Ohm's law: a law that states the relationship between current, voltage and resistance in an electrical circuit

Overheads: regular repeated costs associated with the day-to-day running of a business, for example rent and insurances

Parameters: limits which define the scope of a system, process or activity

Glossary

Partnership agreement: a legally binding contract that sets out terms and conditions for each partner in the business

Passivhaus: ('Passive house' in English) an energy performance standard intended primarily for new buildings, which ensures that buildings are so well constructed, insulated and ventilated that they require little energy for heating or cooling

Philosophy: values and beliefs that act as guiding principles for behaviour

Plant operators: people in control of heavy construction machinery and equipment

Portal frame: a large structural frame made from load-bearing timber and steel beams and columns

Pre-cast: formed into a shape in a factory before being delivered for use on site

Prefabs: buildings manufactured using factory-made components or units that are transported and assembled on site

Preliminaries: pre-construction information that outlines items that are necessary for a contractor to complete the works but are not actually part of the works, for example general plant, welfare facilities and site security

Pre-stressed concrete: a type of concrete that has been compressed during production to improve its strength; this is normally achieved by 'tensioning' (stretching) high-tensile steel wires in the concrete

Prime cost: the actual value of goods and services without any additional costs added, for example profit margins

Principal designer: a designer appointed by the client to take the lead in planning, managing, monitoring and co-ordinating health and safety during the pre-construction phase of a project involving more than one designer

Procurement: the process of agreeing business terms and acquiring goods, products or services from suppliers

Proofread: to check a piece of written communication for errors such as spelling, grammar, punctuation and accuracy before it is shared or published

Proprietor: an individual who owns a business

Public-sector projects: projects funded by the government

Ratio: the amount or proportion of one thing compared to another

Reasonably practicable: a term used in health and safety law to describe realistic steps that should be taken to comply with the law in terms of time, effort and money

Rebar: reinforced steel bar commonly used in concrete to act as a frame to stop it moving and cracking

Render: a pre-mixed layer of sand and cement, similar to mortar, used to make masonry walls flat and prepare them for top coats of finishing plaster

Renewable energy: energy that comes from natural sources or processes that are replenished or replaced, such as water, sun and wind

Renewable resources: resources that can be replaced over time by natural processes, for example wind energy or solar energy

Resistance: the measure of how well a material conducts electricity in ohms (Ω); the lower the value of ohms, the better it conducts

Respiratory protective equipment (RPE): personal protective equipment that protects the user's respiratory system

Responsible person: (in law) usually the employer (assuming control of the workplace), a person with control of the premises or an owner; that person can then nominate other competent people to perform legal responsibilities, for example to manage fire safety

Retrofit: the process of adding new components to older structures

Retrofitting: the process of adding new features and technologies to existing buildings

Risk assessment: a formal process of identifying significant workplace hazards, whom they affect and control measures that could be used to eliminate or reduce risk to an acceptable level

Sabbatical: an extended period of unpaid leave from work, taken in agreement with an employer, often used for holidays, travelling or pursuing interests

Safe system of work: a formal set of procedures that must be followed when hazards cannot be eliminated completely

Safety data sheets: written documents produced by manufacturers and suppliers of hazardous substances that contain important information about how products should be transported, used, stored and safely disposed of after use, any special conditions you should be aware of and how to deal with the substance in an emergency

Self-employed: the state of working for oneself rather than an employer; a self-employed person is responsible for paying their own tax and National Insurance contributions on any earnings

Sentry: a person who supervises workers from the access/egress points of a confined space

Shadow flicker: when the rotating blades of a wind turbine create moving shadows

Shareholders: investors in a company who have purchased at least one share of a company's stock

Site induction: an occasion when all new visitors and workers on a construction site are informed about potential hazards and the control measures used to protect people from harm

Smart technology: computers, smartphones and tablets with software that connects to the internet in order to control, report, monitor and analyse devices and appliances remotely

Snagging: corrective work undertaken by a contractor or their subcontractors that has been identified by the client or their representative

Specification: a detailed description of the materials and working methods that must be used for a project

Spring tides: exceptionally high tides that occur twice monthly at the time of the new moon and the full moon, when the moon's orbit aligns with the sun to create a greater gravitational pull; they are known as spring tides because they act like a spring

State pension: a regular sum of money paid by the state to people of or above official retirement age

Statutory sick pay: payment made by employers to employees when they are too ill to work

Storage heater: an electric heater that stores thermal energy; it heats up internal ceramic bricks when electricity is cheaper at night and then releases heat gradually during the day, acting in the same way as a convection heater

Stress-graded: timber that has been visually or machine-assessed for its strength and stiffness, and certified according to its structural classification to British Standard rules

Subsidence: the sinking of a structure into the ground

Substation: equipment that transforms voltage to a suitable level for consumers

Surface water: water that collects on the ground or above surface structures and buildings, normally in the form of rain

Tendering: the process of inviting bids from contractors to carry out specific projects

Testimonials: statements of recommendation produced by satisfied customers or clients that confirm the quality of a product or service

Thermal: related to heat or temperature

Third-party verification: the process of getting an independent party to confirm that the project meets standards

Tolerances: allowable variations between specified measurements and actual measurements

Toolbox talks: short training sessions arranged at regular intervals at a place of work to discuss health and safety issues; they give safety reminders and inform personnel about new hazards that may have recently arisen

Top-down: proceeding from the most senior to the least senior

Transformers: devices that convert voltages and current, proportionately, to different values

Turbine: a machine that uses a moving stream of air, water, steam or hot gas to turn a wheel and generate power

Visualisations: digital or virtual representations of a structure

Water table: the level below which the ground is saturated with water; this level can rise with rainfall and fall with periods of dry weather

Water and sewerage undertakers: suppliers of wholesome water and sewerage services to properties

Watt: SI unit of power

Weirs: low dams across a river, which increase the force of the water as it flows over the top; sections of a weir can be raised or lowered to regulate the force of the water

Whistleblowing: the act of reporting information about wrongdoing

Wind turbines: a vaned wheel that is rotated by wind to generate electricity

Workers: people who do not have a permanent contract of employment with an employer but are contracted for work or services

Workflow: a sequence of activities needed to complete a work task

Index

A

absence 245
acceleration 45, 55
access equipment 33–7
accident book 20
accidents 2, 7, 18, 20, 262, 282
acoustics principles 77–9
active monitoring 18
Addison Act 1919 188
aesthetics 87
affordability 87–8
aggregates 48, 150
air
   characteristics 73
   quality 141, 145
air source heat pumps 164–5
alloys 48
aluminium 47, 54, 65
anaerobic digestion 156
animation 259
apex 198
appliances
   efficiency ratings 150–1
   waste management 154
appraisals 244
apprenticeships 245
Approved Codes of Practice (ACOPs) 7, 9
Approved Documents 86, 93, 146, 148, 199, 209–11
   see also Building Regulations 2010
arbitration 239
architects 98, 115, 121
area, measuring 45
argon gas 208–9
arm protection 27
arson 22
Art Deco period 188
artificial intelligence (AI) 154, 253–7
asbestos 3, 5, 147
atmospheric pressure 46
attenuation 78
auditing 283
augmented reality (AR) 130, 261
autonomous decision-making 254
autonomous vehicles 191–2

B

baffle 78
basements 193
battered walls 193
beams 70
bedrock 79–80
benchmarking 281–2
bending force 69–70
bill of quantities 119, 132–3, 137
biomass energy 159
biomass heating systems 75, 165
biometric information 253–4
block plan 183
blockboard 50
blockwork 73, 95, 151, 204
boreholes 92, 102
boundaries 209
bricklayers 126
bricks 50, 73, 157, 203
brickwork 89, 151, 194–5, 203–4
bridging 202–3
British Board of Agrément (BBA) 224
British Standards Institute (BSI) 184, 212
brownfield sites 84–5, 91, 92
building control officers (BCOs) 99, 104–5, 115
Building Information Modelling (BIM) 100, 107, 115, 121, 129–31, 212, 218, 222–3, 239, 260, 263
building inspectors 99, 104–5
building line 103
building notice application 104
building preservation notice 152–3
Building Regulations 2010 146
   inspections 99, 104–5
   listed and heritage buildings 93, 152–3
   noise restriction 79
   structural loads 69
   see also Approved Documents
Building Research Establishment Environmental Assessment Method (BREEAM) 140
Building Research Establishment's Home Quality Mark (HQM) 274
building services design engineers 122–3
building surveyors 115
bump caps 27
business
   management 131–2
   models 279–80
   objectives 271–4, 281
   performance assessment 220, 281–2
   plan 271
   structures 111–12, 266–71
   values 275–7
   see also corporate social responsibility (CSR); entrepreneurship
butterfly roof 201

C

capital 280
carbon dioxide 40, 75, 86, 181
carbon emissions 4, 5, 86–7, 114, 141–2, 162
carbon footprint 141–2
carbon monoxide 22, 40
carbon neutral 75
carbonic acid 81
carpenters 125, 190–1
Carriage of Dangerous Goods and Use of Transportable Pressure Equipment Regulations 2009 33
catenary wires 47, 48
cavity walls 126, 202–4
   see also insulation
CE marking 28
cement 49, 144
ceramics 51
certification 224
chasing 70–1
chipboard 50, 52, 73
civil engineering 110
civil engineers 115, 122
Classification, Labelling and Packaging (CLP) Regulation 2010 30
Clean Air Act 1993 145
clerk of works 115, 123
client representatives 115, 123
client specification 222

climate change 60, 82, 114
Climate Change Act 2008 114, 145
closing-off 204
coal 60
cob 187, 189, 202
Code for Sustainable Homes 86
codes of conduct 248, 277
cold roofs 201
collaborative working 232, 277
columns 70
combined heat and power (CHP) systems 168–9
commissioning 123, 223–4
Common Data Environment (CDE) 223
common law 7
common rafters 198
communication 240–5
community, local 87–8
community interest companies (CICs) 270
compact fluorescent lamps (CFLs) 77, 150
Companies House 111, 268
compliance 274
compression 69–70
computer numerical control (CNC) machines 263
computer-aided design (CAD) 100, 115, 121–2, 256, 263
computer-aided manufacturing (CAM) 263
conciliation 239
concrete 49, 73, 89, 150, 196
  testing 52–3
concrete blocks 205
condensation 201
conduction 72
conductors 64–5
confidentiality 227, 248
confined spaces 4, 27, 40–1
  see also excavations
Confined Spaces Regulations 1997 4, 41
conflict-management 239
conservation areas 153
conservation of fuel and power 86–7, 148, 210–11
Construction (Design and Management) (CDM) Regulations 2015 4, 6, 9, 16, 98, 101, 121, 240
construction industry 110–14
construction methods 91, 93–6, 142, 187–90
Construction Skills Certification Scheme (CSCS) 16, 124

container housing 189
continuing professional development (CPD) 128–9, 272
contract managers 123–4
contractors 98, 111, 119–20
contracts 133–5, 241
Control of Asbestos Regulations 2012 3
Control of Noise at Work Regulations 2005 4
Control of Pollution (Oil Storage) (England) Regulations 2001 147
Control of Substances Hazardous to Health (COSHH) Regulations 2002 3, 13–15, 146, 221–2
Control of Vibration at Work Regulations 2005 4
convection 72
convection heaters 72, 75
copper 47, 65
corporate social responsibility (CSR) 91–2, 131, 277–9
corporation tax 267
corporations 112
corrosion 54, 89
costing techniques 176–8, 184, 225
creditors 267
critical path analysis (CPA) 106
culture, organisational 272–3
curved roof 201
customer service 233, 275–6
cut roof 199
cyber security 227, 250

D

damp-proof course (DPC) 105, 195
damp-proof membrane (DPM) 105, 206
data 216
  analysis 221–5, 253, 255
  demographic 279
  legislation 228
  management 226–8
  metadata 216
  sources of 217–20, 221–4, 254, 279
  storage 226
  transfer 256
  uses of 220
  see also internet of things (IoT)
data interoperability 216
Data Protection Act 2018 228
data sheets 14
dead loads 59, 194
debtors 267
decibel 78
decking 202

defamation 249
delivery 257
delivery notes 136–7
demographic data 279
demolition 140, 150
density 46
Department for Business, Energy and Industrial Strategy (BEIS) 161–2
design
  brief 172
  feasibility study 90–1
  impact of 87–92
  planning 105–6
  principles 92–7
  process 101–6, 129–30
  specification 84
  stages 97
  see also architects; computer-aided design (CAD)
desktop survey 101–2
detail drawing 183
digital animation 259
digital engineering 258–61
digital modelling 260–1
  see also Building Information Modelling (BIM); three-dimensional (3D) modelling
Digital Plan of Work (DPoW) 130
digital technology 253–63
discrimination 236
distribution network operators (DNOs) 62–3
Distributor Takeback Scheme (DTS) 155
diversity 235
documentation 132
  bill of quantities 119, 132–3, 137
  contracts 133–5, 241
  delivery notes 136–7
  estimates 135
  purchase orders 137
  quotations 136
  specification 84, 120
  take-off sheets 132–3
  tendering process 98, 119–20
  wiring diagrams 138
  see also drawings
door sets 211
doors 96
dormer windows 200
drains 71, 210
  see also sewerage
draughtsperson 182
drawings 182–3, 243
drinking water 6
driveways 209

drones 192–3
drylining 190–1
drywall *see* plasterboards
ductwork 151
duty holders 2, 4

E

ear protection 26
earth forces 81–2
earth science 79–81
eaves 201
Ecodesign for Energy-Related Products Regulations 2010 99
ecological balance 274
economic downturn 275
economic factors 131
economic sustainability 142
Edwardian period 187
efficiency 58
efflorescence 89
electric shock 22
electric vehicles 169, 211
electrical and electronic equipment (EEE) 154–5
electrical building services (design) engineers 100
Electrical Installation Certificate (EIC) 127
electrical resistance 46
electricians 126–7, 190–1
electricity
  circuit principles 64–9
  distribution 62–3
  generation 60–1, 158
  transmission 61–3
electrolytic corrosion 54
electromotive force 46
elevation drawing 183
email 243
emergencies 21–2
emergency equipment 27
employee benefits 276
employee responsibilities 17–18, 248
employer responsibilities 5–7, 16–18, 29, 248, 276
Employers' Information Requirements (EIR) 129
Employers' Liability (Compulsory Insurance) Act 1969 8
employers' liability insurance 7, 8, 120
Employment Act 2008 236–7
employment rights 236, 245–8
Employment Rights Act 1996 236
encasing services 190–1
encryption 227

energy (work) 46, 57
energy consumption 161–2
energy efficiency 149–51
energy performance certificate (EPC) 85–6, 148
Energy Performance of Buildings Directive 147
Energy-related Products (ErP) Directive 150–1
engineers 115
enterprise asset management (EAM) 219
entrepreneurship 270, 279–81
environment
  effects on materials 54–5
  legislation 4, 145–7
  see also sustainability
Environmental Health Department 115
environmental impact 84–7, 142, 148–9
environmental impact assessment (EIA) 85
Environmental Protection Act 1990 145, 155
environmental risk assessment 102
equality 235–6, 247
Equality Act 2010 236, 247
equilibrium 58
equipment
  health and safety 3
  measuring 180, 184
  mechanical lifting aids 39
  smart technology 256–7
  working at height 33–7
estimates 135
ethics 249, 276, 278
European Convention on Human Rights 237
excavations 39–40
  see also confined spaces
eye protection 26

F

facilities managers 123
fauna 102
feasibility study 90–1
ferrous metal 47, 48, 89, 90
F-gases 156
fibreglass batts 202–3
Fire and Rescue Authority 115
fire safety 21–5, 210
firmware 227
first aid 28–9
first fix 125–6, 190–1
fissures 50, 52

fitness 141
flanking sound 78
flat roofs 200–1
flexible working 248
flooding 80–2
floor plan 183
floor-heating systems 72
floors 206
  insulation 152
  solid 206–7
  suspended 207
flora 102
flow 45
Fluorinated Greenhouse Gases Regulations (2018) 156
foot protection 26
footfall 78
footings 70, 194
force 46, 55–7, 59
forces 69–70
fossil fuels 60, 158, 165
foundations 70–1, 80, 102, 194–8
franchises 270–1
frequency 78
frog bricks 203
frontage line 103
fulcrum 55

G

gable-end roof 201
galvanic corrosion 54
Gantt charts 105–6
gas 60
gas fitters 190–1
gas leaks 22
Gas Safe 75
General Data Protection Regulation (UK GDPR) 2021 228
general principles of prevention 10–11
geofencing 257
Georgian period 187
geotechnical engineering 122
girders 70
glass 65, 73, 90, 143, 150, 208
global warming 81, 156
gold 65
graining 127
granite 73
gravity 58, 82
green buildings 107, 140
green roofs 142–3
greenfield sites 84, 91
greenhouse gases 156
grey-water recycling 143–4
grievance policy 236, 247

Ground Penetrating Radar (GPR) 262
ground source heat pumps 163–4
ground structure 79
gypsum 125, 150

H

haggling 238
hand protection 27
harassment 249
hardcore 48, 52
hardwood 50, 73
hazardous substances 3, 13–15, 29–32, 146–7, 210
Hazardous Waste (England and Wales) Regulations 2005 147
Hazardous Waste Regulations 2005 157
hazards 11, 16–17, 40
head protection 27
health and safety
   employee responsibilities 17–18
   employer responsibilities 5–7, 16–18, 29, 248
   first aid 28–9
   incidents and near misses 20
   inspections 6–7, 18–20
   legislation 2–9
   monitoring 18–20
   procedures 16–17, 21–2
   safe systems of work 4, 5, 10–11
   signage 29–32
   training 16
Health and Safety at Work etc. Act (HASAWA) 1974 2, 6, 16–17, 115, 213
Health and Safety Executive (HSE) 2, 115
   HSE Legal (L) Series 9
   inspections 6–7
Health and Safety (First Aid) Regulations 1981 28
Health and Safety (Safety Signs and Signals) Regulations 1996 29
heat principles
   air changes 73–5
   heat loss 73–5, 151
   heat resistance 73
   heat transfer 72
   heat recovery ventilation (HRV) systems 169
heating methods 75, 149, 164–9
height, working at 3, 20, 33–7
heritage building regulations 93
Heritage Protection Bill 153
hierarchical structure 112

high-pressure mercury (MBFU) lamps 77
high-pressure sodium (SON) lamps 77
hipped-end roof 201
Home Quality Mark 148
horizontal forces 69
hot works 22
hot-water immersion systems 72
human resources department 247
Human Rights Act 1998 237
hydroelectric power 61, 161
hydrology 102

I

illuminance 46
immersive technology 260–1
imposed loads 59, 194
incandescent bulbs 150
inclusion 235
indices 44, 47
infrared radiation 72
infrastructure 87–8, 209, 211, 218
injury 3, 7, 262
   see also first aid
innovation 273, 279
inspections 6–7, 18–20, 219, 224
instantaneous water heaters 72
insulated concrete formwork (ICF) 95
insulation 73, 85
   floors 152
   roofs 85, 151–2, 201–2
   walls 151, 202–3
insulators 64–5
insurance 7, 8, 120
internal pressure 46
International Organization for Standardization (ISO) 212, 274
International System of Units (SI) 44–5, 47, 64
internet of things (IoT) 253–8
interstitial condensation 201
inventory management 117
iron 47
ironmongery 211

J

job roles 98–101, 115–17, 124–7
joiners 125
joists 70
joules 57
just in time (JIT) systems 257–8

K

kerbs 209
key performance indicators (KPIs) 220, 281–2
kinetic energy 57, 159
kinetic lifting 38
K-values 73

L

ladders 33
lamps see lighting
lead time 94, 96
Leadership in Energy and Environmental Design (LEED) 140
lean construction 154
lean-to roof 201
leg protection 26
legislation
   data protection 228
   environmental 4, 145–7
   equality and diversity 236–7
   health and safety 2–9
   non-statutory guidance 7
   waste management 4, 30, 154–7
levers 55–6
life cycle assessment (LCA) 107, 140
lifting see manual handling
Lifting Operations and Lifting Equipment Regulations (LOLER) 1998 19, 20, 37
light-emitting diode (LED) lamps 77, 150
light-emitting diode (LED) technology 75, 77
lighting
   levels 141
   terminology 76
   types of lamp 76–7, 150
lime 48
limestone 81
limited companies 268
limited liability partnership (LLP) 268
lintels 70, 203–4
liquefied petroleum gas (LPG) 167
listed building regulations 93, 152–3
live loads 59, 194
living documents 232
living roofs 142–3
loadings 59, 194, 210
Local Authority Building Control (LABC) 99, 104–5, 115
loft insulation 151–2

Index

loft space 200
lone workers 28
low-emissivity (low-E) glass 208–9
low-pressure mercury (fluorescent) lamps 77
low-pressure sodium (SOX) lamps 77
low-rise buildings 93, 195
luminaires 75
lump-sum contracts 133

M

machine to machine (M2M) communication 253
maintenance 219, 256
malware 227
Management of Health and Safety at Work Regulations 1999 4, 10–11, 41
management team 116, 117
mansard roof 201
manual handling 3, 37–9
Manual Handling Operations Regulations (MHOR) 1992 3, 38
manufacturers 99, 117
manufacturers instructions 213
manufacturers specifications 222
market share 272
masonry
  blockwork 73, 95, 151, 204
  brickwork 89, 151, 194–5, 203–4
  faults in 89, 194
materials 47–52
  calculating quantities 172–3
  common failures 89–90
  delivery 257
  environmental effects on 54–5
  resistivity 65
  source and use of 148–9
  thermal conductivity 73
maternity leave 246–7
mathematical principles 44–5
MDF (Medium Density Fibreboard) 50
measurement
  equipment 180
  methods 175–6
  techniques 179–85
  tolerances 184–5
  units of 44–6, 57, 64, 178–9
mechanical design engineers 100
mechanical loads 55–9
mediation 239
mental health 141, 241, 276
metadata 216
metal halide (HID) lamps 77
metal stud partition walls 206
metals 47–8, 65, 89, 157

method statements 12–13
micro combined heat and power (micro CHP) systems 167–8
micro-hydro systems 166–7
micro-wind turbines 165–6
minerals 51, 52
minimum wage 245
Minor Electrical Installation Works Certificate (MEIWC) 127
mixed reality (MR) 261
mobile bowsers 147
mobile elevating work platforms (MEWPs) 36
mobile scaffold towers 34, 37
modelling *see* digital modelling
modern construction methods 94–6, 188–9
modular components 173–4
modular construction 95, 189–90
mono-pitch roof 201
mortar 49
mould 73
mullioned windows 187
musculoskeletal disorders (MSDs) 38

N

narrow strip foundations 195
National Grid 61–2, 159–61
National Insurance 246, 269
National Living Wage 245
National Minimum Wage 245
National Planning Policy Framework 88
near misses 20, 282
neck protection 27
negotiation skills 237–9
networking 250
new build housing (1990s) 188
Newton metres (Nm) 59
Newton's second law 55, 59
noise 4, 26, 77–9, 210
non-ferrous metal 47, 48
non-statutory guidance 7
non-verbal communication 241–2
notches 70
not-for-profit organisations 270, 272
nuclear power 158–9

O

Ohm's law 64–8
oil 60
oil storage 147
onsite construction methods 187–9
operatives 98, 116, 124–7

organisational culture 272–3
oriented strand board (OSB) 190
overheads 112
overheating 211
over-specification 91

P

pad foundations 197
painters and decorators 127
paints 51, 89
parental leave 247
partition walls 205–6
partnerships 267–8
Party Wall etc. Act 1996 209
passive-aggressive behaviour 240–1
passivhaus 152
payment schedule 134
payslips 246
pebbledash 187
penalty clauses 134
pensions 246
performance assessment 220, 281–2
permits to work 17, 22, 41
permitted development 85, 103
Personal Protective Equipment (Enforcement) Regulations 2018 3, 28
personal protective equipment (PPE) 3, 20, 25–8
Personal Protective Equipment (PPE) at Work Regulations 1992 3, 26
Personal Protective Equipment (PPE) Regulations 2002 20
PESTLE factors 131
PFA (pulverised fuel ash) blocks 202
philanthropy 278
photovoltaic (PV) panels *see* solar photovoltaic (PV) panels
pile foundations 197–8
pipework 151
pitched roof 198–9
planners 99
planning application notice 103–4
planning permission 85, 103–5, 152–3
  process 153
  see also Approved Documents; Building Regulations 2010
plant operators 39, 117, 127
plaster 52, 73, 125
plasterboards 51, 73
plasterers 125, 190
plastics 90, 157
  see also unplasticised polyvinyl chloride (uPVC)
plumbers 126, 190–1

303

plywood 50, 52
podiums 35
point cloud survey 260
political factors 131
pollution 84, 145, 149
portable appliance testing (PAT) 19
portal frame 197
post-occupancy evaluation (POE) 218–19
potential energy 57
power 46, 57–8
power management 255
power stations 158
prefabricated construction 142, 188, 189–90
preliminaries 119–20
Pressure Systems Safety Regulations (PSSR) 2000 20
pre-stressed concrete 208
prime cost 133
principal contractor 9, 237
principal designer 9, 121
private limited companies (Ltd) 268–9
procurement 220
producer compliance scheme (PCS) 155
product data 221
programme of work 105
project planning 105–6, 232–3
Provision and Use of Work Equipment Regulations (PUWER) 1998 3, 20, 36
public liability insurance 8, 120
public limited companies (plc) 269
Publicly Available Specifications (PASs) 212
public-sector projects 212
pulleys 56–7
purchase orders 137

Q

quality 212, 273–4, 282–3
quantities, calculating 176–8
quantity surveyors 115
Queen Anne period 187
quotations 136

R

radiant heaters 72
radiation 72
radio frequency identification (RFID) 257
raft foundations 196–7
rainwater harvesting 168
ratio 181

reactive monitoring 18
rebar 47, 48
recording/reporting incidents 2, 20–1
recycling 150, 154, 156, 157
    see also waste management
redundancy 248
reed beds 144
Regulatory Reform (Fire Safety) Order 2005 21
renewable energy 60–1, 122–3, 159–69, 255
Reporting of Injuries, Diseases and Dangerous Occurrences Regulations (RIDDOR) 2013 2, 21, 38
reservation point 238
residual circuit breaker with overload (RCBO) 69
residual current devices (RCDs) 68–9
resistivity 64
resonance 78
resources allocation 220
respiratory protective equipment (RPE) 25, 27
responsible person 22
rest breaks 246
rest facilities 6
restoration 121, 258
retaining walls 193–4
retention 134
retrofitting 100–1, 114, 258
reverberation 78
rigid foam board 73
risk assessments 3–4, 5, 11–12, 22
risk reduction 262
robotics 192, 262
roof lantern 96–7
roof windows 200
roofs 198–202
    insulation 85, 151–2, 201–2
    shapes 201
    windows 96–7, 200
Royal Institute of British Architects (RIBA) Plan of Work 97, 222
Royal Institution of Chartered Surveyors (RICS) 184
rubber 51, 65
R-values 73–4

S

sabbatical 248
safe systems of work 4, 5, 10–11
safe working load (SWL) 181
safety data sheets 14
safety signs 29–32
scaffolding 34, 37

scale rule 184
scales 182–4
screed 176
second fix 125–7, 191
sectional drawing 183
security 211, 227–8
self-driving vehicles 191–2
self-employment 111–12, 266–7
self-healing concrete 142
semi-detached housing 188
sentry 41
services
    availability of 92
    installing 70
    locating 175
sewerage 71, 146, 210
shear force 69–70
sheet materials 50
SI unit of measurement see International System of Units (SI)
signage
    safety signs 29–32, 240, 242, 244
    visualisations 130, 240, 242–3
    see also symbols
simulation 258–9, 262
site analysis 102
site induction 17
Site Management Safety Training Scheme (SMSTS) 16
site managers 116, 124
site of special scientific interest (SSSI) 102
site plan 183
Site Supervision Safety Training Scheme (SSSTS) 16
site survey 175
site waste management plan (SWMP) 86
small and medium-sized enterprises (SMEs) 269–70
smart cement 144
smart control systems 254–5
smart glass 143
SMART objectives 281
smart products 253–7
smart technology 129
snagging 134
soakaways 144
social media 248, 250
Social Security (Claims and Payments) Regulations 1979 20
social sustainability 142
softwood 50, 73
soils 71, 80
solar gain 96, 149

solar photovoltaic (PV) panels   61, 92, 122–3, 159–61
solar thermal systems   162–3
sole traders   111, 266–7
solid floors   206–7
sound insulation   78–9
sound transmission class (STC) rating   78
specific heat capacity   46
specification   84, 120
spring tides   82
stability of structures   70–1
staging boards   36
stakeholders   231
statutory sick pay (SSP)   245–6
steel   47, 65, 73, 150
steel-frame construction   95
steps   35
storage heaters   72, 75
storm surges   82
straw-bale construction   94
stress   241
stress-grading   205
strip foundations   195–6
structural analysis   258
structural engineers   115, 121
structural insulated panels (SIPs)   95, 190
structural members   70–1
Stuart period   187
subcontractors   112, 116
subsidence   71, 194
substations   62–3
substructure   70, 84, 193–8
sulphate attack   89, 90
superstructure   84, 193, 198–209
suppliers   117
supply chain   114–17
surface water   80–1
surveyors   115
suspended floors   207–8
sustainability   87, 140
  assessment methods   140–1
  objectives   274
  performance measures   148–52
  solutions   142–4, 148
  see also environment
symbols
  hazardous substances   222
  hazardous waste   30–2
  see also signage

T

take-off sheets   132–3
tamping   196

taxation   266–70
teamwork   234–5
tectonic plates   81
temperature   71
tender documents   98, 119–20
tendering process   118–20
tension   69–70
terraced housing   188
terrorist threats   22
testimonials   233
theodolite   180
thermal conductivity   73
thermal efficiency   149
thermal expansion   47, 54–5
thermoplastics   51
third-party verification   140
three-dimensional (3D) modelling   256, 259–60
three-dimensional (3D) printing   192, 262–3
tidal power   161
tilers   126
timber   50, 65, 73, 89, 141, 157, 205
timber frame construction   94
Timber Research and Development Association (TRADA)   141, 199
toilets   6
tolerances   184–5
toolbox talks   17
topography   90–1
torque   55, 57, 59
torsion   69–70
Town and Country Planning Act 1990   153
Town and Country Planning (Environmental Impact Assessment) Regulations 2017   85
trade supervisors   127
traditional construction methods   93–6, 187–8
training   16
transformers   62
transportation   142, 149, 155
trees   71
trestles   36
trusses   70, 199
trust   233
Tudor period   187
tungsten halogen lamp   77
turnover   112

U

ultra-violet (UV) radiation   55
units of measurement   44–6, 57, 64, 178–9

unplasticised polyvinyl chloride (uPVC)   51, 54–5, 65
U-values   74, 96, 181

V

valley roof   201
values   275–7
VAT (value added tax)   111
velocity   45, 55
ventilation   96, 210
verbal communication   240–1, 243
vernacular construction   91
vertical forces   69
vibrating concrete poker   196
vibration   4, 27
Victorian period   187, 188
virtual reality (VR)   260–1
visitors   16–17
visualisations   130, 240, 242–3
voltage   54, 61–2
volume, measuring   45
volumetric construction   95

W

wages   245–6
walkie-talkies   243
walkover survey   102–3
walls
  cavity   126, 202–4
  external   202–4
  insulation   151, 202–3
  internal   205–6
  lintels   70, 203–4
  load-bearing   202
  retaining   193–4
  solid   202
warm roofs   202
washing facilities   6
waste carriers   155
Waste Electrical and Electronic Equipment (WEEE) Regulations 2013   154–5
waste management   4, 30, 84, 86, 142, 147, 149, 154–7, 210
water
  drinking   6
  grey-water recycling   143–4
  leaks   22
  legislation   145–6
  quality   141
  surface   80–1
  usage   149
  wholesome   210
Water Act 2014   145–6

water cycle 81
water source heat pumps 164–5
water table 80, 102–3
watercourses 80–1, 84, 166–7
watertight 94, 96
wattle and daub 187, 189
wave power 61, 161
weather 81–2
weep holes 204
welfare facilities 6

WELL Building Standard 141
whistleblowing 236
whole building 107
whole-body protective equipment 27
wide strip foundations 196
wind, effects of 81
wind power 60, 149, 159, 165–6
wind turbines 159, 165–6
windows
   design elements 96–7

dormer 200
frames and glass 208–9
mullioned 187
roof windows 97, 200
wiring diagrams 138
Work at Height Regulations 2005 3, 20
working at height 3, 20, 33–7
Working Time (Amendment) Regulations 2007 246
written communication 241–2

Photo credits

The Publishers would like to thank the following for permission to reproduce copyright material.

p.1 © Eleventh Hour Photography/Alamy Stock Photo; p.7 © Health and Safety Executive; p.17 © Bogdan Vasilescu – Fotolia; p.18 © Seventyfour/stock.adobe.com; p.19 © Matjoe/Shutterstock.com; p.20 © Health and Safety Executive; p.21 t © Bertrand Cadilhac/Shutterstock.com, b © Grand Warszawski/stock.adobe.com; p.22 © Kumar Sriskandan/Alamy Stock Photo; p.23 © wellphoto – Fotolia; p.24 t © Jenny Thompson/stock.adobe.com, c © Alan Stockdale/Fotolia.com, b © Hartphotography/stock.adobe.com; p.25 t © Mr.Zach/Shutterstock.com; c © Jijomathai/stock.adobe.com, b © Mr.nutnuchit phutsawagung/EyeEm/stock.adobe.com; p.27 © Kamolrat/stock.adobe.com; p.28 l © Ross Swift/Alamy Stock Photo; r © Poco_bw/iStock/Getty Images; p.29 l © St John Ambulance, rt © Yongyuth Chanthabutr/stock.adobe.com, rb © Daseaford/stock.adobe.com; p.30 t © Viktorijareut/stock.adobe.com, c1 © Viktorijareut/stock.adobe.com, c2 © Ricochet64/stock.adobe.com, b © Ricochet64/stock.adobe.com; p. 31 all © Ricochet64/stock.adobe.com except last row © olando/stock.adobe.com; p.32 all © Ricochet64/stock.adobe.com; p.35 © Euro Towers Ltd; p.36 lt © Southern Plant & Tool Hire Ltd, rt © andersphoto/stock.adobe.com, rb © Sergejus Bertasius/123RF.com; p.37 © Totz_photo/Shutterstock.com; p.39 l © Africa Studio/stock.adobe.com, © Sveta/stock.adobe.com, r © shime – Fotolia.com; p.40 l © Linus/stock.adobe.com, r © EdNurg/stock.adobe.com; p.41 © Chitsanupong/stock.adobe.com; p.43 © tong2530/stock.adobe.com; p.52 t © Matthew Ashmore/stock.adobe.com, b © YOR/stock.adobe.com; p.53 lt © Walter/stock.adobe.com, lb © Bannafarsai/stock.adobe.com, r © Astrosystem/stock.adobe.com; p.54 © Dmitriy Sladkov/stock.adobe.com; p.59 © SPP/stock.adobe.com; p.73 © Burdun/stock.adobe.com; p.79 © SEMPATAP S.A.S.; p.80 © Kadmy/stock.adobe.com; p.81 © VectorMine/stock.adobe.com; p.83 © Chlorophylle/stock.adobe.com; p.84 l © Sveta/stock.adobe.com, r © Candy1812/stock.adobe.com; p.85 l © Silent Corners/stock.adobe.com, r © Alekss/stock.adobe.com; p.86 l © UrbanImages/Alamy Stock Photo, r © Phil Wills/Alamy Stock Photo; p.87 © Alicia G. Monedero/Shutterstock.com; p.88 © Duncan/stock.adobe.com; p.89 lt1 © Baharlou/stock.adobe.com, lt2 © Ian Shipley IND/Alamy Stock Photo, c © Hoda Bogdan/stock.adobe.com, lb1 © Evening_tao/stock.adobe.com, lb2 © Petert2/stock.adobe.com; p.90 t © Fasphotographic/stock.adobe.com, b © Kim sayer/Alamy Stock Photo; p.91 l © Pelvidge/stock.adobe.com, r © Beataaldridge/stock.adobe.com; p.92 © Rdonar/stock.adobe.com; p.93 © FotoVoyager/E+/Getty Images; p.94 c © Christian Delbert/stock.adobe.com, lb1 © Oleg/stock.adobe.com, lb2 © Kingspan Insulation Ltd; p.95 lt1 © Bildlove/stock.adobe.com, lt2 © Timmins Engineering & Construction Ltd, lc © Simon Turner/Alamy Stock Photo, lb1 © Kingspan Insulation Ltd, lb2 © Kara/stock.adobe.com; p.96 © Brizmaker/stock.adobe.com; p.97 © Laurence Berger/Shutterstock.com; p.99 l © Primagefactory/123RF.com, r © KomootP/stock.adobe.com; p.101 © Ronstik/stock.adobe.com; p.102 c © Pellinni/stock.adobe.com, b © John Copland/Shutterstock.com; p.103 © Frank/stock.adobe.com; p.104 l © Avpics/Alamy Stock Photo, r © Torbay Council, Torbay, Devon, UK; p.105 © Sculpies/stock.adobe.com; p.109 © HENADZ/stock.adobe.com; p.110 lc © ceebeestock/stock.adobe.com, lb © Dan Kosmayer/stock.adobe.com, r © Simon Turner/Alamy Stock Photo; p.113 © Avpics/Alamy Stock Photo; p.114 l © Sonyachny/stock.adobe.com, rt © Scotframe/Saint-Gobain Construction Products UK Ltd, rb © Shutterstock/Anton Gvozdikov; p.117 tl© Lightfield Studios/stock.adobe.com, bl © Monkey Business/stock.adobe.com; p.118 © Cineberg/stock.adobe.com; p.121 © Sergey Nivens/stock.adobe.com; p.122 lt © Cyberstock/Alamy Stock Photo, lb © Travelview/stock.adobe.com, r © Microgen/stock.adobe.com; p.123 l © Diyana Dimitrova/123RF.com, r © Kaspars Grinvalds/stock.adobe.com; p.124 t © Rido/stock.adobe.com, b © Kings Access/stock.adobe.com; p.125 lt © Monkey Business/stock.adobe.com, lb © Kzenon/stock.adobe.com, r © Alexander/stock.adobe.com; p.126 l © Ladanifer/stock.adobe.com, r © Monkey Business/stock.adobe.com; p.127 t © Collins Photography/stock.adobe.com, b © Highwaystarz/stock.adobe.com; p.129 © Aisyaqilumaranas/Shutterstock.com, r © HENADZ/stock.adobe.com; p.130 © Zapp2photo/stock.adobe.com; p.131 © konstruktor1980/stock.adobe.com; p.132 © HM Government published by the RIBA; p.134 l © The Joint Contracts Tribunal Ltd, r © KMPZZZ/stock.adobe.com; p.139 © Polifoto/stock.adobe.com; p.141 © Jandrie

Lombard/stock.adobe.com; p.143 l © Elke Hotzel/stock.adobe.com, r © Smart Glass VIP; p.145 © Inna/stock.adobe.com; p.146 © maximilian_100/stock.adobe.com; p.147 © John Borda/stock.adobe.com; p.149 © Khamkhor/stock.adobe.com; p.150 l © Pioneer111/stock.adobe.com; r © You can more – Fotolia.com; p.155 © Ungvar/stock.adobe.com; p.161 © Blueringmedia/stock.adobe.com; p.164 © Peter Gudella/Shutterstock.com; p.169 © Steve Lagreca/Shutterstock.com; p.171 © Kadmy/stock.adobe.com; p.172 l © Monkey Business/stock.adobe.com, r © Robert Kneschke/stock.adobe.com; p.173 © Hybrid Images/Cultura Creative RF/Alamy Stock Photo; p.174 l © DragonImages/stock.adobe.com, r © Michaeljung/stock.adobe.com; p.175 lt1 © Zero Creatives/Cultura Creative RF/Alamy Stock Photo, lt2 © hansenn/stock.adobe.com, lb1 © Fotolia RAW/stock.adobe.com, lb2 © Andy Dean/stock.adobe.com; p.176 t © Philippe Devanne/stock.adobe.com, b © Avalon/Construction Photography/Alamy Stock Photo; p.177 © AnnaStills/stock.adobe.com; p.178 l © SteF/stock.adobe.com, r © StockPhotosArt/stock.adobe.com; p.180 l © Mint Images Limited/Alamy Stock Photo, r © Carsten Rehder/dpa picture alliance/Alamy Stock Photo; p.181 © Scharfsinn86/stock.adobe.com; p.182 l © Brozova/stock.adobe.com, r © Ionia/stock.adobe.com; p.184 l © Bokic Bojan/Shutterstock.com, r © NorGal/stock.adobe.com; p.186 © ceebeestock/stock.adobe.com; p.188 l © esdras700/stock.adobe.com; r © Ron Ellis/Shutterstock.com; p.189 tl © Peter Turner Photography/Shutterstock.com, bl © Miriam Heppell/Alamy Stock Photo, rt © oatautta/stock.adobe.com; p.190 © Peter Jordan_NE/Alamy Stock Photo; p.191 © Andrey_Popov/Shutterstock.com; p.192 l © Built Robotics, rt © Construction Robotics, rb © PERI AG; p.193 l Gorodenkoff/stock.adobe.com, r © Sveta/stock.adobe.com; p.194 t © eelnosiva/stock.adobe.com, b © Jason/stock.adobe.com; p.196 l © Readin Wilson/Shutterstock.com, r © bandung/stock.adobe.com; p.197 r © Ali/stock.adobe.com; p.200 tl © Moving Moment/stock.adobe.com, bl © Wirestock/stock.adobe.com, rt © U. J. Alexander/stock.adobe.com, rb © Marko Beric/Alamy Stock Photo; p.204 lt © Keystone Lintels Ltd, rb © ARC Building Solutions Ltd; p.206 © jpldesigns/stock.adobe.com; p.207 rt © helgidinson/stock.adobe.com, rb © David Darkin/Shutterstock.com; p.208 l © Simple Shots/Ron Rovtar Photography/Alamy Stock Photo, rt © len44ik/stock.adobe.com; p.215 © Denisismagilov/stock.adobe.com; p.216 © Gorodenkoff/stock.adobe.com; p.217 t © Goodluz/stock.adobe.com, b © Chris Pearsall/Alamy Stock Photo; p.218 © Karel Miragaya/123RF.com; p.219 l © JoeFoxBelfast/Radharc Images/Alamy Stock Photo, r © Addictive Stock Creatives/Alamy Stock Photo; p.221 © Cookiecutter/stock.adobe.com; p.222 © WoGi/stock.adobe.com; p.223 © Construction Photography/Avalon/Alamy Stock Photo; p.224 © Andrey Popov/stock.adobe.com; p.225 © Atstock Productions/stock.adobe.com; p.226 © Blue Planet Studio/stock.adobe.com; p.227 t © Asadykov/stock.adobe.com, b © Gorodenkoff/stock.adobe.com; p.228 © Funtap/stock.adobe.com; p.230 © Rido/stock.adobe.com; p.231 t © Halfpoint/stock.adobe.com, b © Kzenon/stock.adobe.com; p.233 l © Syda Productions/stock.adobe.com, rt © Wordley Calvo Stock/stock.adobe.com; p. 234 l © Justin Kase zsixz/Alamy Stock Photo, r © BigPixel Photo/Shutterstock.com; p.235 l © Gorodenkoff/stock.adobe.com, rt © Frank/stock.adobe.com, rb © cineberg/stock.adobe.com; p.236 © TommyStockProject/stock.adobe.com; p.237 © pressmaster/stock.adobe.com; p.240 l © trgowanlock/stock.adobe.com, r © Bangkok Click Studio/Shutterstock.com; p.241 © eccolo/stock.adobe.com; p.242 © insta_photos/stock.adobe.com; p.244 © pressmaster/stock.adobe.com; p.245 © Monkey Business/stock.adobe.com; p.246 © Kevin Wheal/Alamy Stock Photo; p.247 © Anoushkatoronto/stock.adobe.com; p.248 © iQoncept/Shutterstock.com; p.249 © Alex Kosev/Shutterstock.com; p.252 © Zephyr_p/stock.adobe.com; p.253 l © Nuamfolio/stock.adobe.com, r © Arie Storm/Alamy Stock Photo; p.254 l © Simon Turner/Alamy Stock Photo, r © Andrey Popov/stock.adobe.com; p.255 l © Simon Turner/Alamy Stock Photo, r © Hans-Peter Merten/Mauritius images GmbH/Alamy Stock Photo; p.256 © Cyberstock/Alamy Stock Photo; p.257 l © hansenn/stock.adobe.com, r © Justin Kase zsixz/Alamy Stock Photo; p.258 © Bits and Splits/stock.adobe.com; p.259 l © Yon Marsh/Alamy Stock Photo, r © Murattellioglu/stock.adobe.com; p.260 l © Estan Cabigas/Alamy Stock Photo, r © Escapejaja/stock.adobe.com; p.261 © Gorodenkoff/stock.adobe.com; p.262 l © Yoshio Tsunoda/Nippon News/Aflo Co. Ltd./Alamy Stock Photo, r © Steven May/Alamy Stock Photo; p.263 l © Solkafa/Shutterstock.com, r © Alotofpeople/stock.adobe.com; p.265 © Hipproductions/stock.adobe.com; p.266 © highwaystarz/stock.adobe.com; p.267 © MK photograp55/Shutterstock.com; p.268 © Isabelle Plasschaert/Alamy Stock Photo; p.269 © Chris leachman/Alamy Stock Photo; p.271 © bernardbodo/stock.adobe.com; p.272 © Rido/stock.adobe.com; p.273 © Seventyfour/stock.adobe.com; p.275 © Andrew Paterson/Alamy Stock Photo; p.276 l © MilanMarkovic78/Shutterstock.com, r © Isuzek/E+/Getty Images; p.277 © Albert Pego/Shutterstock.com; p.279 © goodluz/stock.adobe.com; p.280 © fizkes/stock.adobe.com

Photo credits

The Publishers would like to thank City & Guilds for permission to reproduce photos and artworks from the following textbooks:

Carpentry & Joinery for the Level 1 Diploma (6706)

Site Carpentry and Architectural Joinery for the Level 2 Apprenticeship (6571), Level 2 Technical Certificate (7906) & Level 2 Diploma (6706)

Site Carpentry & Architectural Joinery for the Level 3 Apprenticeship (6571), Level 3 Advanced Technical Diploma (7906) & Level 3 Diploma (6706)

Bricklaying for the Level 1 Diploma (6705)

Bricklaying for the Level 2 Technical Certificate & Level 3 Advanced Technical Diploma (7905), Level 2 & 3 Diploma (6705) and Level 2 Apprenticeship (9077)

Plumbing Book 1 for the Level 3 Apprenticeship (9189), Level 2 Technical Certificate (8202) & Level 2 Diploma (6035)

Plumbing Book 2 for the Level 3 Apprenticeship (9189), Level 3 Advanced Technical Diploma (8202) and Level 3 Diploma (6035)

Electrical Installations Book 1 for the Level 3 Apprenticeship (5357), Level 2 Technical Certificate (8202) & Level 2 Diploma (2365)

Electrical Installations Book 2 for the Level 3 Apprenticeship (5357), Level 3 Advanced Technical Diploma (8202) & Level 3 Diploma (2365)

Level 2 NVQ Diploma in Plumbing and Heating

Level 3 NVQ Diploma in Plumbing and Heating 6189